Exploring the Northern Rocky Mountains

edited by

Colin A. Shaw
Department of Earth Sciences
Montana State University
Bozeman, Montana 59717, USA

Basil Tikoff
Department of Geoscience
University of Wisconsin
1215 W. Dayton Street
Madison, Wisconsin 53706, USA

Field Guide 37

3300 Penrose Place, P.O. Box 9140 ▪ Boulder, Colorado 80301-9140, USA

2014

Published by The Geological Society of America, Inc.
3300 Penrose Place, P.O. Box 9140, Boulder, Colorado 80301-9140, USA
www.geosociety.org

Printed in U.S.A.

Library of Congress Cataloging-in-Publication Data

Exploring the northern Rocky Mountains/edited by Colin A. Shaw, Department of Earth Sciences, Montana State University, Bozeman, Montana, Basil Tikoff, Department of Geoscience, University of Wisconsin-Madison.
pages cm. — (Field guide; 37)
Includes bibliographical references.
ISBN 978-0-8137-0037-3 (pbk.)
1. Geology—Rocky Mountains Region. 2. Geology—Idaho. 3. Geology—Montana. I. Shaw, Colin A. (Colin Arthur) editor of compilation. II. Tikoff, Basil, editor of compilation.
QE13.R54E97 2014
557.86′5—dc23

2014008704

Cover: Early morning alpenglow highlights the summit and west ridge of Granite Peak beneath a full moon. At 12,799 feet (3901 m), Granite Peak is the highest peak in Montana and crowns the Beartooth Range in the northern Greater Yellowstone geo-ecosystem. The peak lies within the Absaroka–Beartooth Wilderness area about 12 miles (20 km) northeast of the Cooke City/Silvergate entrance to Yellowstone National Park and is accessible only by foot or on horseback. The peak consists of Archean quartzofeldspathic gneiss intruded by Precambrian mafic dikes. The uppermost reaches of a small remnant glacier clinging to the north-facing headwall of Avalanche Lake basin are visible in the lower right of the photograph. Photo by David R. Lageson, 21 August 2005.

10 9 8 7 6 5 4 3 2 1

Contents

The Geological Society of America
Field Guide 37
2014

Foreword

Colin A. Shaw*
David R. Lageson*
Department of Earth Sciences, Montana State University, Bozeman, Montana 59717, USA

Basil Tikoff*
Department of Geoscience, University of Wisconsin, Madison, Wisconsin 53706, USA

Jeff D. Vervoort*
School of the Environment, Washington State University, Pullman, Washington 99164, USA

The northern Rocky Mountain Region encompasses a mosaic of overlapping tectonic provinces recording the accumulated effects of tectonic, magmatic, sedimentary, and geomorphic processes spanning more than three billion years of Earth history. The remarkable geologic diversity and good exposure that characterize the region make it an unsurpassed outdoor classroom for geology field camps from around the country and a focus for countless field trips. Over the years, this long tradition of field-based education and scientific discourse "on the outcrop" has spawned a rich literature of published and unpublished field guides (e.g., Beaver, 1985; Childs and Lageson, 1993; Czamanske and Zientek, 1985; Fournier et al., 1994; Lewis and Berg, 1988). This volume brings together a diverse collection of nine field guides, providing a timely update to the field-based literature on southwestern Montana and neighboring areas of Idaho and Wyoming. The guides cover geologic features ranging in age from Archean to recent, representing a broad sampling of the geology of the region.

Archean orthogneiss and supracrustal metamorphic rocks of the Wyoming craton (Fig. 1) are exposed in the cores of a number of Laramide arches in southwest Montana (e.g., Houston et al., 1993; Mogk et al., 1988; Mueller and Frost, 2006; Mueller et al., 1998). This domain of Archean basement is juxtaposed against the poorly exposed, northeast-trending Great Falls tectonic zone (e.g., O'Neill and Lopez, 1985) and locally overlapped by the Mesoproterozoic Belt basin (Harrison, 1972; Ross et al., 1963; Ross and Villeneuve, 2003). An economically important aspect of the multi-stage Precambrian geologic history that produced these rocks is explored in the Chapter 8 field guide (Underwood et al.), which focuses on hydrothermal metasomatism of deformed Archean dolomitic marbles coeval with extension along the margin of the Belt basin. This episode produced substantial deposits of talc that are currently mined in the Gravelly and Ruby Ranges.

Subduction of the Farallon plate beneath the western margin of North America during the Mesozoic and Paleogene resulted in the growth of the continent by accretion of exotic terranes, as well as by the emplacement of calc-alkaline batholiths into older continental crust (e.g., Dickinson, 2004). The field guide related to Chapter 1 (Lewis et al.) explores processes of continental growth along the Mesozoic margin as it traces a transect from exotic terranes in westernmost Idaho across the western Idaho shear zone and through the Idaho batholith into western Montana (Fig. 1).

*E-mails: colin.shaw1@montana.edu; lageson@montana.edu; basil@geology.wisc.edu; vervoort@wsu.edu.

Shaw, C.A., Lageson, D.R., Tikoff, B., and Vervoort, J.D., 2014, Foreword, *in* Shaw, C.A., and Tikoff, B., eds., Exploring the Northern Rocky Mountains: Geological Society of America Field Guide 37, p. v–viii, doi:10.1130/2014.0037(00). For permission to copy, contact editing@geosociety.org.

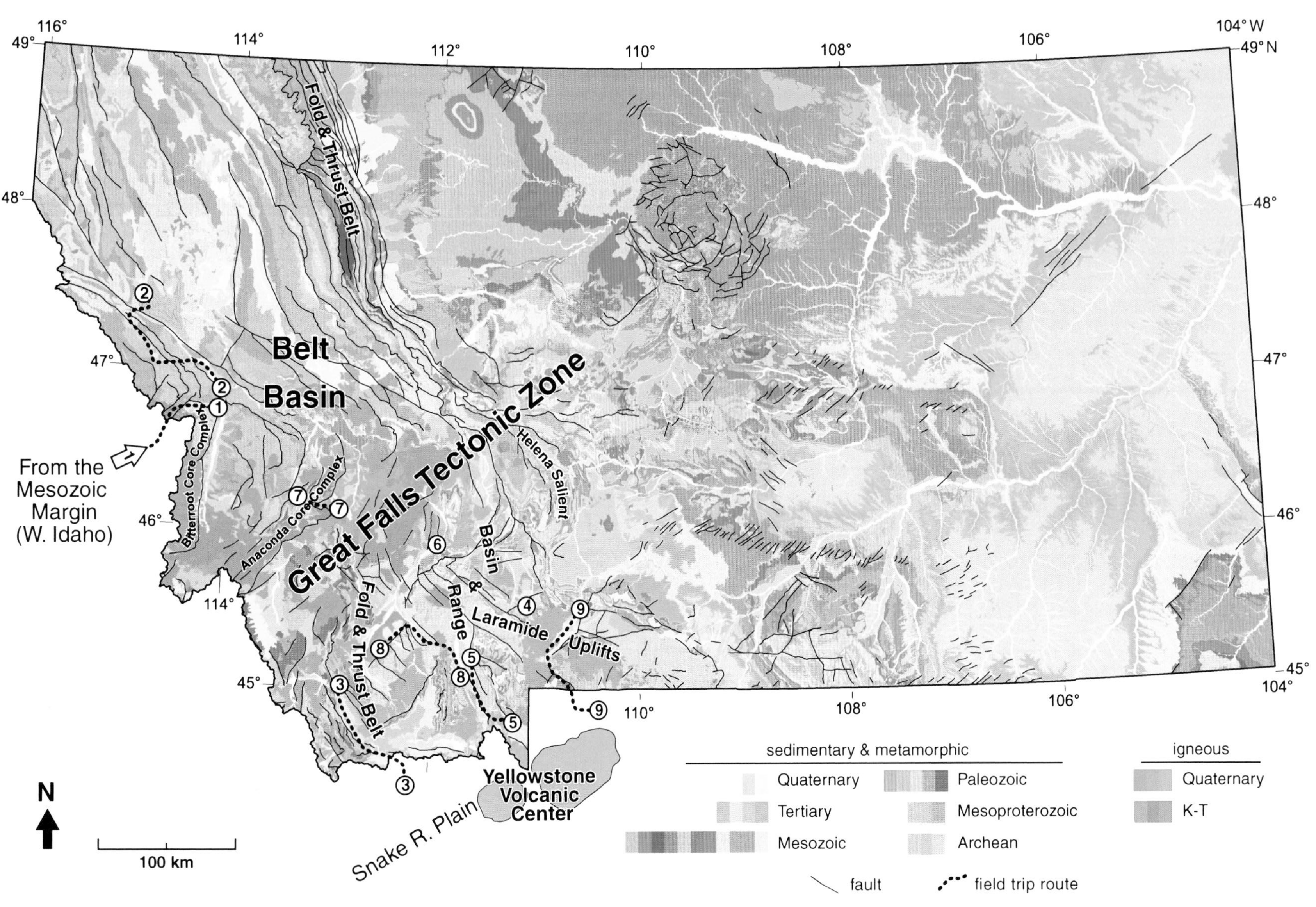

Figure 1. Generalized geologic map of Montana (Vuke et al., 2007) showing approximate routes of field trips. Numbered end-points indicate the chapter number: (1) Lewis et al.—transect across an accretionary margin, (2) Smith and Hanson—sedimentary record of glacial Lake Missoula, (3) Ruleman et al.—neotectonics of Yellowstone Tectonic Parabola, (4) Sears—tracking a big Miocene river, (5) Sugden et al.—Hyalite Canyon soil lithosequence, (6) Oyer et al.—Golden Sunlight Mine, (7) Kalakay et al.—Anaconda core complex, (8) Underwood et al.—talc deposits in SW Montana, and (9) Pierce et al.—Quaternary geology of northern Yellowstone area. Yellow colors indicate Quaternary and Recent deposits, tan and light brown represent Paleogene and Neogene sedimentary strata, shades of green represent Mesozoic sedimentary strata, shades of blue represent Paleozoic strata, brown represents rocks of the Mesoproterozoic Belt Supergroup, and gray represents Archean and Paleoproterozic crystalline basement. Major faults are shown as thin black lines. R. stands for River. Base map modified from Figure 4 of the booklet accompanying the *Geologic Map of Montana* (Vuke et al., 2007).

Mesozoic to Paleogene convergence along the plate margin resulted in continental shortening extending hundreds of kilometers eastward from the plate margin (Fig. 1). Laramide-style basement-cored arches dominate the landscape east of the fold-and-thrust belt as far as central Montana, and the thrust front bulges eastward where tectonic inversion of the Belt basin produced the Helena salient (Harlan et al., 1988; Schmidt et al., 1988). The field trip related to Chapter 6 (Oyer et al.) visits the Golden Sunlight Mine, where gold-silver mineralization was localized in a late Cretaceous latite porphyry breccia pipe that pierces the southern margin of the Helena salient. Late orogenic collapse (Paleogene) of the thrust stack within the fold-and-thrust belt produced the Bitterroot and Anaconda core complexes, the latter of which is the focus of Chapter 7 (Kalakay et al.).

By the middle Miocene, Sevier Laramide shortening had given way to Basin and Range extension in southwestern Montana (Lageson and Stickney, 2000). Classic normal fault range-front morphology marks moderate to steep north-south–trending faults that bound distinctive down-dropped basins like the Gallatin and upper Madison Valleys. Chapter 4 (Sears) examines evidence for a proposed north-draining paleoriver that may have found a path through these basins during the Miocene.

The eastward relative migration of the Yellowstone volcanic center and associated eruption of voluminous basalts on the Snake River Plain brought renewed tectonic and magmatic activity to the region, culminating in three climactic caldera-forming eruptions between ca. 2.1 Ma and 0.64 Ma (Christiansen, 2001; Pierce and Morgan, 1992; Smith and Braile, 1994). Neotectonic and geomorphic effects of Yellowstone volcanism are the focus of Chapter 3 (Ruleman et al.) and, in part, Chapter 9 (Pierce et al.).

Pleistocene glaciers effected the penultimate sculpting of the landscape of the northern Rockies before streams resumed their predominant role upon retreat of the ice. The Cordilleran ice sheet advanced as far as north-central Montana, while alpine glaciers filled the upper reaches of many valleys and icecaps covered high plateaus in the Yellowstone region (Fullerton et al., 2004; Locke, 1990). Chapter 9 (Pierce et al.) includes discussion of glacial features in the upper Yellowstone River drainage. Chapter 2 (Smith and Hanson) showcases geomorphic and sedimentologic evidence for the repeated filling and draining of the famous glacial Lake Missoula (e.g., Pardee, 1942). Chapter 5 (Sugden et al.) examines the influence of bedrock lithology on soil development in the spectacular partly glaciated Hyalite Canyon south of Bozeman.

REFERENCES CITED

Beaver, P.C., 1985, Geology and Mineral Resources of the Tobacco Root Mountains and Adjacent Region: Guidebook of the 10th Annual Tobacco Root Geological Society Field Conference: Bozeman, Montana, Northwest Geology, 87 p.

Childs, J.F., and Lageson, D.R., 1993, Economic and Regional Geology of the Tobacco Root–Boulder Batholith Region, Montana: Northwest Geology, Tobacco Root Geological Society, 129 p.

Christiansen, R.L., 2001, The Quaternary and Pliocene Yellowstone Plateau Volcanic Field of Wyoming, Idaho, and Montana: U.S. Geological Survey Professional Paper, 145 p.

Czamanske, G.K., and Zientek, M.L., 1985, The Stillwater Complex, Montana; Geology and Guide: Montana Bureau of Mines and Geology Special Publication 92, 396 p.

Dickinson, W.R., 2004, Evolution of the North American Cordillera: Annual Review of Earth and Planetary Sciences, v. 32, p. 13–45, doi:10.1146/annurev.earth.32.101802.120257.

Fournier, R.O., Christiansen, R.L., Hutchinson, R.A., and Pierce, K.L., 1994, A Field-Trip Guide to Yellowstone National Park, Wyoming, Montana, and Idaho; Volcanic, Hydrothermal, and Glacial Activity in the Region: U.S. Geological Survey Bulletin, 46 p.

Fullerton, D.S., Colton, R.B., Bush, C.A., and Straub, A.W., 2004, Map Showing Spatial and Temporal Relations of Mountain and Continental Glaciations on the Northern Plains, Primarily in Northern Montana and Northwestern North Dakota: U.S. Geological Survey Scientific Investigations Map 2843, scale 1:1,000,000, 1 sheet.

Harlan, S.S., Geissman, J.W., Lageson, D.R., and Snee, L.W., 1988, Paleomagnetic and isotopic dating of thrust-belt deformation along the eastern edge of the Helena salient, northern Crazy Mountains Basin, Montana: Geological Society of America Bulletin, v. 100, p. 492–499, doi:10.1130/0016-7606(1988)100<0492:PAIDOT>2.3.CO;2.

Harrison, J.E., 1972, Precambrian Belt basin of Northwestern United States: Its geometry, sedimentation, and copper occurrences: Geological Society of America Bulletin, v. 83, p. 1215–1240, doi:10.1130/0016-7606(1972)83[1215:PBBONU]2.0.CO;2.

Houston, R.S., ed., and 11 others, 1993, The Wyoming province, *in* Reed, J.C., Jr., Bickford, M.E., Houston, R.S., Link, P.K., Rankin, D.W., Sims, P.K., and Van Schmus, W.R., eds., Precambrian: Conterminous U.S.: Boulder, Colorado, Geological Society of America, North American Geology, v. C-2, p. 121–170.

Kalakay, T.J., Foster, D.A., and Lonn, J.D., 2014, this volume, Polyphase collapse of the Cordilleran hinterland: The Anaconda metamorphic core complex of western Montana—The Snoke symposium field trip, *in* Shaw, C.A., and Tikoff, B., eds., Exploring the Northern Rocky Mountains: Geological Society of America Field Guide 37, p. 145–159, doi:10.1130/2014.0037(07).

Lageson, D.R., and Stickney, M.C., 2000, Seismotectonics of Northwest Montana, *in* Schalla, R.A., and Johnson, E.H., eds., Montana/Alberta Thrust Belt and Adjacent Foreland, Volume I: Billings, Montana Geological Society, p. 109–126.

Lewis, R.S., Schmidt, K.L., Gaschnig, R.M., LaMaskin, T.A., Lund, K., Gray, K.D., Tikoff, B., Stetson-Lee, T., and Moore, N., 2014, this volume, Hells Canyon to the

Bitterroot front: A transect from the accretionary margin eastward across the Idaho batholith, *in* Shaw, C.A., and Tikoff, B., eds., Exploring the Northern Rocky Mountains: Geological Society of America Field Guide 37, p. 1–50, doi:10.1130/2014.0037(01).

Lewis, S.E., and Berg, R.B., 1988, Precambrian and Mesozoic Plate Margins: Montana Bureau of Mines and Geology Special Publication 96, 195 p.

Locke, W.W., 1990, Late Pleistocene glaciers and the Climate of western Montana, U.S.A.: Arctic and Alpine Research, v. 22, p. 1–13, doi:10.2307/1551716.

Mogk, D.W., Mueller, P.A., and Wooden, J.L., 1988, Archean tectonics of the North Snowy Block, Beartooth Mountains, Montana: The Journal of Geology, v. 96, p. 125–141, doi:10.1086/629205.

Mueller, P.A., and Frost, C.D., 2006, The Wyoming Province: A distinctive Archean craton in Laurentian North America: Canadian Journal of Earth Sciences, v. 43, p. 1391–1397, doi:10.1139/e06-075.

Mueller, P.A., Wooden, J.L., Nutman, A.P., and Mogk, D.W., 1998, Early Archean crust in the northern Wyoming province: Evidence from U–Pb ages of detrital zircons: Precambrian Research, v. 91, p. 295–307, doi:10.1016/S0301-9268(98)00055-2.

O'Neill, J.M., and Lopez, D.A., 1985, Character and regional significance of Great Falls tectonic zone, east-central Idaho and west-central Montana: American Association of Petroleum Geologists Bulletin, v. 69, p. 437–447.

Oyer, N., Childs, J., and Mahoney, J.B., 2014, this volume, Regional setting and deposit geology of the Golden Sunlight Mine: An example of responsible resource extraction, *in* Shaw, C.A., and Tikoff, B., eds., Exploring the Northern Rocky Mountains: Geological Society of America Field Guide 37, p. 115–144, doi:10.1130/2014.0037(06).

Pardee, J.T., 1942, Unusual currents in glacial Lake Missoula, Montana: Geological Society of America Bulletin, v. 53, p. 1569–1599.

Pierce, K.L., and Morgan, L.A., 1992, The track of the Yellowstone hot spot; volcanism, faulting, and uplift, *in* Link, P.K., Kuntz, M.A., and Platt, L.B., eds., Regional Geology of Eastern Idaho and Western Wyoming, Geological Society of America Memoir 179, p. 1–53.

Pierce, K.L., Licciardi, J.M., Krause, T.R., and Whitlock, C., 2014, this volume, Glacial and Quaternary geology of the northern Yellowstone area, Montana and Wyoming, *in* Shaw, C.A., and Tikoff, B., eds., Exploring the Northern Rocky Mountains: Geological Society of America Field Guide 37, p. 189–203, doi:10.1130/2014.0037(09).

Ross, C.P., Skipp, B.A., and Rezak, R., 1963, The Belt Series in Montana, with a Geologic Map and a Section on Paleontologic Criteria: U.S. Geological Survey Professional Paper 346, 122 p.

Ross, G.M., and Villeneuve, M., 2003, Provenance of the Mesoproterozoic (1.45 Ga) Belt basin (western North America): Another piece in the pre-Rodinia paleogeographic puzzle: Geological Society of America Bulletin, v. 115, p. 1191–1217, doi:10.1130/B25209.1.

Ruleman, C.A., Larsen, M., and Stickney, M.C., 2014, this volume, Neotectonics and geomorphic evolution of the northwestern arm of the Yellowstone Tectonic Parabola: Controls on intra-cratonic extensional regimes, southwest Montana, *in* Shaw, C.A., and Tikoff, B., eds., Exploring the Northern Rocky Mountains: Geological Society of America Field Guide 37, p. 65–87, doi:10.1130/2014.0037(03).

Schmidt, C.J., O'Neill, J.M., and Brandon, W.C., 1988, Influence of Rocky Mountain foreland uplifts on the development of the frontal fold and thrust belt, southwestern Montana: Geological Society of America, v. 171, p. 171–201, doi:10.1130/MEM171-p171.

Sears, J.W., 2014, this volume, Tracking a big Miocene river across the Continental Divide at Monida Pass, Montana/Idaho, *in* Shaw, C.A., and Tikoff, B., eds., Exploring the Northern Rocky Mountains: Geological Society of America Field Guide 37, p. 89–99, doi:10.1130/2014.0037(04).

Smith, L.N., and Hanson, M.A., 2014, this volume, Sedimentary record of glacial Lake Missoula along the Clark Fork River from deep to shallow positions in the former lakes: St. Regis to near Drummond, Montana, *in* Shaw, C.A., and Tikoff, B., eds., Exploring the Northern Rocky Mountains: Geological Society of America Field Guide 37, p. 51–63, doi:10.1130/2014.0037(02).

Smith, R.B., and Braile, L.W., 1994, The Yellowstone hotspot: Journal of Volcanology and Geothermal Research, v. 61, p. 121–187, doi:10.1016/0377-0273(94)90002-7.

Sugden, J.C., Hartshorn, A.S., Dixon, J.L., and Montagne, C., 2014, this volume, A slice through time: A Hyalite Canyon soil lithosequence, *in* Shaw, C.A., and Tikoff, B., eds., Exploring the Northern Rocky Mountains: Geological Society of America Field Guide 37, p. 101–114, doi:10.1130/2014.0037(05).

Underwood, S.J., Childs, J.F., Walby, C.P., Lynn, H.B., Wall, Z.S., Cerino, M.T., and Bartlett, E., 2014, this volume, The Yellowstone and Regal talc mines and their geologic setting in southwestern Montana, *in* Shaw, C.A., and Tikoff, B., eds., Exploring the Northern Rocky Mountains: Geological Society of America Field Guide 37, p. 161–187, doi:10.1130/2014.0037(08).

Vuke, S.M., Porter, K.W., Lonn, J.D., and Lopez, D.A., 2007, Geologic Map of Montana: Montana Bureau of Mines and Geology, GM-62, scale 1:500,000, 2 sheets.

The Geological Society of America
Field Guide 37
2014

Hells Canyon to the Bitterroot front: A transect from the accretionary margin eastward across the Idaho batholith

Reed S. Lewis*
Idaho Geological Survey, University of Idaho, 875 Perimeter Drive, MS3014, Moscow, Idaho 83844, USA

Keegan L. Schmidt*
Division of Natural Science, Lewis-Clark State College, 500 8th Avenue, Lewiston, Idaho 83501, USA

Richard M. Gaschnig*
Department of Geology, University of Maryland, College Park, Maryland 20742, USA

Todd A. LaMaskin*
Department of Geography and Geology, University of North Carolina Wilmington, 601 South College Road, Wilmington, North Carolina 28403, USA

Karen Lund*
U.S. Geological Survey, Box 25046, Denver Federal Center, Denver, Colorado 80225, USA

Keith D. Gray*
Wichita State University, Department of Geology, 1845 Fairmount Street, Wichita, Kansas 67260, USA

Basil Tikoff*
Tor Stetson-Lee*
Department of Geoscience, University of Wisconsin, 1215 W. Dayton Street, Madison, Wisconsin 53706, USA

Nicholas Moore*
Department of Geography and Geology, University of North Carolina Wilmington, 601 South College Road, Wilmington, North Carolina 28403, USA

ABSTRACT

This field guide covers geology across north-central Idaho from the Snake River in the west across the Bitterroot Mountains to the east to near Missoula, Montana. The regional geology includes a much-modified Mesozoic accretionary boundary along the western side of Idaho across which allochthonous Permian to Cretaceous arc complexes of the Blue Mountains province to the west are juxtaposed against

*E-mails: reedl@uidaho.edu; klschmidt@lcsc.edu; gaschnig@umd.edu; lamaskint@uncw.edu; klund@usgs.gov; k.gray@wichita.edu; basil@geology.wisc.edu; stetsonlee@wisc.edu; nom2943@uncw.edu.

Lewis, R.S., Schmidt, K.L., Gaschnig, R.M., LaMaskin, T.A., Lund, K., Gray, K.D., Tikoff, B., Stetson-Lee, T., and Moore, N., 2014, Hells Canyon to the Bitterroot front: A transect from the accretionary margin eastward across the Idaho batholith, *in* Shaw, C.A., and Tikoff, B., eds., Exploring the Northern Rocky Mountains: Geological Society of America Field Guide 37, p. 1–50, doi:10.1130/2014.0037(01). For permission to copy, contact editing@geosociety.org.

autochthonous Mesoproterozoic and Neoproterozoic North American metasedimentary assemblages intruded by Cretaceous and Paleogene plutons to the east. The accretionary boundary turns sharply near Orofino, Idaho, from north-trending in the south to west-trending, forming the Syringa embayment, then disappears westward under Miocene cover rocks of the Columbia River Basalt Group. The Coolwater culmination east of the Syringa embayment exposes allochthonous rocks well east of an ideal steep suture. North and east of it is the Bitterroot lobe of the Idaho batholith, which intruded Precambrian continental crust in the Cretaceous and Paleocene to form one of the classical North American Cordilleran batholiths. Eocene Challis plutons, products of the Tertiary western U.S. ignimbrite flare-up, intrude those batholith rocks. This guide describes the geology in three separate road logs: (1) The Wallowa terrane of the Blue Mountains province from White Bird, Idaho, west into Hells Canyon and faults that complicate the story; (2) the Mesozoic accretionary boundary from White Bird to the South Fork Clearwater River east of Grangeville and then north to Kooskia, Idaho; and (3) the bend in the accretionary boundary, the Coolwater culmination, and the Bitterroot lobe of the Idaho batholith along Highway 12 east from near Lewiston, Idaho, to Lolo, Montana.

INTRODUCTION

This guide is intended to provide an overview of the geology from the Snake River in westernmost Idaho eastward across the Bitterroot lobe of the Idaho batholith to near Missoula, Montana. It consists of three separate road logs: (1) White Bird, Idaho, west into Hells Canyon; (2) White Bird to the South Fork Clearwater River east of Grangeville and then north to Kooskia, Idaho; and (3) Highway 12 from near Lewiston, Idaho, east to Lolo, Montana. The first of these was compiled by LaMaskin, Schmidt, and Moore. The second road log was compiled by Schmidt and Lewis. The Highway 12 portion of this guide was modified from an unpublished road log by Reed S. Lewis, Thomas P. Frost, Keegan L. Schmidt, Robert J. Fleck, and Peter B. Larson, which was originally produced for a Goldschmidt Conference premeeting field trip in May 2005. Contributors for this 2014 version of the Highway 12 road log were Lewis, Lund, Gaschnig, Gray, Tikoff, and Stetson-Lee.

As we view the geologic past, we will pass points of historical interest related to the mythology and history of the Nez Perce Indians, the Lewis and Clark expedition of 1805–1806, and the Nez Perce War of 1877. Notes regarding these historical features are found throughout the guide.

REGIONAL GEOLOGY

Island-Arc Terranes and the Arc-Continent Boundary

The close affinity of rocks in west-central Idaho to those of modern volcanic arcs was suggested by Hamilton (1963), who compared the chemical composition of volcanic rocks in the Riggins region with those from the Aleutian islands of Alaska. With the advent of plate tectonic theory, multiple terranes representing late Paleozoic to Mesozoic island-arc complexes accreted to the ancestral Precambrian core of North America (Laurentia) were recognized in western Idaho, northeastern Oregon, and southeastern Washington (Figs. 1 and 2; Jones et al., 1977; Brooks and Vallier, 1978; Davis et al., 1978; Dickinson, 1979). In a classic Sr-isotope and geochronologic study, Armstrong et al. (1977) demonstrated that plutons intruded into these arc terranes are isotopically distinct from those intruded into Laurentian crust. These terranes, which comprise the Blue Mountains province (Silberling et al., 1987), consist of island-arcs and basins that formed both in proximity to the Laurentian margin (Olds Ferry, Izee, and Baker terranes) and in a distal intra-oceanic setting (Wallowa terrane; Dickinson, 1979; Hillhouse et al., 1982; Stanley and Whalen, 1989; Vallier, 1995; LaMaskin et al., 2008, 2011; Schwartz et al., 2010).

Polyphase structural analyses in northeastern Oregon (Avé Lallemant, 1995) document Late Jurassic amalgamation of the Wallowa terrane to the other terranes in the Blue Mountain province prior to Early Cretaceous (?) collision with the Laurentian margin. Local structural geometries and age data, in combination with plate motion data, are interpreted by some to indicate dextral transpressive motion between arc-related and Laurentian blocks at regional and local scales (Lund and Snee, 1988; Tikoff et al., 2001; Giorgis and Tikoff, 2004). Others favor normal convergence on the basis of Late Jurassic to Early Cretaceous (ca. 150–125 Ma) plate motions (Engebretson et al., 1985; Blake et al., 1988) and high-strain linear-planar fabrics, which in the Riggins region show no evidence of transpressional strain (Gray et al., 2012; Gray, 2013). Figure 3 summarizes major Late Jurassic to Eocene tectonic and magmatic events in the region.

Road Log 1 focuses on the Wallowa terrane exposed in Hells Canyon. Collectively, rocks of the Wallowa terrane record four major stages of development. Least understood is the formation of Wallowa terrane basement, which consists of Permian intrusive rocks and subordinate Permian or Pennsylvanian amphibolitic gneiss, and includes the Cougar Creek complex at Pittsburg Landing (Walker, 1986; Vallier, 1995; Kurz et al., 2012).

Subsequent Permian to Late Triassic arc-related magmatism and sedimentation are recorded by the Seven Devils Group (Vallier, 1977). The Seven Devils Group includes felsic and mafic volcanogenic rocks of the Permian Windy Ridge and Hunsaker Creek formations, overlying subaqueous mafic volcanic rocks of the Triassic Wild Sheep Creek Formation, and partly coeval volcaniclastic submarine-fan deposits of the Doyle Creek Formation (Nolf, 1966; Goldstrand, 1994; Vallier, 1998). These units record deposition in a forearc and (or) intra-arc basin during Wallowa-arc magmatism. Younger Late Triassic to Early Jurassic carbonate platform-to-basin sedimentation and possible synorogenic basin formation is deduced from the Martin Bridge and Hurwal formations (Follo, 1994; Vallier, 1977; Dorsey and LaMaskin, 2007). A major regional unconformity bevels the Seven Devils Group, Martin Bridge, and Hurwal formations, evidence of Late Triassic and Early Jurassic syn-magmatic uplift and tilting of the Wallowa terrane (Avé Lallemant et al., 1985; Dorsey and LaMaskin, 2007, 2008; Kurz et al., 2012). Finally, Upper Jurassic to Earliest Cretaceous basin filling is represented by the Coon Hollow Formation (Morrison, 1963; Vallier, 1977, 1995; Goldstrand, 1994; LaMaskin et al., 2008).

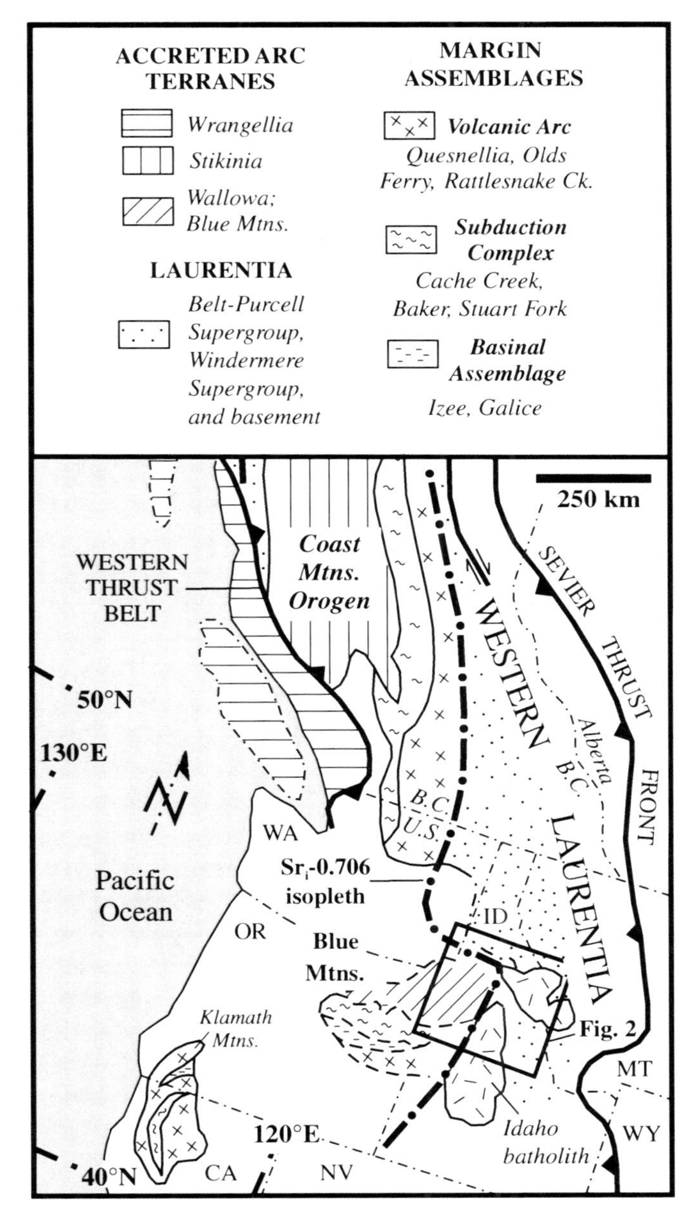

Figure 1. Generalized tectonic map of the North American Cordillera showing allochthonous arc terranes and Laurentian margin. Modified from McClelland et al. (2000) and Gray et al. (2012). B.C.—British Columbia; CA—California; ID—Idaho; MT—Montana; NV—Nevada; OR—Oregon; WA—Washington; WY—Wyoming.

East of the Wallowa terrane proper are poorly dated amphibolite-facies metasedimentary assemblages of presumed late Paleozoic or Mesozoic age that include the Pollock Mountain amphibolite south of Riggins, Idaho (Selverstone et al., 1992; Manduca et al., 1993; Blake et al., 2009), the Riggins Group to the north (Hamilton, 1963; Onasch, 1987; Lund et al., 1993; Lund, 1995), and the Orofino series farther to the north near Orofino (Anderson, 1930; Hietanen, 1962; Snee et al., 2007). Most authors agree that at least some of the assemblages may be part of the Wallowa terrane but they may also include other terranes and (or) back-arc basin deposits. Eastern parts of these assemblages are intruded by magmatic epidote-bearing plutons emplaced at ~8–11 kbar (Manduca et al., 1992; Selverstone et al., 1992; Getty et al., 1993) and dated by U-Pb zircon methods at ca. 120–110 Ma (Manduca et al., 1993; Lee, 2004; McClelland and Oldow, 2007; Unruh et al., 2008). Paucity of detrital zircon makes these assemblages difficult targets for U-Pb analysis and numerous attempts have failed. However, recent work has produced results from a few analyses. Lund et al. (2008) reported detrital zircon results from a paragneiss (Swiftwater gneiss) in the Coolwater Ridge structural culmination on the Middle Fork of the Clearwater drainage basin, which they interpreted as indicating a Late Cretaceous protolith that received detritus of both Wallowa arc and Laurentian origin. Farther south along the South Fork of the Clearwater River and Slate Creek, two samples of Riggins Group paragneiss produced similar detrital zircon age spectra with single, nearly identical peaks at ca. 200 Ma and no grains older than Permian (Schmidt et al., 2013). These authors proposed that the Riggins Group likely originated as part of the Wallowa terrane or a related basinal assemblage.

The boundary between island-arc terranes and Laurentia, which has been termed the Salmon River suture zone (Lund and Snee, 1988), Salmon River suture (Lund et al., 2008), or western Idaho suture zone (Strayer et al., 1989; Fleck, 1990; Fleck and Criss, 2007), extends from southwestern Idaho north to the Orofino area where it turns westward into Washington state (Figs. 1 and 2). In this field guide, we employ the term "arc-continent boundary" to describe the location of the change in country rocks from arc derived in the west to Laurentian crust in the east. In the Owyhee Mountains in southwestern Idaho and north near McCall, Idaho, much of the arc-continent boundary is obscured by tabular plutons with ages of ca. 118–90 Ma (Lund and Snee,

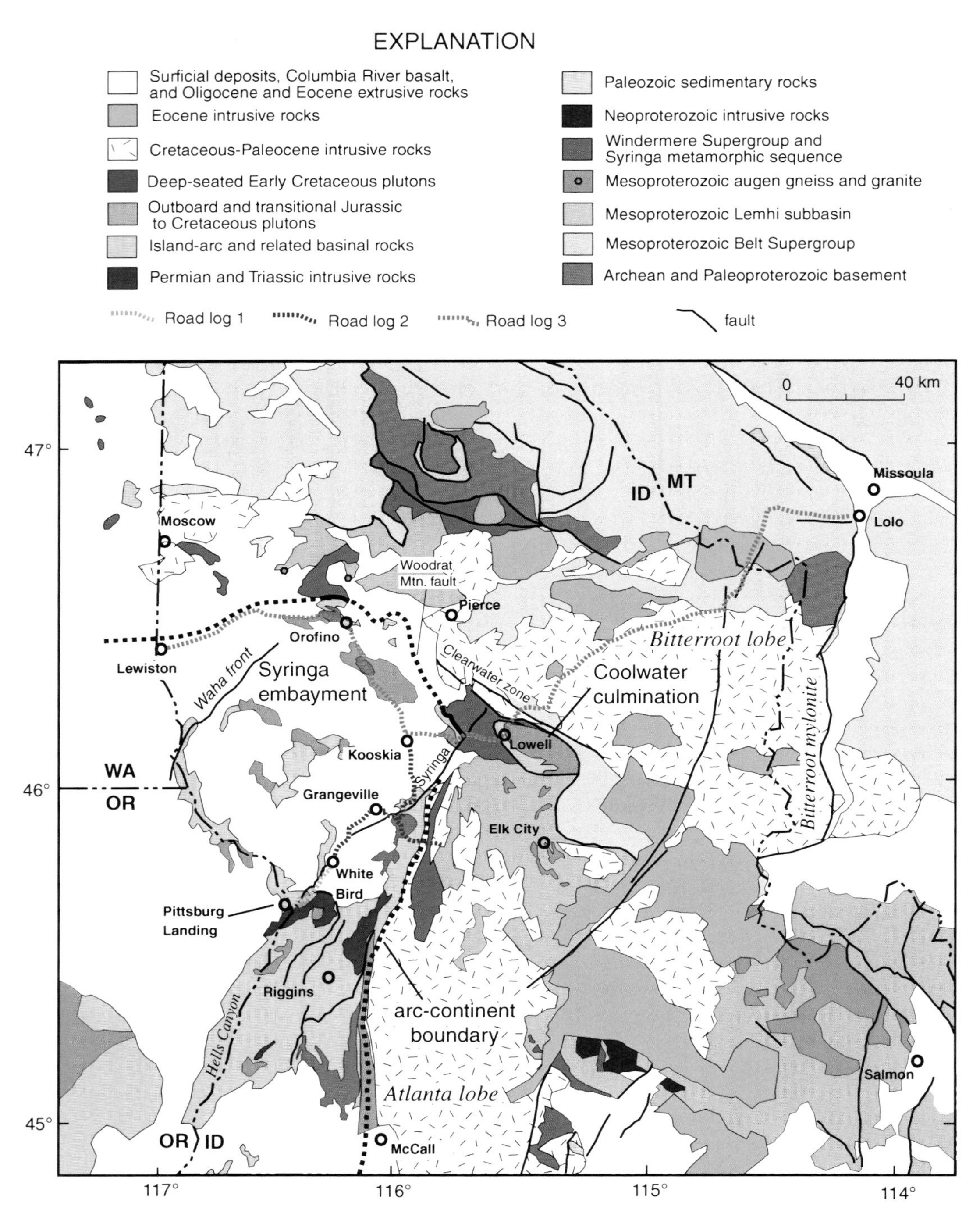

Figure 2. Geologic map of north-central Idaho and vicinity showing field trip routes, the arc-continent boundary (bold dashed line), the Coolwater culmination, and the southern (Atlanta) and northern (Bitterroot) lobes of the Idaho batholith. Modified from Lewis et al. (2012). ID—Idaho; MT—Montana; OR—Oregon; WA—Washington.

1988; Manduca et al., 1993; Snee et al., 1995; Lund, 2004; Giorgis et al., 2008; Benford et al., 2010). A relatively narrow zone (10–15 km) of ductile deformation, the western Idaho shear zone (McClelland et al., 2000), is superposed on these plutons and minor zones of country rocks. Within the zone, pervasive fabrics are characterized by steeply east-southeast–dipping foliation and down-dip–plunging lineation (Lund and Snee, 1988; Manduca et al., 1993; Braudy, 2013), although reconstructing Miocene-present normal fault movement restores the fabrics to a subvertical orientation (Tikoff et al., 2001; Giorgis et al., 2006). Workers

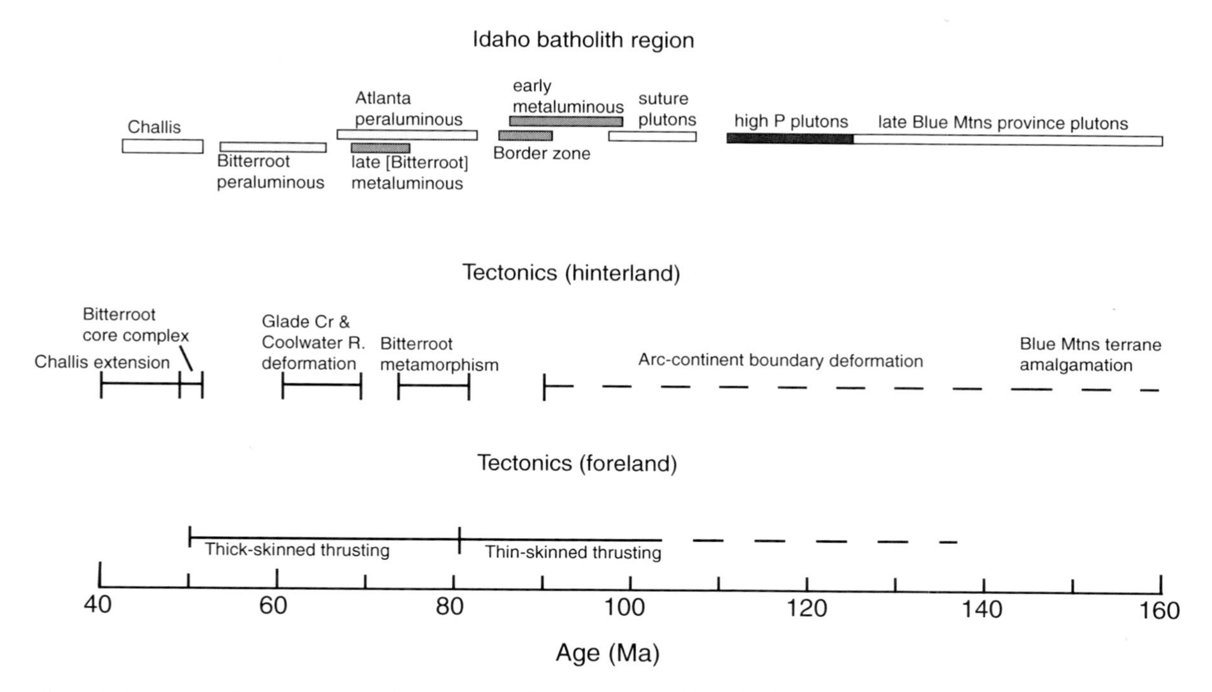

Figure 3. Timeline of magmatism and major tectonic events in the Idaho batholith region and Wallowa terrane from the Late Jurassic through the Eocene. Modified from Gaschnig et al. (2010).

in the McCall area and southward attribute these fabrics to dextral transpression, based on consistent rotational fabrics indicating dextral shear (Giorgis and Tikoff, 2004; Braudy, 2013).

From the Riggins area north to the South Fork of the Clearwater River, Late Cretaceous plutons along the arc-continent boundary contain northerly striking gneissic foliation and moderate to steeply plunging mineral stretching lineation (Myers, 1982; Lund, 1995; Lund et al., 1993). Fabric orientations in adjacent units (layered oceanic and continental rocks and outlying plutons) are typically more varied, but still exhibit high strains that extend >15 km away from the boundary (Aliberti, 1988; Gray, 2013). From west-to-east across the boundary, strain markers show no marked asymmetry (e.g., triaxially deformed lithic clasts, mantled porphyroclasts, boudinaged felsic veins) and kinematic indicators are rare. However, inclusion trail patterns in syntectonic garnet of the Riggins Group and Pollock Mountain amphibolite support top-to-the-west shear (McKay et al., 2011). Some workers argue that the western Idaho shear zone extends north from McCall into this area as a spatially overlapping but temporally distinct orogenic belt (McClelland et al., 2000; Giorgis et al., 2008; Blake et al., 2009). Others favor a protracted contractional history whereby regional synmetamorphic structures (Gray and Oldow, 2005) accumulated over a pre-118 Ma to ca. 92 Ma interval without an overprinting orogen-scale ductile shear zone (Gray et al., 2012). East of Grangeville, the arc-continent boundary is offset by the Mount Idaho shear zone and the Syringa fault (Fig. 2) in a right-lateral sense. The plutons east of Grangeville, and the structures west of them, are described in Road Log 2.

Syringa Embayment and Coolwater Culmination

East of Kooskia, the trend of the arc-continent boundary changes from north to northwest, then changes again farther north near Orofino to west (Fig. 2), forming a "corner" (Armstrong et al., 1977; Fleck and Criss, 2007; Mohl and Thiessen, 1995) referred to as the Syringa embayment (Lund et al, 2008). Proposed origins for the bend in the boundary include Mesozoic rifting (Davis et al., 1978), oroclinal bending (Schmidt et al., 2003, 2009c), northeast-directed subduction (Strayer et al., 1989), truncation of the north-south trend by sinistral strike-slip faulting on a northwest trend (McClelland and Oldow, 2004, 2007; Snee et al., 2007), and a restraining bend at the northern end of the dextral transpressional arc-continent boundary to the south (Lund et al., 2008). Recent geologic mapping around the embayment (Lewis et al., 2005a, 2007a) shows that the tectonostratigraphic assemblages and plutonic belts that are apparent along the north-trending segment of the arc-continent boundary to the south continue around the Syringa embayment as far as basement exposures can be traced before they disappear completely beneath Columbia River Basalt Group cover rocks to the west in Washington. The embayment is bordered to the north and northeast by a complex structural belt consisting of numerous high-strain zones. The older deformation, metamorphism, and rock cooling history of this belt relate closely to the structural and metamorphic history to the south, but the Syringa embayment also contains a much younger deformational and cooling history that apparently has no counterpart to the south. This complex history is explored further on field trip 3.

The northwest-trending segment of the Syringa embayment includes the Woodrat Mountain fault (Fig. 2), a steeply east-northeast–dipping mylonitic shear zone that juxtaposes the arc-related Orofino series and continental Neoproterozoic Syringa metamorphic sequence. Outboard and southwest of the Woodrat Mountain fault is a broad mylonitic shear zone with northeast-dipping foliation and steeply pitching lineation developed in gneissic rocks of the Orofino metamorphic series and plutons that intrude it. The southwestern margin of this shear zone is termed the "Ahsahka thrust," which displays top-to-southwest shear sense (Strayer et al., 1989; Davidson, 1990; Snee et al., 2007). Both of these major structures can be traced around the bend at Orofino and project to the west where they are covered by strata of the Columbia River Basalt Group. Whether or not the bend is a primary feature or reflects later folding or faulting is not agreed upon, but will be discussed further on field trip 3.

Inboard and northeast of the Woodrat Mountain fault along the northwest-trending segment of the Syringa embayment are several northwest-striking faults and shear zones of considerably younger age than the outboard structures described above. These shear zones are part of a strong northwest-southeast structural grain that is present along the southwest edge of the Bitterroot lobe of the Idaho batholith and which deform ca. 73 Ma plutonic rocks in the westernmost exposures of the Bitterroot lobe (McClelland and Oldow, 2007). This deformation is less well developed (or perhaps absent?) in the ca. 66–54 Ma main body of the Bitterroot lobe. This northwest grain continues northwest to the area southwest of Pierce, where rocks are covered by Columbia River basalt. Yates (1968) recognized this pronounced northwest trend, which he termed the "trans-Idaho discontinuity." It has been suggested that the trans-Idaho discontinuity reactivated a Neoproterozoic transform fault of the nascent Cordilleran margin (Dickinson, 2004). More recent work has shown that west-northwest–striking faults such as the Glade Creek fault and Brown Creek Ridge shear zone may be segments of a regional possibly left-lateral strike-slip structure called the "Clearwater zone" by Sims et al. (2005) and Lund et al. (2008). The Clearwater zone is interpreted as a transpressional Late Cretaceous to Paleocene structure formed in response to east-west stress in the Syringa embayment (Lund et al., 2008), possibly remobilizing a structure from the Neoproterozoic Laurentian rift margin (Lund et al., 2010). Alternatively, the north-striking arc-continent boundary and western Idaho shear zone may have been truncated and offset by younger structures of the Orofino shear zone (McClelland and Oldow, 2004, 2007).

Within Laurentian rocks well east of the bend from north-northeast–striking to northwest-striking arc-continent boundary, the antiformal Coolwater culmination (Fig. 2; Lund et al., 2008) exposes Cretaceous paragneiss derived from multiple sources, including arc terranes and Precambrian Laurentia. Lund et al. (2008) proposed that paragneiss with a Cretaceous depositional age, and the ca. 86 Ma Coolwater orthogneiss that intruded it, formed a wedge of oceanic rocks that were inserted into the Laurentian margin between 98 and 73 Ma (based on ages of deformed plutons) and that the wedge split supracrustal Laurentian rocks from their basement. Crustal thickening, melting, and intrusion within the wedge, and subsequent folding to form the Coolwater culmination, continued until ca. 61 Ma (Lund et al., 2008). This folded crustal wedge is a major anomaly along the arc-continent boundary that, for 400 km to the south, manifests as a steeply east-dipping to vertical feature. The model proposed by Lund et al. (2008) envisions that during dextral oblique translation and clockwise rotation against Laurentia, the Blue Mountains province impinged into a restraining bend of the Syringa embayment. In this model, ensuing orthogonal contraction in the embayment produced a crustal wedge of oceanic rocks that delaminated Laurentian crust.

Recent work on the Ahsahka thrust along the east-west segment of the Syringa embayment indicates that it is similar in age to the north-south–striking western Idaho shear zone near McCall. Stetson-Lee et al. (2013) described reverse-sinistral transpressive kinematics preserved along this segment of the Ahsahka thrust and suggested that the Syringa embayment is an orogenic syntaxis, a feature created by Late Cretaceous plate motion toward the apex of the bend at Orofino. Paleomagnetic work on three ca. 90 Ma plutons intruded along the east-west–oriented arc-continent boundary west of Orofino indicates that all of these plutons experienced ~30° of clockwise rotation since cooling through their blocking temperatures between 90 and 84 Ma as constrained by hornblende $^{40}Ar/^{39}Ar$ closure ages (Byerly et al., 2013). The similarity of these paleomagnetic results with those from the McCall area suggests that the entire region of present western Idaho, including the arc-continent boundary and Blue Mountains province, has rotated ~30° clockwise since the Late Cretaceous.

Continental North America (Laurentia)

All Precambrian rocks immediately north and west of the Bitterroot lobe of the Idaho batholith have undergone amphibolites-facies metamorphism (e.g., Hietanen, 1962), rendering correlation with low-grade Precambrian strata in the region difficult. Peak conditions of 6–8 kb and 650–750 °C reported from north of the Bitterroot lobe (House et al., 1997; Foster et al., 2001) are probably in part a result of thrust stacking in the hinterland of the Sevier thrust belt, but may also record Mesoproterozoic and Neoproterozoic metamorphism documented northwest of the Bitterroot lobe (Zirakparvar et al., 2010; Nesheim et al., 2012).

Our present understanding is that the package of continental rocks decreases in age from the Mesoproterozoic Belt Supergroup exposed near Missoula (and equivalent rocks to the south such as around Elk City; Fig. 2), to Neoproterozoic Windermere-equivalent rocks west of Elk City (Syringa metamorphic sequence). Detrital zircon analyses help distinguish these two major rock groups (Lund et al., 2008; Lewis et al., 2007c, 2010), and indicate that although discontinuously exposed, Neoproterozoic metasedimentary rocks wrap around the bend in the boundary near Orofino (Fig. 2). A distinguishing constituent of the Neoproterozoic strata is feldspar-poor quartzite that contrasts with the feldspathic quartzite of the Belt Supergroup.

Idaho Batholith

The Idaho batholith, one of the largest of the Cordilleran batholiths, is well exposed in the steep canyons of northern Idaho. The Idaho batholith region is one of the first areas in which Rb/Sr isotopic zonation was shown to correspond to basement boundaries (Armstrong et al., 1977; Fleck and Criss, 1985; Criss and Fleck, 1987). The batholith was also the target of pioneering work in oxygen isotope systematics in granitic terranes (Taylor and Margaritz, 1978; Criss and Taylor, 1983). Whole-rock geochemical work has delineated several geochemical suites within the batholith and Challis plutons (Hyndman, 1984; Lewis and Kiilsgaard, 1991; Kiilsgaard et al., 2001; Lewis and Frost, 2005). Nevertheless, mapping of much of the northern (Bitterroot) lobe has only been at the reconnaissance level, so the ages and distribution of many intrusive units are poorly known. U-Pb dating has been hampered by inherited Proterozoic zircons (Fig. 4; Gaschnig et al., 2013) as well as inherited zircons from earlier intrusive phases of the batholith (Gaschnig et al., 2010), but with the advent of in situ geochronology (Fig. 5), a better chronologic understanding is developing (Foster and Fanning, 1997; Lund et al., 2008; Gaschnig et al., 2010, 2011).

The batholith proper formed during multiple stages (Fig. 3). The first included the intrusion of metaluminous tonalitic and megacryst-bearing plutons ca. 100–85 Ma, mostly in south-central Idaho, but including megacrystic phases on the west side of the Bitterroot lobe. This was followed by a voluminous pulse of peraluminous granite to granodiorite ca. 80–67 Ma and formed the southern (Atlanta) lobe of the batholith (Gaschnig et al., 2010). The metaluminous rocks contain a juvenile mantle-derived component and likely formed in a continental arc setting, whereas the peraluminous rocks reflect a shift to crustal melting driven by regional crustal thickening (Gaschnig et al., 2011). Formation of the main mass of the Bitterroot lobe began ca. 70 Ma with intrusion of metaluminous granodiorite and tonalite plutons and was followed by voluminous peraluminous granodiorite and granite ca. 66–54 Ma (Chase et al., 1978; Toth and Stacey, 1992; Foster and Fanning, 1997; Gaschnig et al., 2010). The large contribution of crustal melting, localized distribution, and younger age of Bitterroot lobe magmatism suggests a possible link to the focused compression and deformation responsible for the formation of the Coolwater culmination (Gaschnig et al., 2011).

Igneous rocks of the Blue Mountains have juvenile isotopic characteristics approaching that of mid-ocean ridge basalt, which reflects the province's oceanic island-arc heritage (Kurz, 2010; Schwartz et al., 2011). In contrast, Idaho batholith and Challis rocks all have considerably more evolved isotopic signatures (Fig. 6), reflecting the involvement of Precambrian continental crust in their genesis (e.g., Fleck and Criss, 1985; Manduca et al., 1992; Gaschnig et al., 2011). Both the Atlanta and Bitterroot peraluminous suites are predominantly products of crustal melting, although different crustal lithologies and (or) terranes were involved, which is reflected in their differing Nd isotopic compositions (Gaschnig et al., 2011). Challis plutons differ north and south of the latitude of Riggins based on Pb isotopic distinctions. The highly varied Sr-Nd isotopic compositions of the Challis plutons likely reflect mixing between mantle input and multiple crustal components.

Challis Magmatic Event and Bitterroot Lobe Extension

The Challis magmatic event, part of the Tertiary western U.S. ignimbrite flare-up, ensued during Eocene time (ca. 51–43 Ma) across much of northern and central Idaho when granodiorite and granite plutons and their associated volcanics were emplaced (Gaschnig et al., 2010). Northwest-southeast extension resulted in northeast-striking dike swarms and normal faults, and the formation of the Bitterroot mylonite zone (Fig. 2), a moderately east-dipping

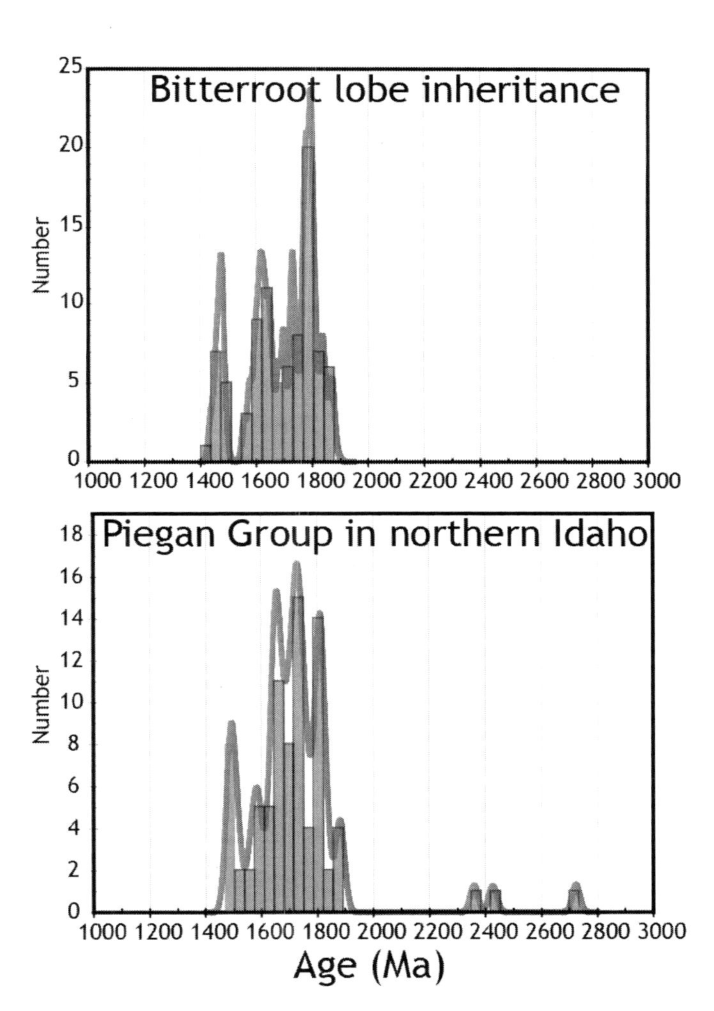

Figure 4. Histograms comparing the age distribution of inherited Precambrian zircons in the Bitterroot lobe (Gaschnig et al., 2013) to the detrital zircon distribution of the Piegan Group of the Belt Supergroup (Lewis et al., 2007c). The similarity strongly suggests that Belt sediments were incorporated into Bitterroot magmas.

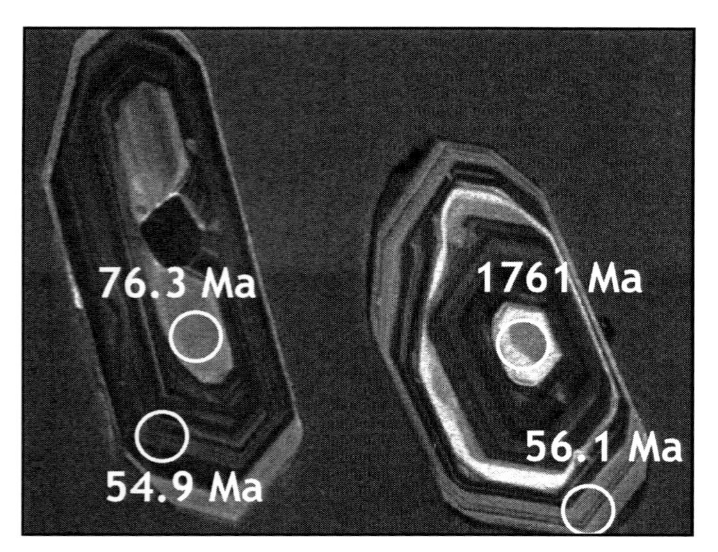

Figure 5. Scanning electron microscope (SEM)-cathodoluminescence image of Idaho batholith zircons, after Gaschnig et al. (2010). Circles are 30 µm spots analyzed for LA-ICP-MS (laser-ablation inductively coupled plasma-mass spectrometry) U-Pb geochronology, with ages given. This is a Bitterroot peraluminous suite sample from the crest of the Bitterroot Range on the Idaho-Montana border. The rim ages are among the youngest in the Bitterroot peraluminous suite, but the presence of both Cretaceous and Proterozoic inherited cores is typical of the suite's zircons.

north-south zone of extension with lineations that trend 100–110° and which began developing ca. 53 Ma (Hyndman, 1980; Foster and Fanning, 1997; Foster et al., 2001, 2007; House et al., 2002). Right-lateral strike-slip movement along west-northwest–striking fault zones aided in transfer of the extensional motion across the region (Doughty and Sheriff, 1992; Lewis et al., 2007b).

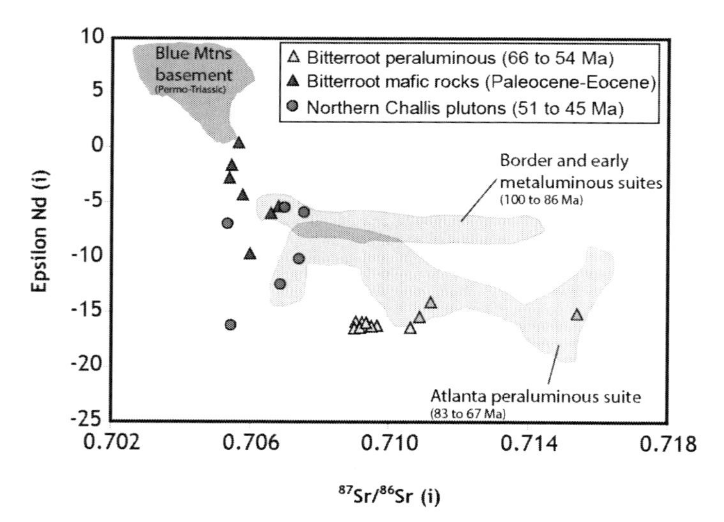

Figure 6. Sr and Nd isotopic ratios of some of the igneous rock units encountered on the trip. Blue Mountains field is from Kurz (2010); the Atlanta lobe field and other data are from Gaschnig et al. (2011).

Miocene and Younger Structures

In the Miocene, the eruption of the Columbia River flood basalts from north-northwest–trending dikes inundated much of the Pacific Northwest (e.g., Hooper and Swanson, 1990). In west-central Idaho, the resulting basalt sequence covers large sections of the Wallowa terrane (Swanson et al., 1979a, 1979b; Camp, 1981), particularly in the area of the Syringa embayment (Fig. 2). During early stages, the Imnaha and Grande Ronde units were emplaced at very high eruption fluxes and west-central Idaho was inundated by lava that rapidly constructed a plateau on an erosional surface that had considerable topographic relief.

Several large-scale structural features were active during basalt flow emplacement and were strongly controlled by older structures in underlying basement rocks. The Lewiston Hill structure immediately north of Lewiston is delineated by the arc-continent boundary line on Figure 2 and consists of south-verging folds and thrust faults along the westward projection of the arc-continent boundary near Orofino (Kauffman et al., 2009). The folding first occurred during emplacement of the Grande Ronde units (Alloway et al., 2013; Reidel et al., 2013). The southern side of the east-northeast–trending Lewiston basin is the Waha front, a structural zone of northeast-trending folds and steeply dipping faults that were also active during extrusion of the Grande Ronde flows (Reidel et al., 1992) and are centered on an older northeast-striking reverse fault in basement rocks that probably also accommodated dextral strike-slip displacement (Limekiln structure; Kauffman et al., 2009). Farther south near Grangeville, the northeast-striking Mount Idaho fault accommodated reverse faulting and folding along the trace of an older northeast-striking reverse shear zone that also accommodated dextral strike-slip faulting in basement rocks (Mount Idaho shear zone; Schmidt et al., 2007). South of White Bird Hill, a series of north-striking normal faults related to Neogene Basin and Range extension follows the Salmon River corridor. These faults are strongly controlled by basement structures that are associated with the arc-continent boundary (Giorgis et al., 2006). Most bend into and merge with the Mount Idaho fault, but some continue northward as north-striking faults onto the Camas Prairie.

ROAD LOG 1—WHITE BIRD–PITTSBURG LANDING (HELLS CANYON)

This road log starts near White Bird, Idaho, along the Salmon River. The route crosses over the high divide to the west into the Pittsburg Landing area along the Snake River in Hells Canyon (Fig. 7). There, Triassic and Jurassic sub-greenschist grade volcanic and sedimentary rocks have been overthrust by mid-Permian to Late Triassic partially migmatized crystalline basement rocks of the Cougar Creek complex along the northeast-striking brittle-ductile Klopton Creek fault (Fig. 8; White and Vallier, 1994; White, 1994; Kurz et al., 2012). The Klopton Creek fault is a 100–200-m-wide zone of mylonite that displays consistent northwest-vergent thrust shear sense indicators. Kurz

(2001) determined a dextral-oblique thrust fault mechanism from fracture pattern analysis of brittle structures that overprint the mylonite along the fault. Timing of faulting is bracketed by the Late Jurassic age of the youngest deformed rocks and undeformed Miocene basalt that overlies the fault. Oldest rocks in the footwall are mostly andesite to basalt of the Middle (?) to Late Triassic Wild Sheep Creek Formation. Late Triassic rocks of the Doyle Creek Formation that are coeval with the Wild Sheep Creek Formation are represented by quartzolithic grainstones, *Halobia*-ammonoid mudstone and/or wackestone, and calcareous quartzolithic sandstone of the "Kurry unit" (Figs. 8, 9; White and Vallier, 1994; LaMaskin et al., 2008; Schmidt et al., 2009a). The Kurry unit is overlain by the informally named "red tuff unit" (White and Vallier, 1994) along a slightly discordant unconformity (Figs. 10A, 10B; White and Vallier, 1994; Schmidt et al., 2009a). The red tuff unit is dominated by pink lithic sandstone and monomict volcanic conglomerate; welded tuff is present in its upper parts. U-Pb zircon geochronology of the uppermost welded tuff yields a weighted mean $^{206}Pb/^{238}U$ age of 196.82 ± 0.06 Ma (Tumpane, 2010; Northrup et al., 2011). This is an Early Jurassic (Sinemurian) age and is considerably older than the previously accepted Middle Jurassic (Bajocian) age (White and Vallier, 1994).

The Coon Hollow Formation is a fluvial-to-marine transgressive sequence that overlies the red tuff on an unconformity with slight discordance (Figs. 8, 9, 10C, 10D, 10E). The lower fluvial unit consists of polymict (predominantly but not exclusively volcanic and volcaniclastic) brown conglomerate and distinctive yellow-weathering, plant-fossil–bearing mudstone. Black shale and mudstone of the marine unit gradationally overlie fluvial deposits and contain ammonoid, pelecypod, and gastropod fossils. An important structure, the Kurry Creek fault places older strata of the Coon Hollow Formation over younger Coon Hollow strata (White and Vallier, 1994; Schmidt et al., 2009a). An additional unit, the "Oregon turbidite unit" of the Coon Hollow Formation (White and Vallier, 1994), structurally overlies the fluvial Coon Hollow unit above a southeast-dipping thrust fault on the Oregon side of the Snake River (Schmidt et al., 2009a). These rocks may be age-correlative with distinctive chert-bearing rocks in the upper plate of the Kurry Creek fault in Idaho at Stop 1-4B (Fig. 10F; cf. Schmidt et al., 2009a). The Coon Hollow Formation has been assigned to the Wallowa terrane and is therefore considered allochthonous relative to North America (Vallier, 1977, 1995). It is typically interpreted to have been deposited during a Middle–early Late Jurassic interval (Bajocian–Oxfordian; ca. 170.3 ± 1.4 to 157.3 ± 1.0 Ma) on the basis of sparse marine fauna at disparate locations (Morrison, 1963; Stanley and Beauvais, 1990; White and Vallier, 1994). More recent U-Pb zircon geochronology in the Pittsburg Landing area from a hornblende-phyric lithic lapilli tuff located ~40 m above the base of the fluvial unit indicates a maximum depositional age of 159.62 ± 0.10 Ma for the Coon Hollow Formation (Tumpane, 2010; Northrup et al., 2011).

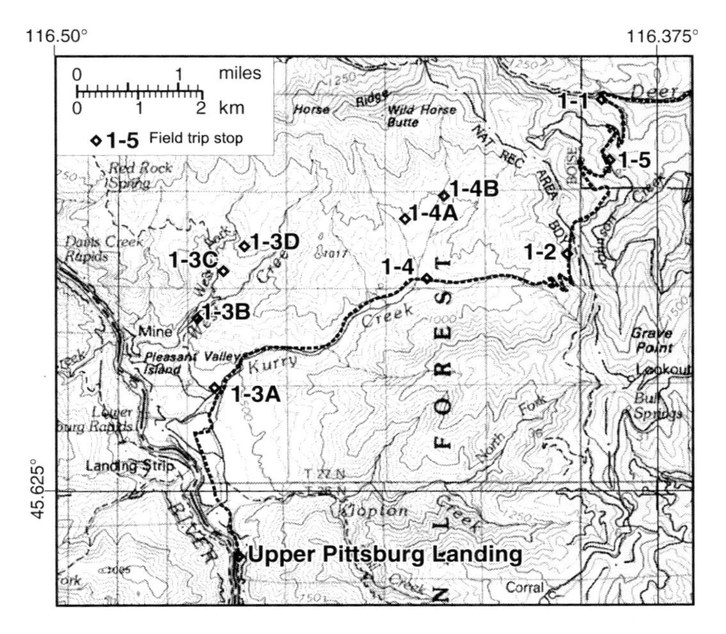

Figure 7. Field trip stops for Road Log 1 in and east of the Pittsburg Landing area.

The Klopton Creek fault can be traced northeast across the major ridge into the Salmon River canyon into the Hammer Creek fault (Garwood et al., 2008; Schmidt et al., 2009a, 2009b). Even farther northeast, the Klopton–Hammer Creek fault system appears to join the Mount Idaho shear zone and Syringa fault (Fig. 2), which offsets the arc-continent boundary and associated fault-bounded assemblages in a right-lateral sense as detailed in Day 2 of this guide. The Hammer Creek and Mount Idaho structures have been reactivated as multiple Neogene normal faults that cut the Miocene Columbia River basalt sequence along the north-south–trending zone of extension that follows the Salmon River.

In the following log, "Mile" represents road mileage from turnoff for Hammer Creek Recreation area on Highway 95, 1.3 miles south of White Bird, Idaho. See Figure 7 for locations of field trip stops, which are also given below in decimal degrees using WGS84 datum. Figure 8 shows the geology of the area in and near Pittsburg Landing west of Stop 1-2.

Mile 0.0: U.S. Highway 95, at intersection with Old Highway 95, signed as the turnoff for Hammer Creek Recreation Area, 1.3 miles south of White Bird, Idaho. Turn off Highway 95 (right if southbound; left if northbound) onto Old Highway 95 (also known as River Road). Reset odometer to zero and proceed northwest along Old Highway 95.

Mile 0.9: Turn left onto Doumecq Road and cross bridge over Salmon River.

Mile 1.0: Turn left onto Deer Creek Road (USFS Road 493). Mile markers along Deer Creek Road are zeroed at this intersection and, thus, are one mile less than mileage given in this road log.

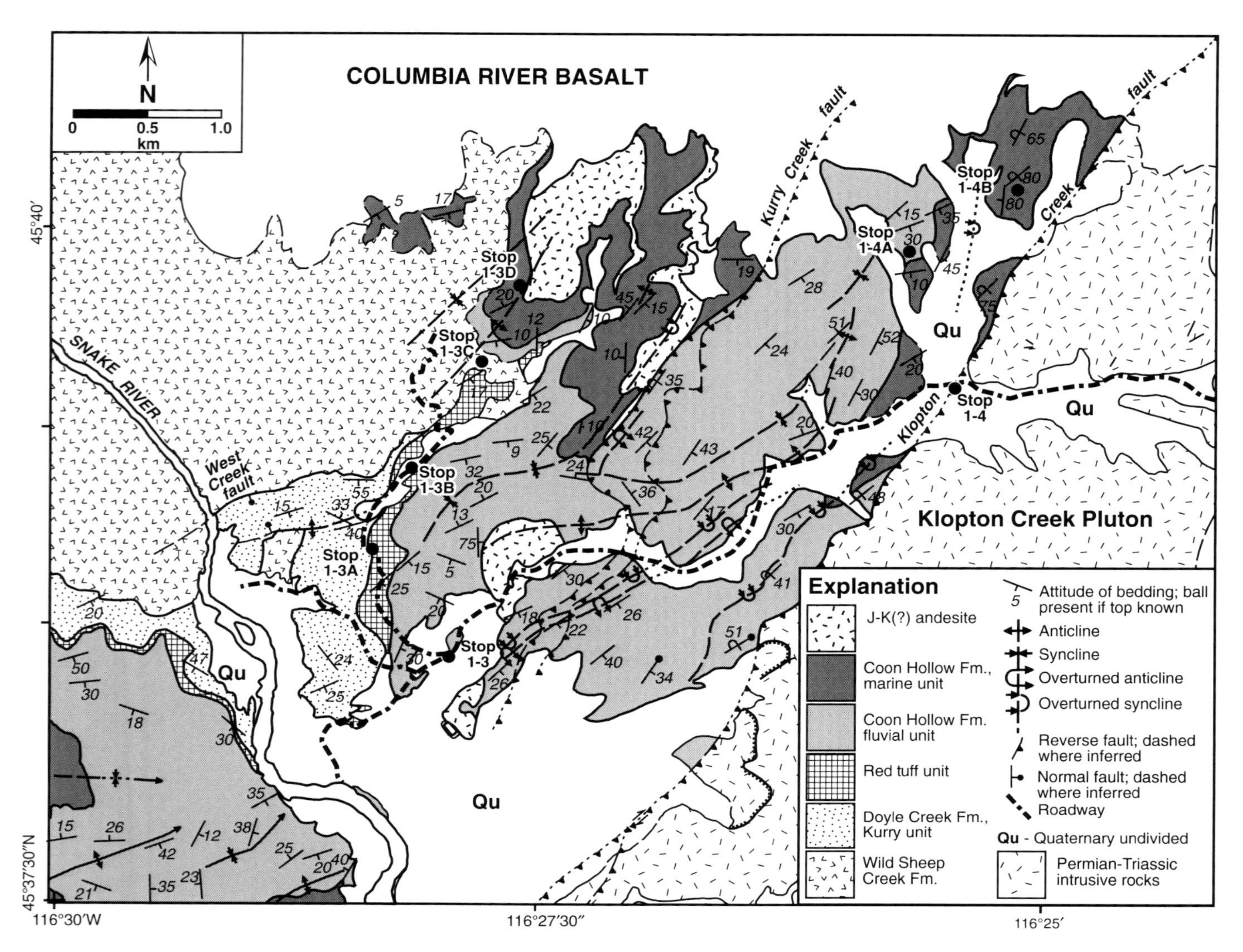

Figure 8. Geologic map of the Pittsburg Landing area showing field trip stops. Modified from Schmidt et al. (2009a).

Mile 2.7: Silty sedimentary interbed in Columbia River Basalt Group on right.

Mile 2.9: Unconsolidated river terrace gravels of the Salmon River 620 feet above present river elevation.

Mile 3.1: Columbia River basalt and sedimentary units in roadcut on right. Roadcuts from here to mile 8.3 were mapped as the Deer Creek sediments (Coffin, 1967). Sediments consist largely of silt, sand, pebbles, and ash, above Grande Ronde R2 and below the basalt of Grangeville (part of Saddle Mountains Basalt of Columbia River Basalt Group; Schmidt et al., 2009b). Most of this sequence is strongly deformed by normal faulting, fracturing, and mass wasting.

Mile 4.9: Sheared and brecciated Columbia River basalt exposed in roadcuts and on the quarry face on right mark the trace of a major north-northeast–striking, steeply east-southeast–dipping normal fault that projects through the notch in ridge to north.

Mile 5.6: Deer Creek maintenance shop on left. The road to Pittsburg Landing is kept open year-round, mainly for river access to the boat ramp at Lower Pittsburg Landing.

Mile 6.2: Poe Creek.

Mile 7.6: Harry Robinson Ranch on left.

Mile 8.0: Junction with Cow Creek Saddle Road (USFS Road 672).

Mile 8.9: First switchback. Outcrops and roadcut on the right consist mostly of polymict conglomerate interbedded with sandstone and black shale, mapped as part of the marine unit of the Coon Hollow Formation (Schmidt et al., 2009a) in the footwall block of the Klopton Creek fault.

Mile 9.0: Pullout on left.

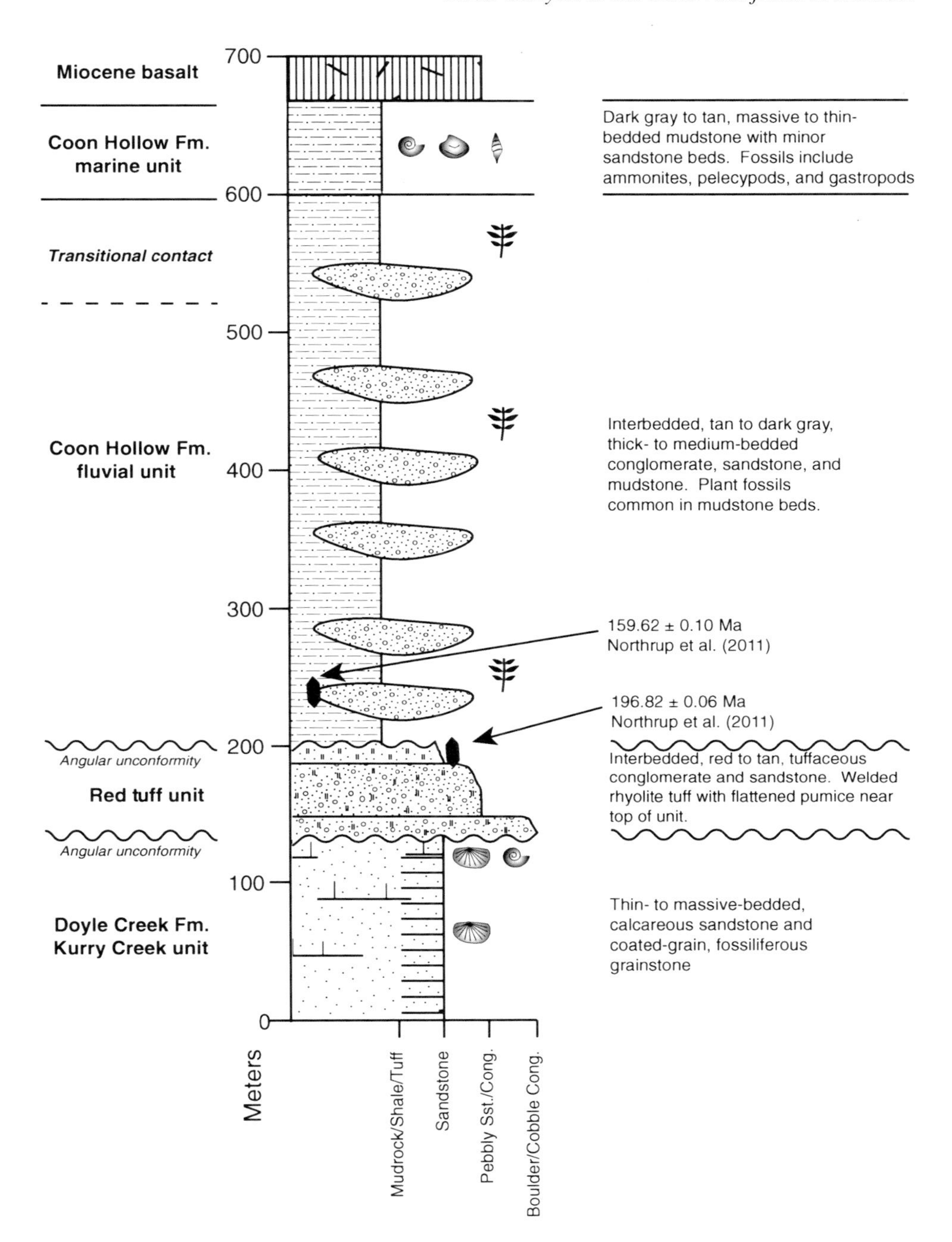

Figure 9. Generalized stratigraphic column of rock units in the Pittsburg Landing area.

STOP 1-1: Klopton Creek Fault (45.6838 °N, 116.3865 °W)

This is the only roadcut exposure of the Klopton Creek fault in the area, and a poor one at that. The deeply weathered roadcut across from the pullout consists of hornblende diorite and hornblende biotite quartz diorite of the Permian–Triassic Cougar Creek complex that have experienced both ductile and brittle shearing in the hanging wall (Schmidt et al., 2009a). Farther down the road to the north are outcrops of black shale, sandstone, and conglomerate of the Middle to Late Jurassic Coon Hollow Formation passed at mile 8.9. Southwest from here, the Klopton Creek fault continues beneath undeformed Miocene Columbia River Basalt Group on the ridge above Pittsburg Landing. To the northeast, it becomes the Hammer Creek thrust fault in the Salmon River valley west of White Bird. In that area, Late Triassic to Early Jurassic volcanic and sedimentary rocks of the red tuff unit make up the footwall assemblage and the hanging wall is mostly hornblende diorite.

Mile 11.0: Junction with USFS Road 420 to left. Stay to the right, following signs to Pittsburg Landing.

Figure 10. Field photographs of lithofacies within the Pittsburg Landing area. (A) Nested fluvial channels in the red tuff unit. (B) Close-up view of granule–cobble conglomerate in the red tuff unit. (C) Field view of Coon Hollow Formation, Jcc unit. Fluvial channel composed of yellow-brown polymict conglomerate is cut into yellow mudstone containing plant fossils. (D) Typical plant fossils found in the Coon Hollow Formation. This is *Phlebopteris tracyi*, a fern-like foliage (Ash, 1991). (E) Ripple-cross–laminated sandstone from the Coon Hollow Formation Jcc unit. Note outsized clast to the right. White lines denote truncated ripple-cross–laminated bed-sets. (F) Chert-clast granule conglomerate from the Coon Hollow Formation Jcsm (?) unit in the upper plate of the Kurry Creek fault.

Mile 12.1: Pullout on right at overlook with picnic bench and signage.

STOP 1-2: Overview of Pittsburg Landing (45.6606 °N, 116.3938 °W)

This location affords a spectacular view into Hells Canyon (Fig. 11A). Pittsburg Landing is one of the few places where the confining walls of Hells Canyon recede (a result of the presence of soft sedimentary rocks) and the canyon broadens to permit road access to the Snake River. The high skyline is dominated by well-bedded, brown weathering Miocene basalt that forms an ~1-km-thick regional cover. At this stop, we are standing on the Imnaha Basalt, near the base of the Columbia River Basalt Group. In most of Hells Canyon, the Snake River has eroded through the basalt into volcanogenic rocks of the Blue Mountains province. Most of the valley on the Idaho side of the river is underlain by Mid–Late Triassic to Late Jurassic–Early Cretaceous volcanic and sedimentary rocks of the Wallowa terrane. The trace of the Klopton Creek fault trends northwest from Upper Pittsburg Landing at the southern end of the subdued topography across the low hills below us and through a saddle just north of this

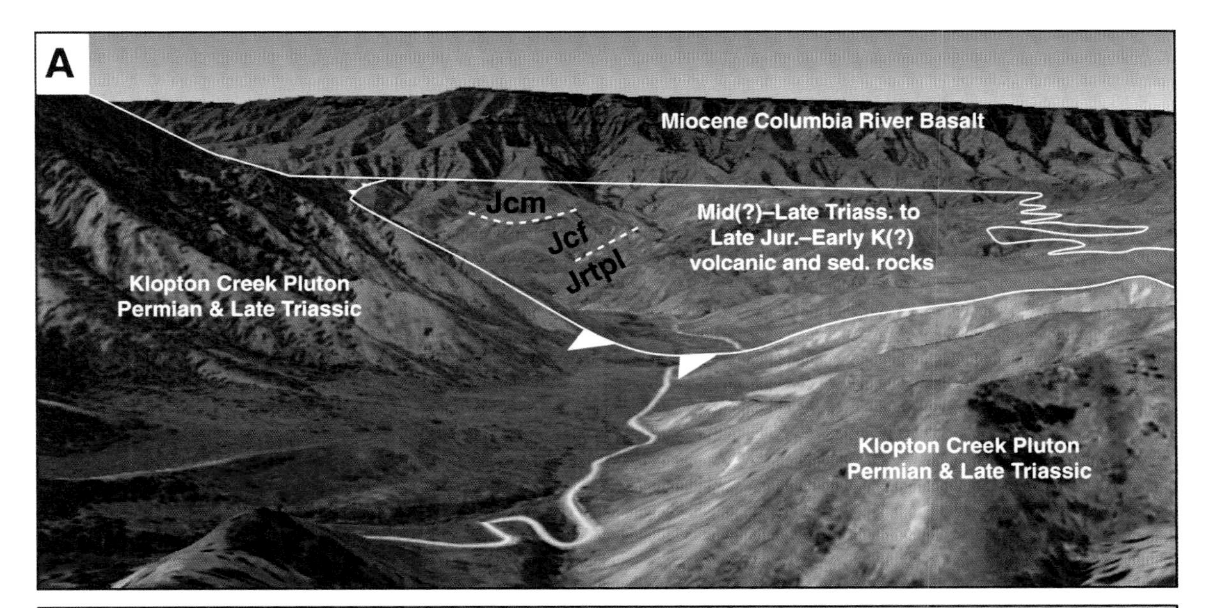

Figure 11. Interpreted views of the distribution of geologic units in the Pittsburg Landing area. (A) Google Earth® image from Stop 1-2. Viewer is on the upper plate of the Klopton Creek fault looking west. Permian to Late Triassic intrusive rocks are thrust over younger Triassic–Jurassic and Cretaceous (?) dominantly sedimentary rocks. The high skyline is dominated by volcanic flow deposits of the Columbia River Basalt Group. (B) View from Stop 1-4A. Viewer is on the lower plate of the Klopton Creek fault looking southeast. Here, Permian to Late Triassic intrusive rocks are thrust over younger Late Jurassic to Early Cretaceous (?) rocks of the Coon Hollow Formation. Exposures of lithologies in the footwall of the fault are dominated by chert-grain sandstone and conglomerate. In the immediate foreground, a transitional contact between the fluvial Jcc and marine Jcsm unit is well exposed.

stop (Schmidt et al., 2009a). A major splay of the Klopton Creek fault, the Kurry Creek fault, cuts across the middle of the low hills and repeats footwall rock of the Klopton Creek fault. Across the Snake River in Oregon, either this structure or a splay of it thrusts the late Middle Jurassic "Oregon turbidite unit" (White and Vallier, 1994) over the Coon Hollow fluvial unit. Quaternary cover along the river makes cross-river correlations difficult. Directly below us, hanging-wall rocks consist of Permian and Triassic quartz diorite and diorite plutonic rocks, and chlorite-epidote schist and gneiss wall rock screens.

Mile 13.4: At sharp left-hand turn in road, sheared diorite and quartz diorite in the hanging wall of Klopton Creek fault.

Mile 14.7: Crossing of Klopton Creek fault covered by alluvium. Rocks from here to next stop are mostly sedimentary sequence in footwall (Fig. 8).

Mile 16.4: Crossing Kurry Creek fault. Trace of the fault continues northeast from here. An upright anticline on nearby hill just above Kurry Creek alluvium is in the hanging wall.

Mile 17.0: Rock wall on right is porphyritic andesite stock of Jurassic or Cretaceous age that intruded fluvial Coon Hollow sequence (Fig. 8). The andesite contains plagioclase and minor hornblende phenocrysts in a strongly recrystallized chlorite and epidote groundmass. These intrusions are restricted to the footwall sequence of the Kurry Creek fault.

Mile 17.4: Pullout on right at junction with BLM four-wheel drive road and park your vehicle. Note that the four-wheel drive road is drivable when it is dry, but it becomes a slippery mess when wet.

STOP 1-3: Complex Relationships among Triassic and Jurassic Volcanic and Sedimentary Assemblages in the Footwall of the Kurry Creek Fault (45.6405 °N, 116.4667 °W)

Hike 1.0 mile north along road into West Creek, the next major drainage (Fig. 7). We are in float of the Kurry unit of the Doyle Creek Formation of Vallier (1977) as we begin this hike, and then pass back and forth across an unconformity between underlying Late Triassic calcareous sandstone and fossiliferous limestone (*Halobia*- and ammonoid-bearing) of the Kurry unit and overlying Early Jurassic volcanic, volcaniclastic, and epiclastic rocks of the "red tuff" of White and Vallier (1994). Stop at channel in red tuff (same as Northrup et al., 2011, their stop 2.3) where the small road makes a sharp left-hand bend.

Stop 1-3A: Red Tuff Unit, Triassic and Jurassic Volcanic and Sedimentary Sequences in West Creek, Lower Plate of Kurry Creek Fault (45.6471 °N, 116.4735 °W)

Here are excellent exposures of the basal and middle portions of the red tuff unit. Rocks at this location represent a largely reworked volcaniclastic succession. Lithologies vary from medium-grained lithic arenite to boulder conglomerate. Large fluvial channel features are visible and cross-stratification features are ubiquitous. Welded tuff units are accessed by climbing the ridgeline to the east. The red tuff unit is a volumetrically small portion of the Pittsburg Landing section, but Early Jurassic (Sinemurian) red volcanic-volcaniclastic rocks are dramatically thicker to the north-northeast in the Hammer Creek area of the Salmon River (Garwood et al., 2008).

Paleomagnetic results from seven sites in these welded tuff outcrops (R.F. Burmester, 2013, written commun.) yield a tilt-corrected mean direction that is similar in inclination to but 5–25° clockwise of that expected for the 200 Ma paleomagnetic field for this location calculated from the J1 cusp of North America's (Euler) polar wander path (Beck and Housen, 2003). A straightforward interpretation is that the red tuff unit records no net change in latitude, but 15° more clockwise than counterclockwise rotation, relative to North America since then.

The contact with the overlying Coon Hollow Formation is placed differently by different workers. LaMaskin and Moore suggest here that the contact is marked by the first appearance of characteristic yellow mudstone and sandstone. Previous work by Schmidt et al. (2009a) and Northrup et al. (2011) placed the contact at the base of the first conglomerate interval. Another possibility is that a third, previously unrecognized unit is present between the red tuff and the Coon Hollow Formation.

Continue walking north along four-wheel drive road to left-hand corner.

Stop 1-3B: View of Contact between Wild Sheep Creek Formation and Younger Units, Triassic and Jurassic Volcanic and Sedimentary Sequences in West Creek, Lower Plate of Kurry Creek Fault (45.6509 °N, 116.4709°W)

We are standing in the middle of the red tuff unit. Looking directly to the west, we see the contact between craggy, high-relief outcrops of Wild Sheep Creek Formation to the north, and low hills of the Kurry unit to the south. The contact between the Wild Sheep Creek Formation and overlying units has been a point of controversy since it was first recognized by Vallier (1974, 1977). White's (1994) map shows the contact as the normal-sense West Creek fault cutting his lower and upper fluvial units (JTRsv, Jcc, Jcms of Schmidt et al., 2009a; red tuff and fluvial-to-marine units here). His "schematic diagram," however, shows his lower and upper fluvial units above an angular unconformity. White and Vallier (1994) interpreted the contact as a normal fault beginning in Oregon and continuing northeast across the Snake River into Idaho, terminating ~1.9 km to the east of the Snake River in an area south of a prominent conical hill where massive beds of Wild Sheep Creek Formation seem clearly to extend across the proposed fault. This "problem area" seems to have confounded interpretations in the area for many mappers, and investigation of these rocks is a major objective at the next stop.

Walk ~0.4 miles north along the road to a sharp bend at 45.6545 °N, 116.4662 °W. Leave the road and climb to the low ridgeline. Walk ~0.25 miles to the next stop.

Stop 1-3C: Contact between Wild Sheep Creek Formation and Younger Units (45.6579 °N, 116.4651 °W)

We are at the problem area (Figs. 12A, 12B). Observe massive beds of Wild Sheep Creek Formation apparently extending across the proposed fault, and black, vitreous rocks of possible pseudotachylyte origin. Schmidt et al. (2009a) interpreted the contact to be a long-lived buttress unconformity and show the West Creek fault as a small structure ending just ~0.55 km (0.34 mi) to the east of the Snake River (Fig. 8). Thus, portions of the Kurry unit and all of the overlying Mesozoic units are interpreted to onlap a depositional substrate of Wild Sheep Creek Formation. Key to this interpretation, White and Vallier (1994) and Schmidt et al. (2009a) observed massively bedded, wedge-shaped, 1–3-m-thick breccia and underlying soft-sediment deformation of the Kurry unit, and interpreted these as channel deposits.

Recent mapping by Moore and LaMaskin (Fig. 12B) suggests that the West Creek fault may be a structure with considerable offset and a long strike length, extending ~4.0 km (2.5 mi) to the east of the Snake River until being covered by Miocene basalt. In this interpretation, the following units are truncated by the West Creek fault: (1) Wild Sheep Creek (Footwall block), (2) Kurry unit of Doyle Creek Formation, (3) red tuff, (4) Coon Hollow Formation, fluvial unit, and (5) Coon Hollow Formation, marine unit. In this interpretation, the massive breccias of White and Vallier (1994) and Schmidt et al. (2009a) may be syn-depositional slide blocks generated along an active West Creek fault, modern mass-wasting slide blocks, or complicated post-depositional, fault-generated structures, among other interpretations.

Assuming the West Creek fault is a large structure, it could be either a normal fault with Wild Sheep Creek Formation rocks in the footwall and younger units in the hanging wall, or a reverse fault with top-to-the-northwest motion. Moore and LaMaskin suggest that the problem area is either a complex horst structure, or a small structural window. In addition, the lithologies north and south of the problem area were mapped by Schmidt et al. (2009a) within the red tuff unit; however, another interpretation is that yellow-brown, polymict conglomerate and yellow mudstone here are part of the Coon Hollow Formation. A specific goal of this field trip is to make further observations about this complex map area.

Walk north, 0.1 miles to black shale exposures on the west, south, and east flanks of a prominent conical hill.

Stop 1-3D: Late Jurassic Marine Unit, Coon Hollow Formation, Lower Plate of Kurry Creek Fault (45.6616 °N, 116.4610 °W)

Note the gradual transition from yellow-brown conglomerates and mudstones of the Coon Hollow fluvial unit, to black shale of the marine unit. The marine unit has yielded a sparse number of fossil ammonites, pelecypods, and gastropods at this

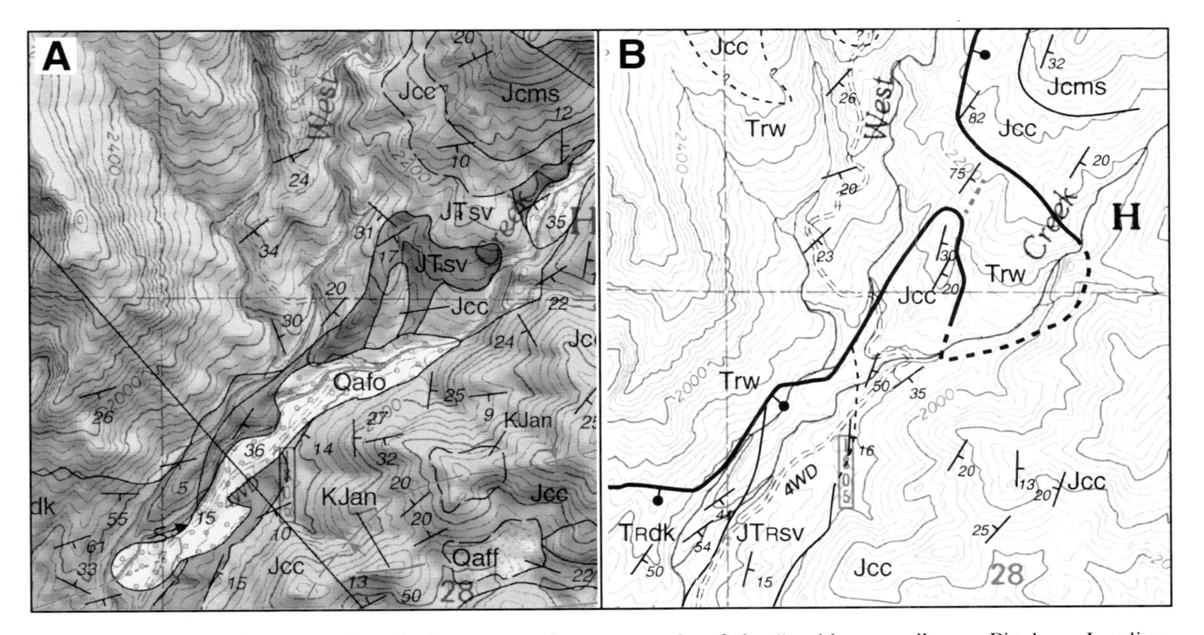

Figure 12. Detailed images and geologic maps at the same scale of the "problem area" near Pittsburg Landing. (A) Schmidt et al. (2009a). Qafo—older alluvial fan deposits; Qls—landslide deposits; KJan—andesitic sills and stocks; Jcms—marine mudstone, sandstone, and conglomerate of Coon Hollow Formation; Jcc—fluvial conglomerate, sandstone, and mudstone of Coon Hollow Formation; JTRsv—sedimentary and volcanic rocks; TRwbc—Big Canyon unit of Wild Sheep Creek Formation. (B) Preliminary mapping of Moore and LaMaskin. Fault shown is interpreted to be the northeast extension of the West Creek fault. Jcms—marine Coon Hollow Formation; Jcc—fluvial Coon Hollow Formation; Trw—Wild Sheep Creek Formation.

locality. Because our biostratigraphic investigations here are in their infancy, the authors would appreciate donation of any fossils found here during the trip or in subsequent visits.

Return to vehicle. The road log for the side trip to see the Cougar Creek intrusive complex at Upper Pittsburg Landing immediately follows. The road log for returning back to Highway 95 continues on subsequent pages.

SIDE TRIP to Upper Pittsburg Landing

Continue on USFS Road 493 toward Lower Pittsburg Landing (Fig. 7).

Mile 17.7: Turn left at junction with Upper Pittsburg Landing road. Road is signed for both Upper Pittsburg Landing and Ranger Station.

Mile 17.8: Road crosses Kurry Creek.

Mile 17.9: Junction with road to Ranger Station on left. From this junction to the parking area at Upper Pittsburg Landing the road crosses a huge Pleistocene Bonneville flood gravel bar. Large boulders on the terrace characterize this unit. Such deposits extend from river elevation below us at 1150 feet to a maximum of 1700 feet in elevation in Pittsburg Landing (O'Connor, 1993). The Bonneville flood was a catastrophic release of water down the Snake River from pluvial Lake Bonneville at ca. 17.4 calibrated ka before present (Janecke and Oaks, 2011). In contrast, the sand and silt bar across the river and below us is a later Pleistocene Missoula flood deposit caused by one of the larger jökulhlaup discharges from ice dams that formed in the Clark Fork and Columbia River systems in northern Idaho and Washington. These repeated floods backed up the Snake River from the Lower Columbia River at Wallula Gap in central Washington and reached a maximum of 1200 feet in elevation in the Snake River canyon.

Mile 18.6: Pullout into parking area on left for viewing petroglyphs chipped into patina formed on Bonneville flood boulders. These anthropogenic features are clearly younger than the Bonneville flood, but an intriguing issue concerning ice age floods in Idaho and Washington is that, as understood at present, there is an overlap of radiocarbon ages for flood events and evidence for earliest human occupation in the major river valleys. Low, regularly spaced grassy hills in foreground on the left are Bonneville flood megaripples. Their unusual scale makes them difficult to make out at this close range and a much better vantage point is from across the river looking back (Fig. 13).

Mile 19.0: Stock and overflow parking for Upper Pittsburg Landing trailhead in pullout on left.

Mile 19.4: Park at trailhead for Upper Pittsburg Landing.

OPTIONAL STOP: Cougar Creek Complex in Hanging Wall of Klopton Creek Fault

Hike the trail along the east side of the Snake River southward from Upper Pittsburg Landing. The contrast from the open terrain of Pittsburg Landing to the narrow rocky canyon is striking and the dramatic geomorphic change here marks the trace of the poorly exposed Klopton Creek fault. The Cougar Creek complex in the hanging wall of the fault is considerably more competent than the softer sedimentary sequences of the footwall. After ~0.25 miles, there are good outcrops of the Cougar Creek complex along the steep canyon wall.

Figure 13. Megaripples developed on Bonneville bar near Upper Pittsburg Landing. View to east across river from Oregon side toward Idaho side. Flood flow was from right to left (south to north). Optional stop to look at Cougar Creek complex is just off photo on right.

The Permian and Triassic Cougar Creek complex consists of dikes, sills, and stocks of gabbro, basalt, diorite, quartz diorite, trondhjemite, and tonalite (Walker, 1986; Vallier, 1995; Northrup et al., 2011; Kurz et al., 2012). The complex is interpreted as basement to the Seven Devils Group and includes rocks of probable volcanic protolith that may predate the Seven Devils Group. Layering and foliation within the canyon are predominantly vertical. Most of the tonalite rocks are porphyritic and contain light gray to bluish quartz phenocrysts as much as 1 cm across. The Middle Permian to Early Triassic tonalites (Kurz et al., 2012) are extensively mylonitized in the canyon. Mylonitic foliation is nearly vertical and lineation is nearly horizontal. Kinematic analysis of the mylonite indicates dominantly left-lateral shear sense (Kutz and Northrup, 2008), which predates deformation on the Klopton Creek fault.

Return on trail to Upper Pittsburg Landing and drive 2.0 miles back to Stop 1-3 at junction with BLM four-wheel drive road. **Reset odometer to zero and continue back up road toward Highway 95.**

Mile 2.6: Pullout on right.

STOP 1-4: Parking for Access to Stops 1-4A and 1-4B (45.6572 °N, 116.4242 °W)

Hike north along drainage and travel northwest to low hills (Fig. 7).

Stop 1-4A: Coon Hollow Formation, Fluvial and Marine Unit, Upper Plate of Kurry Creek Fault (45.6658 °N, 116.4274 °W)

We are now in the upper plate of the Kurry Creek fault (Figs. 7, 8, 11B). Rocks mapped here represent both the fluvial and marine units of the Coon Hollow Formation. Here, the transition from fluvial to marine rocks is a cyclical interbedded contact. Note multiple transitions from yellow-brown conglomerates and mudstones of the Coon Hollow fluvial unit, to black shale of the marine unit. Here, the transition consists of 5–10 parasequence-scale cycles of yellow-brown conglomerate and yellow mudstone; moving upsection, the mudstone becomes increasingly gray to black, and the proportion of shale in each cycle increases.

Hike northeast to the unusually rounded hill—the "Hobbit Hill"—in middle of drainage.

Stop 1-4B: Coon Hollow Formation, Chert-Bearing Beds, Marine Unit, Upper Plate of Kurry Creek Fault (45.6693 °N, 116.4191 °W)

In the hills just beneath the Klopton Creek fault (not exposed here), we see clastic rocks with a distinctly different composition than those of the Coon Hollow fluvial unit. Lithologies here contain chert clasts and range from fine sandstone to boulder conglomerate as well as interbedded black shale (Fig. 10F). These chert-rich clastic rocks appear to represent the uppermost rocks of the Coon Hollow Formation marine unit; however, they are not exposed in the footwall block of the Kurry Creek fault. Identical beds are present at the top of the upper flysch succession at the type locality in Hells Canyon 50 km to the northwest (Goldstrand, 1994).

These rocks may be overturned in a footwall syncline to the Klopton Creek fault; however, unequivocal top indicators (sedimentary structures that provide facing direction) have been difficult to find. Part of this field trip is to observe the chert-bearing units for such indicators. Ammonites, pelecypods, and gastropods have been found in black shale and fine sandstone north of the hill. A very limited collection of ammonites and bivalves has been tentatively identified by T. Poulton of the Geological Survey of Canada as from the Late Bathonian to Early Callovian Stages (167–165 Ma; late Middle Jurassic; 2009, written commun.). These ages are similar to late Middle Jurassic ages on ammonites from the "Oregon turbidite unit," which has provided a rich ammonite fauna of probable Callovian age, ca. 166–163 Ma. Again, potential cross-river correlations have been historically difficult. Investigation is under way to further constrain the depositional ages of upper- and lower-plate successions in Idaho and Oregon. For reasons noted above, the authors would appreciate obtaining any fossils that might be found at this locality during the trip or in subsequent visits.

Return to vehicle and continue driving east, out of Pittsburg Landing.

Mile 6.9: Pullout on right.

STOP 1-5: Overview of Upper Deer Creek (45.6746 °N, 116.3850 °W)

We are now back in the hanging wall of the Klopton Creek fault (Fig. 14). The fault continues northeast on the north side of Deer Creek. Below us, flows of the Columbia River Basalt Group are clearly out of place with respect to the geology we have examined in the Pittsburg Landing area. A system of mostly north-striking Neogene normal faults occurs along the north-flowing segment of the Salmon River, and the westernmost of these structures, the Deer Creek detachment, bends to the northeast below to join the trace of the Klopton Creek fault. This normal fault carries the Miocene basalt sequence in its hanging wall, which is juxtaposed with Mesozoic basement rocks in its footwall (also the footwall of the Klopton Creek fault). The Deer Creek area became a major extensional basin in late Miocene time and many roadcuts are through light-colored sand and mud deposits in the basalts. Some of the quartz-rich sand deposits were probably ancestral Salmon River alluvium. Mapping in this part of the Salmon River canyon has produced many spectacular examples of Neogene extensional structures reactivating Mesozoic contractional structures.

Continue on Deer Creek Road 10.5 miles, retracing the route back to junction with Highway 95.

END OF ROAD LOG 1.

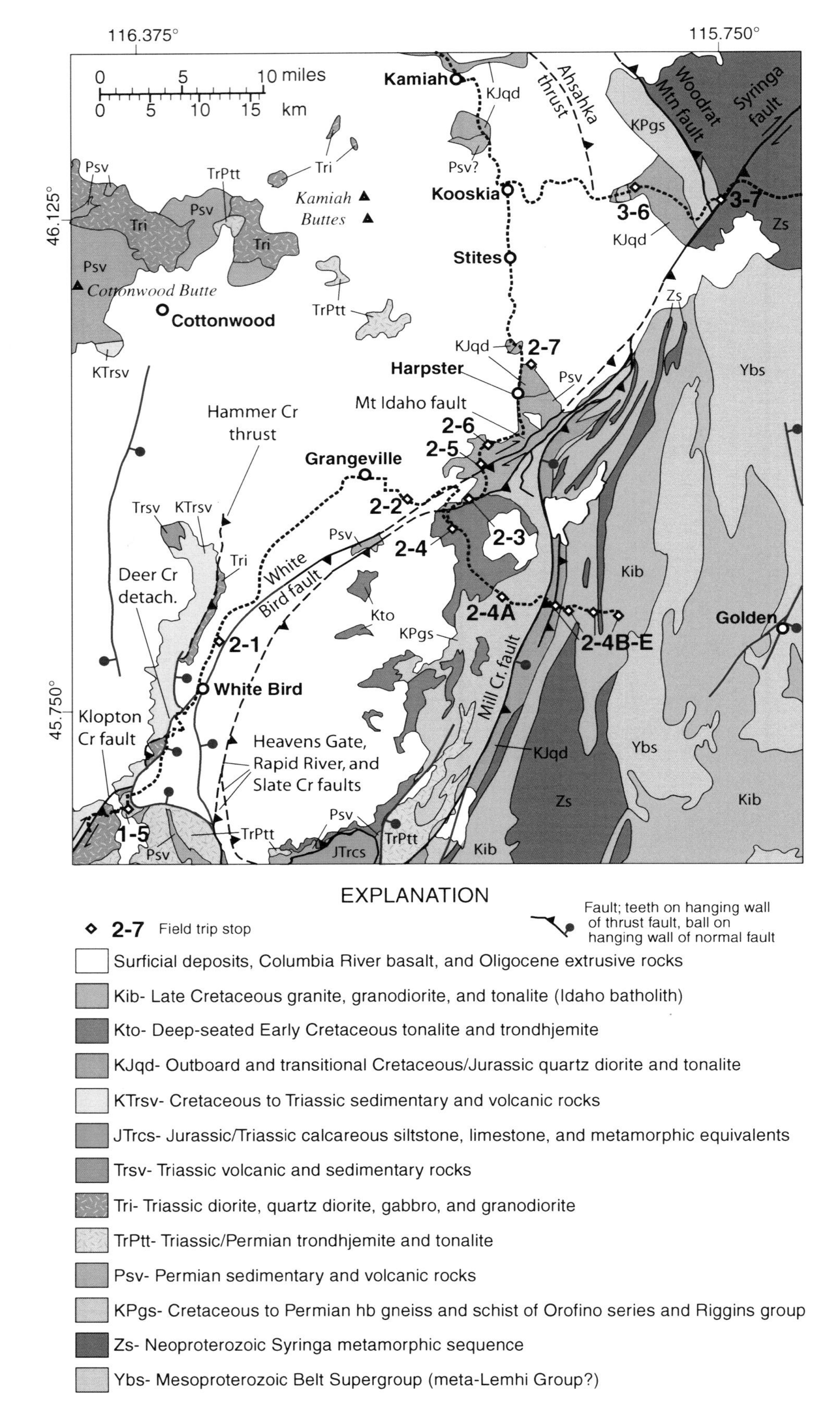

Figure 14. Field trip stops and simplified geology for Road Log 2 in the Grangeville area. Geology compiled from numerous Idaho Geological Survey maps referenced in the text. Final stop of trip 1 (1-5) also shown.

ROAD LOG 2—WHITE BIRD–GRANGEVILLE–SOUTH FORK CLEARWATER RIVER–KOOSKIA

This section of the field trip begins at White Bird Hill with a discussion of Neogene extensional reactivation of Cretaceous contractional structures along and west of the arc-continent boundary (Figs. 2 and 14). We continue northeast to the canyon of the South Fork of the Clearwater River, following the trace of the northeast-striking Mount Idaho shear zone. Along the South Fork we will examine the Early Cretaceous Blacktail pluton, a deeply emplaced tonalite-trondhjemite pluton, then continue north to look at quartz porphyries and chert-epiclastic sequences in the likely Permian portion of the Seven Devils Group of the Wallowa terrane. We finish this segment by examining magma mingling relationships in the Late Jurassic Harpster pluton, which intruded the eastern part of the Wallowa terrane. An optional side trip up the South Fork of the Clearwater River is included because the canyon affords a spectacular transect across the arc-continent boundary.

Warning: Numerous stops along busy highways. Watch for traffic! In the following log, "Mile" represents road mileage from turnoff for Hammer Creek Recreation area on Highway 95, 1.3 miles south of White Bird, Idaho. See Figure 14 for locations of field trip stops.

Mile 0.0: U.S. Highway 95, at intersection with Old Highway 95, signed as the turnoff for Hammer Creek Recreation Area, 1.3 miles south of White Bird, Idaho. **Reset odometer to zero and proceed north on Highway 95.**

Mile 0.4: Turnoff for Hoots Café on right. The trace of the informally termed "Hoots fault," a north-striking Neogene normal fault that forms part of the north-south–trending zone of extension along the Salmon River, is just west of the highway here. After the local quadrangle mapping by the Idaho Geological Survey was completed (Schmidt et al., 2009b), the proprietor of the café was approached with a complimentary copy of the map. She purportedly "didn't give a hoot" about the geologic map or any Neogene faults that passed near her establishment. However, faults farther south are active, with an M3.6 event near Riggins, Idaho, on 26 May 1987 and numerous events over M3.0 near McCall, Idaho, in the past few decades.

Mile 1.3: Turnoff for the town of White Bird on right. Highway 95 crosses White Bird canyon on an impressive bridge and starts up White Bird Hill, following the trace of the White Bird Hill fault, a north-northeast–striking normal fault (Garwood et al., 2008). White Bird canyon, which lies to the southeast in the hanging wall of the fault, contains a second normal fault with even more displacement (White Bird fault). The Hoots fault joins these structures on the north side of the bridge.

Mile 4.7: Turn right into large pullout at U.S. National Park Service White Bird battlefield memorial.

STOP 2-1: Overview of White Bird Fault and Salmon River Extensional Corridor (45.8018 °N, 116.2874 °W, WGS84 datum)

We are standing above the trace of the east-southeast–dipping White Bird fault, and one of its splays, the White Bird Hill fault, which bend from north-striking to the south to northeast-striking to the north of us and join the northeast-striking Mount Idaho fault (Fig. 15; Garwood et al., 2008; Kauffman et al., 2008). Exposed in the footwall across the highway to the west of us are nearly horizontal basalt strata. The lower 20 m (65 ft) of basalt containing well-developed columns belong to the Imnaha Formation, the lowermost unit in the Columbia River Basalt Group. It is overlain by more hackly jointed basalt of the Grande Ronde R1 unit (lowest reversed magnetostratigraphic unit). The ridge top above is capped by the first normal magnetostratigraphic unit (N1). The hanging wall exposed on the gently west-northwest–dipping slope across White Bird Canyon to the southeast of us consists of a dip slope of mostly Grande Ronde N1 and overlying Grande Ronde R2 units (Fig. 16). We interpret the White Bird and White Bird Hill structures as listric normal faults (Fig. 17), which collectively have a normal offset of over 350 m (1150 ft) (Garwood et al., 2008).

A consequence of the White Bird Hill fault at this location is the development of large landslide deposits that underlie the White Bird battlefield below us (Fig. 17). The landslides are actively forming in Miocene sediments that likely were deposited in the hanging wall of the normal fault system. Kirkham and Johnson (1929) reported an earthquake near here on 21 November 1927. They also described the formation of a set of north-northwest-trending fissures in part of the landslide deposits that were first noticed on 25 November 1927, following two months of unusually high precipitation. They measured complex active displacement across some of these fissures over the course of most of the following year and speculated about whether the fissures were due to faulting or mass wasting.

The Mount Idaho fault to the northeast of this stop is a northeast-striking, southeast-dipping reverse fault that offsets Columbia River Basalt Group units along the northwest side of Mount Idaho (Schmidt et al., 2007; Kauffman et al., 2008). This relatively young structure probably roots in an older structure (termed the "Mount Idaho shear zone") that diverges from the north-striking Hammer Creek thrust and carries reverse dextral displacement in basement rocks between the Salmon and South Fork of the Clearwater Rivers (southwest extension of Syringa fault in Fig. 2). We interpret the Mount Idaho fault as a late brittle segment of a continuous northeast-striking mylonite shear zone of Jurassic to Paleocene age that carries reverse dextral displacement from the Klopton Creek fault at Pittsburg Landing (see Road Log 1) along the southern part of the Hammer Creek fault along the Salmon River to the Mount Idaho shear zone (Stop

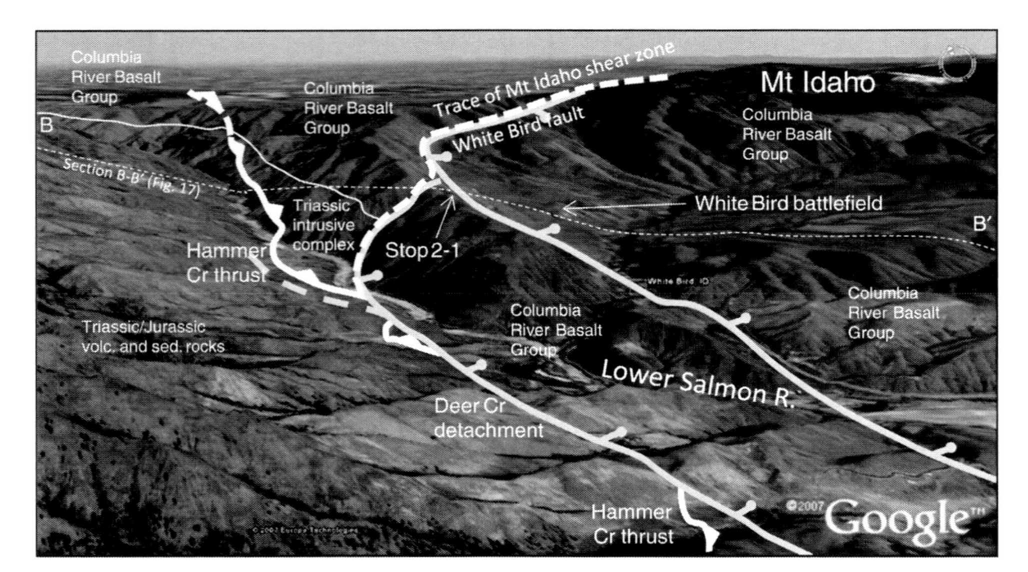

Figure 15. Annotated oblique overview of White Bird Hill area. View to north-northeast. See text for description. From GoogleEarth®.

2-3 below) along the South Fork of the Clearwater River to the Syringa fault near Stop 3-7 (Road Log 3) on the Middle Fork of the Clearwater River.

The Mount Idaho fault is an important tectonic boundary in north-central Idaho. To the south of us, the Salmon River flows due north from Riggins to White Bird as it follows the north-south–trending narrow zone of extension that connects to the Neogene Basin and Range province to the south. Along this Salmon River zone, the extensional structures reactivate older fabrics and faults associated with Mesozoic contraction along the arc-continent boundary. The Mount Idaho fault serves as an accommodation structure along which normal faults in the extensional corridor to the south of us bend and merge (Fig. 15). The Salmon River cuts a narrow canyon as it leaves the extensional corridor, follows the Hammer Creek fault north, then flows west 80 km (50 miles) to join the Snake River in Hells Canyon. The Camas Prairie plateau to the north of Mount Idaho represents a very different structural province. There, extensional structures are less common and northwest-trending gentle folds are prevalent in Columbia River basalt strata.

Historical interest: The low hills toward the southern end of the White Bird battlefield consist of basalt of Grangeville, part of the Saddle Mountains Basalt that occurs near the top of the Columbia River basalt sequence. These basalt knobs are interpreted as slump blocks (Garwood et al., 2008), and tactical control of them was critical in the decisive Nez Perce victory in the opening battle of the Nez Perce war on 17 June 1877. On the morning of the 17th, a U.S. cavalry regiment and volunteers led by Captain David Perry descended southward into White Bird Canyon from the Camas Prairie. They had been dispatched from Fort Lapwai two days before in response to recent raids on settlers along the Salmon River by Nez Perce warriors of the White Bird and Chief Joseph bands. The Chief Joseph band had been forced off of their lands to the west in the Wallowa country by a U.S. military contingent led by General O.O. Howard earlier in the spring of 1877, and they were

Figure 16. View to southeast from top of ridge above Stop 2-1, showing trace of White Bird fault and the large hanging wall panel that is tilted toward the fault. Region at bottom of hill in sunlight is the White Bird battlefield. Salmon River canyon in distance turns to right behind shaded slope.

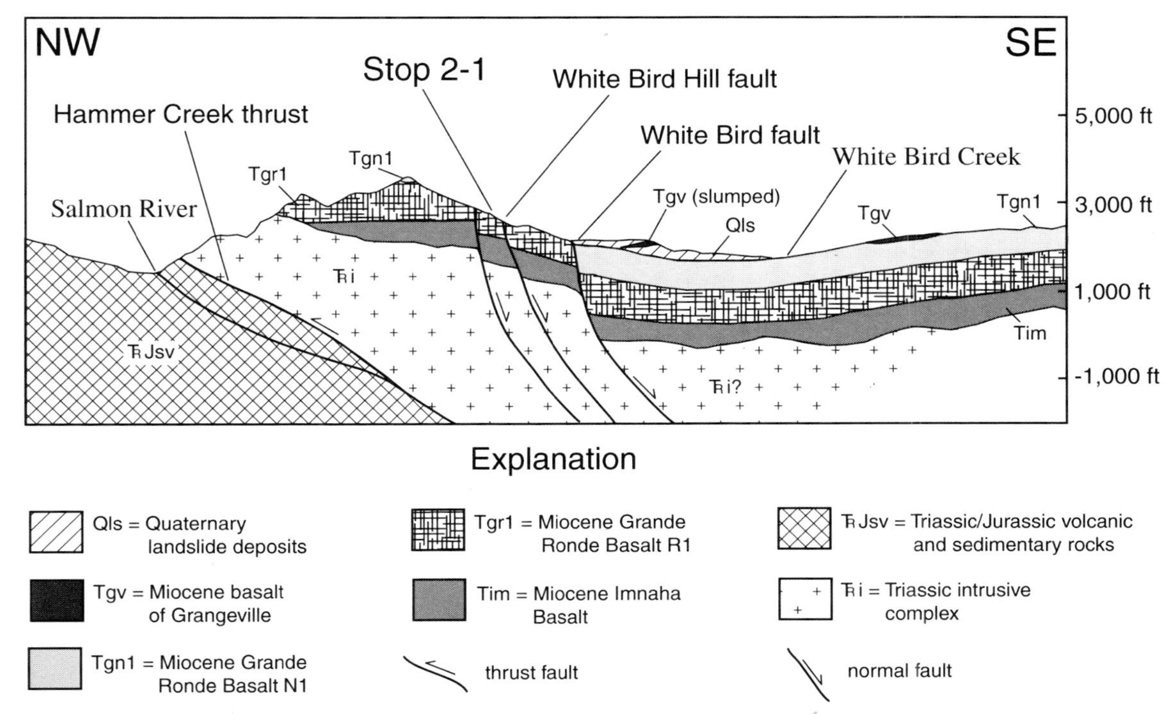

Figure 17. Cross section drawn through Stop 2-1 from northwest to southeast showing major normal faults (including the White Bird Hill fault) that offset units of the Columbia River Basalt Group and presumably sole into the Hammer Creek thrust fault. Modified from Garwood et al. (2008).

making their way north to the Nez Perce Reservation at Lapwai. Learning of Captain Perry's advance from Fort Lapwai, the Nez Perce had retreated to an encampment along White Bird Creek near the present town of White Bird. They sent a peace party to meet the advancing soldiers as they descended into White Bird canyon, but they were also prepared for an attack. For unknown reasons, Perry's force fired upon the peace party, initiating the battle of White Bird on the battlefield below us. Although outnumbered and out-armed, the Nez Perce were more skilled marksmen and horsemen than the soldiers. The Nez Perce were also more tactically skilled and their control of the basalt slide blocks during the brief battle that ensued led to the rout of Captain Perry's force back up White Bird Canyon, leaving a sizeable stock of arms and ammunition behind them on the battlefield. Despite the initial Nez Perce victory at White Bird, the remainder of the Nez Perce war was a fighting retreat by Chief Joseph's band as they fled east and then north toward Canada, pursued by a much larger U.S. military force led by General Howard (Hendrickson and Laughy, 1999, p. 215–216).

Mile 5.8: Pillow basalt lava flow in Grande Ronde R1 unit.

Mile 7.7: Junction with Old Highway 95 in the saddle of the ridge. Old Highway 95 was constructed in 1915 and traverses an impressive number of switchbacks to climb to this elevation from White Bird. The new highway leaves the trace of the White Bird Hill fault at this notch to climb up to White Bird Summit.

Mile 8.0: Sedimentary interbeds on Grande Ronde N1-R1 contact. Interbeds are commonly clay-rich in the Columbia River basalts and tend to fail on steeply cut slopes.

Mile 10.7: White Bird Hill Summit, elevation 4245 ft.

Mile 12.1: Turnoff for Tolo Lake, a traditional Nez Perce rendezvous site. It was also the site of a major discovery of Pleistocene Columbian mammoth and bison when a silt-removal project was undertaken in 1994. The remains of a complete mammoth were excavated and are displayed in Grangeville.

Mile 17.5: Intersection with Highway 13 (West Main Street) in Grangeville, Idaho. Turn right and proceed east through Grangeville. The Tolo Lake mammoth is displayed 0.5 miles farther north on Highway 95 on the right-hand side of the road.

Mile 17.8: Pass through the only traffic light in Idaho County, one of the largest counties in the country and larger than the state of New Jersey. Grangeville is the county seat.

Mile 18.7: Junction with Mount Idaho Grade Road. Turn right.

Mile 19.3: Turn left at old drive-in movie theater to stay on Mount Idaho Grade Road. The Grangeville-Salmon Road continues straight.

Mile 20.6: Junction with View Road. Pullout on right.

STOP 2-2: Overview of Camas Prairie (45.9102 °N, 116.0864 °W)

We are looking north and northwest across the Camas Prairie from the base of Mount Idaho. Lava flows of the Miocene Columbia River Basalt Group and Pliocene–Pleistocene loess underlie most of the prairie. Most high points are steptoe buttes with outcrops of the Wallowa terrane. The prominent tree-covered hill to the northwest is Cottonwood Butte, originally mapped as Permian–Triassic Seven Devils Group undivided (Schmidt et al., 2005). We now think that it consists mostly of sedimentary and volcanic rocks of the Permian Hunsaker Creek Formation (lower Seven Devils Group) intruded by Triassic quartz diorite and diorite plutons. The southeast shoulder of the butte is underlain by porphyritic lithic-crystal tuff dated at 199 ± 8 Ma (U-Pb zircon, LA-ICPMS methods, Darin Schwartz and Jeff Vervoort, 2011, written commun.) that is correlative with the Triassic–Jurassic "red tuff" at Pittsburg Landing (Stop 1-3A, Road Log 1). St. Gertrude's Monastery near Keuterville at the base of Cottonwood Butte is constructed of the ca. 200 Ma tuff.

The treeless hills in the foreground to the north of us are Kamiah Buttes, underlain by Oligocene andesite and rhyolite lavas and tuffs of the Kamiah volcanics (Jones, 1982; Kauffman et al., 2009). The treed skyline from the north to the east of us is on the continental side of the arc-continent boundary and consists mostly of Neoproterozoic metasedimentary rocks intruded by Late Cretaceous granite and granodiorite plutons of the Idaho batholith. We will continue to the east, dropping into the canyon of the South Fork of the Clearwater River, which provides one of the major transects across the arc-continent boundary.

Mile 21.1: Historic hamlet of Mount Idaho. Founded in 1862, the town served nearby gold-mining areas. This was the original county seat for Idaho County from 1875 to 1902.

Mile 24.7: Deeply weathered boulders mark the unconformity beneath the Columbia River Basalt Group as we descend into the South Fork Clearwater Canyon. Outcrops below here are sheared chlorite-epidote schist of volcanic and sedimentary protolith tentatively assigned to the Seven Devils Group of Vallier (1977) by Schmidt et al. (2007).

Mile 25.3: Pullout on left.

STOP 2-3: Mylonitized Permian Tonalite in the Mount Idaho Shear Zone (45.9099 °N, 116.0228 °W)

Roadcut exposes mylonitized tonalite (or trondhjemite) and diorite. A sample collected from this outcrop yielded a U-Pb zircon (LA-ICPMS, laser-ablation inductively coupled plasma-mass spectrometry) age of 255 ± 8 Ma (Richard Gaschnig and Jeff Vervoort, 2011, written commun.). These rocks are identical in lithology, structural position, and age to the Permian–Triassic Cougar Creek complex in the hanging wall of the Klopton Creek fault exposed at upper Pittsburg Landing (Vallier, 1995; Kurz et al., 2012; Stop 1-1, Road Log 1). Well-developed mylonite in these rocks contains northeast-striking, moderately to steeply southeast-dipping foliation and shallowly northeast- to east-plunging lineation. Abundant S-C and extension crenulation fabrics indicate consistent dextral, southeast-side-up sense of shear parallel to lineation across this zone (Fig. 18). The footwall assemblage of this ~2-km-wide shear zone consists of the epiclastic and quartz porphyry assemblage that we will visit at Stops 2-5 and 2-6. The hanging wall assemblage is hornblende gneiss, schist, and marble of the Riggins Group intruded by Early Cretaceous tonalite and trondhjemite of the Blacktail pluton.

The Mount Idaho shear zone continues beneath basalt cover to the southwest, where it is exposed again in an erosional window at the base of Mount Idaho. North-striking shear zones located to the west of the arc-continent boundary between Riggins and Slate Creek to the south of us are either cut by, or merge with, the Mount Idaho shear zone under basalt cover rocks southwest of Grangeville. As noted at Stop 2-1, we interpret the Mount Idaho shear zone to merge with the Hammer Creek and Klopton Creek faults farther southwest. The shear zone continues northeast of us, promptly disappearing beneath more basalt cover on the other side of the South Fork canyon. We link this segment of the shear zone with the Syringa fault, which is along-strike to the northeast where pre-Miocene basement rocks are exposed along the Middle Fork of the Clearwater River (near Syringa, MP-90, Road Log 3). This composite shear zone appears to offset the arc-continent boundary by tens of km in oblique reverse right-lateral sense. It postdates Jurassic (and possibly Early Cretaceous?) sedimentary rocks of the Coon Hollow Formation at Pittsburg Landing and appears to predate Paleocene intrusive rocks of the Bitterroot lobe of the Idaho batholith to the northeast. To the north of this shear zone, the strike of the arc-continent boundary changes, first to the northwest, then to the west around the Syringa embayment near Orofino.

Figure 18. Outcrop photo of mylonite in Mount Idaho shear zone along the South Fork Clearwater River northeast of Stop 2-3; view to northeast, showing southeast-side-up (reverse) shear sense.

Mile 25.6: Basalt and sedimentary interbed debris in large slump.
Mile 27.0: Outcrop of tonalite and trondhjemite of Early Cretaceous Blacktail pluton. We have crossed into the hanging wall of the Mount Idaho shear zone.
Mile 28.1: Junction with Highway 14. Turn right, then take immediate left into parking area next to river.

STOP 2-4: Tonalite and Trondhjemite of the Blacktail Pluton (45.8886 °N, 116.0353 °W)

The Blacktail pluton, originally mapped by Myers (1982), consists of biotite tonalite cut by dikes and pods of fine-grained to pegmatitic, muscovite-garnet trondhjemite. Tonalite contains unusually large euhedral biotite grains as wide as 2 cm that locally define a foliation. Subhedral epidote, interpreted to be magmatic in origin, is common. Criss and Fleck (1987) determined low Sr_i values (<0.704) for this pluton. These rocks were originally described as trondhjemite by Myers (1982), but only the muscovitic cross-cutting dikes classify as such on normative Ab-An-Or diagrams (after Barker, 1979). A U-Pb zircon age of 111.0 ± 1.6 Ma was determined by McClelland and Oldow (2007) from a tonalite sample collected nearby. The pluton is covered by Miocene basalt to the southwest, but is exposed in uncommon erosional windows exposing hornblende gneiss intruded by biotite tonalite. Similar rocks crop out along the north side of Slate Creek 33 km (20 mi) south of here, where Unruh et al. (2008) obtained a U-Pb zircon age of 113.1 ± 0.6 Ma. Thus, the Blacktail pluton likely continues to the south beneath Miocene cover to Slate Creek and is probably on the order of 400 km^2. It is part of a suite of tonalite and trondhjemite plutons that occur along and west (outboard) of the arc-continent boundary. In common to these plutons is primary epidote indicative of high pressure (>8 kb) crystallization (Zen and Hammarstrom, 1984; Zen, 1988). The 119–116 Ma Sixmile Creek pluton along the Clearwater River also forms part of this suite (MP-56.0, Road Log 3).

At this stop, the large rocks ringing the parking lot are mostly biotite tonalite intruded by trondhjemite dikes. Cross Highway 14 to the outcrop ("Bubba" graffiti marks the spot) to examine garnet-muscovite trondhjemite. The road log for a side trip across the arc-continent boundary on the South Fork of the Clearwater River follows.

SIDE TRIP across the Arc-Continent Boundary along the South Fork of the Clearwater River

Reset odometer to zero. Proceed east from Stop 2-4 and cross bridge over the South Fork. This portion of the field trip covers a region mapped by Myers (1982) and Hoover (1986).

Mile 0.8: Cotter Bar Picnic Area on right.
Mile 2.7: McAllister Picnic Area on right.
Mile 4.9: First outcrops of hornblende gneiss past the eastern margin of the Blacktail pluton.
Mile 5.1: Pullout on right at entrance to an unnamed camping area.

STOP 2-4A: Hornblende Gneiss, Schist, and Marble of the Riggins Group or Orofino Series (45.8349 °N, 115.9827 °W)

This unit is characterized by fine- to medium-grained hornblende gneiss that grades into hornblende ± biotite ± chlorite ± garnet schist, muscovite-plagioclase-quartz ± biotite ± epidote schist, epidote-plagioclase quartzite, and marble. Hornblende gneiss contains subordinate epidote, zoisite, chlorite, quartz, and plagioclase. The protolith for this assemblage was probably dominated by sedimentary carbonate-clastic and volcaniclastic lithologies. This assemblage occurs along and west of the arc-continent boundary and was intruded by the Early Cretaceous tonalite-trondhjemite plutonic suite containing magmatic epidote.

At this outcrop, well-layered, shallowly dipping, garnet-quartz schist and garnet-biotite quartzite predominate. Wilford (2012) obtained a Lu/Hf garnet age of 117.5 ± 3.6 Ma from this locality. This age suggests peak metamorphism of this assemblage shortly before intrusion of deep-seated tonalite and trondhjemite suite plutons. We attempted detrital zircon U-Pb LA-ICPMS analysis on the same sample, but had trouble obtaining zircons. However, two other samples from the Riggins Group assemblage collected along Slate Creek and the Grangeville-Salmon Road south of here yielded sufficient zircon populations. The two samples produced nearly identical detrital zircon age spectra with well-defined peaks at ca. 200 Ma. (Schmidt et al., 2013), and indicate a sedimentary protolith ranging in age from a maximum of Early Jurassic (youngest detrital zircon ages) to a minimum of Early Cretaceous (Blacktail pluton age). We interpret the protolith of this unit as either Wallowa terrane rocks or a related basinal assemblage. Thus, the assemblage attained peak pressures ~8 kb by ca. 118 Ma during approximately coeval intrusion by tonalite-trondhjemite plutons.

Mile 5.3: Cross Nelson Creek.
Mile 6.1: Entrance to Castle Creek Campground on right.
Mile 6.4: Entrance to South Fork Campground on right.
Mile 6.6: Roadcut on left shows vertical layering in Riggins Group or Orofino series marble. Remains of extensive placer operations are evident across river on right.
Mile 8.1: Junction on right of U.S. Forest Service Road 309 that continues up Mill Creek.
Mile 8.3: Pullout on right just past Meadow Creek Campground on left.

STOP 2-4B: Sheeted Orthogneiss Suite along the Arc-Continent Boundary (45.8280 °N, 115.9278 °W)

Steeply dipping sheets of biotite tonalite gneiss, migmatite, and biotite schist comprise most of this outcrop, which is representative of the mid-Cretaceous plutonic suite that intrudes the arc-continent boundary and within which the Sr_i 0.704/0.706 transition occurs. Hoover (1986) described porphyritic granodiorite and quartz monzodiorite gneiss near here that contains K-feldspar porphyroclasts. These rocks are similar to K-feldspar megacrystic orthogneiss in the ca. 105 Ma Little Goose plutonic complex near McCall, Idaho (Manduca et al., 1993), which was involved in late-stage dextral transpression along the western Idaho shear zone (Giorgis et al., 2008). Unruh et al. (2008) dated a sample of foliated granodiorite from near this location that yielded a mean age of 99.4 ± 3.2 Ma from concordant fractions (U-Pb zircon, ID-TIMS).

Mile 9.1: USFS Road 484 takes off to the left.
Mile 9.3: Pullout on right at lone pine tree.

STOP 2-4C: Hornblende-Biotite Tonalite Pluton East of 0.704/0.706 Line (45.8247 °N, 115.9121 °W)

Steeply foliated hornblende-biotite tonalite containing large hornblende crystals characterizes this zone just east of the Sr_i 0.704/0.706 boundary. This unit has been dated at 90.2 ± 2.7 Ma (U-Pb zircon, LA-ICPMS) by McClelland and Oldow (2007). This lithology and its age are similar to the 90 Ma Payette River tonalite near McCall that preserves final deformation on the western Idaho shear zone (Giorgis et al., 2008).

Mile 10.7: Pullout on left, just past suspension bridge, in trail-head parking area for Johns Creek, and walk back to outcrops opposite bridge.

STOP 2-4D: Neoproterozoic Biotite Quartzite of the Syringa Metamorphic Sequence (45.8254 °N, 115.8854 °W)

This outcrop is feldspathic garnet-biotite quartzite intruded by pegmatite dikes. The outcrop is on strike with more quartz-rich quartzite that typifies the Syringa metamorphic sequence. U-Pb detrital zircon analyses from a sample of the Syringa quartzite at Wall Point, 16 km (10 mi) north of here, yielded mostly 1150–1050 Ma grains but included two significantly younger grains (<700 Ma) consistent with a Neoproterozoic age for the unit (Lewis et al., 2010). The quartzite at Wall Point is similar to Neoproterozoic quartzite in the Gospel-Hump Wilderness to the south (Umbrella Butte Formation; Lund et al., 2003), which is interpreted as the metamorphosed equivalent of the Neoproterozoic Windermere Supergroup. Wilford (2012) obtained a Lu/Hf garnet age of 89.9 ± 1.8 Ma from this outcrop. This age suggests peak metamorphism during the waning stages of deformation along the arc-continent boundary.

Mile 11.8: View of prominent exfoliating granite dome up river on left marks the western margin of granodiorite plutons of the Idaho batholith.
Mile 12.1: Pullout on right, just past Cougar Creek trailhead.

STOP 2-4E: Late Cretaceous Granodiorite of the Atlanta Lobe of the Idaho Batholith (45.8222 °N, 115.8595 °W)

Muscovite-bearing biotite granodiorite in outcrop is typical of the ca. 87–67 Ma peraluminous suite of the Atlanta lobe of the Idaho batholith (Lund, 2004; Gaschnig et al., 2010), which includes biotite granodiorite and muscovite-biotite granite. This region along the South Fork of the Clearwater River is the northern extent of the voluminous plutonic suite of the Idaho batholith.

Turn around and return to junction of Highway 14 with Mount Idaho grade road. Reset odometer to zero and continue northwest on Highway 14 toward Kooskia, Idaho.

Mile 1.4: Eroded toe of large slump that Mount Idaho Grade Road traverses after Stop 2-3 above.
Mile 2.5: Pullout on right has large boulders of fresh biotite tonalite and garnet-muscovite trondhjemite of the Blacktail pluton blasted from roadcuts along Highway 14. A worthy stop for fresh rock!
Mile 2.7: Contact between the northern margin of the Blacktail pluton and hornblende gneiss and marble of the Riggins Group or Orofino series. The southeastern side of the Mount Idaho shear zone is a short distance beyond this outcrop.
Mile 4.7: Fault gouge in outcrop on left (Fig. 19) marks the northwest side of the Mount Idaho shear zone. Chlorite-epidote schist predominates in the shear zone south of here and lower grade greenstone is the common lithology north of here. This part of the Mount Idaho shear zone has been reactivated by brittle faulting, which also deforms the Miocene Columbia River Basalt Group.
Mile 4.9: Mouth of Mill Creek.

Figure 19. Fault gouge along the Mount Idaho fault where it crosses the South Fork of the Clearwater River. View is looking southwest. See Schmidt et al. (2007) for geologic map of area.

Mile 5.1: Pullout on right at private bridge that crosses the South Fork.

STOP 2-5: Permian (?) Mineralized Quartz Porphyry of the Dewey Mine (45.9363 °N, 116.0080 °W)

The iron-stained outcrop on the left side of the road consists of pyrite-bearing quartz porphyry and greenstone. Both the quartz porphyry and greenstone are sheared in places. The quartz porphyry pluton is highly altered to quartz, sericite, and iron oxide, and hosts Cu-Au mineralization at the Dewey Mine on Mill Creek above the houses visible on the highway to the south. Although not dated here, Cu-Au mineralization in the Wallowa terrane is typically associated with Permian quartz porphyry plutons.

Mile 6.3: Pullout at the second, smaller of two pullouts on right. Walk 0.1 mile northeast along road to colorful roadcuts at the beginning of a long right-hand curve in the highway.

STOP 2-6: Chert and Epiclastic Sequence, Likely Permian Hunsaker Creek Formation (45.9508 °N, 115.9976 °W)

Interbedded coarse-grained graywacke, argillite, siliceous argillite, siltstone, and chert are well exposed in the long roadcut, and large chert clasts are present locally (Fig. 20). Nearby volcanic flow and intrusive rocks (mostly sills) are commonly plagioclase ± hornblende porphyritic. Epidote, albite, and chlorite alteration indicate lower greenschist metamorphic conditions. We correlate this assemblage with the Permian Hunsaker Creek Formation of the lower part of the Seven Devils Group (Vallier, 1977) based on the presence of chert, absence of carbonate, and prevalence of epiclastic units in these rocks. This association is corroborated by the quartz porphyry at the Dewey Mine (Stop 2-5), 1.5 km (0.9 mi) south of here.

Figure 20. Chert clast in likely Permian graywacke (Hunsaker Creek Formation?) at Stop 2-6.

Mile 8.6: Junction of Highway 14 with Highway 13. Go straight (north) on Highway 13, which stays along the river toward Kooskia. Turning left on Highway 13 would take you back to Grangeville. Rocks in roadcuts are still part of Hunsaker Creek Formation.

Mile 8.8: Junction with Lightning Creek Road on right after crossing bridge. Myers (1982) mapped a structural transect from here across the arc-continent boundary to the east. His fieldwork coincided with the heyday of logging in this area, which provided unprecedented roadcut exposures as well as hair-raising logging truck encounters. The spectacular roadcuts have long ago succumbed to weathering and his detailed map delineates what we can no longer see on the ground.

Mile 10.2: Junction with Sears Creek Road on right. Outcrops along the highway consist of quartz diorite of the Harpster pluton and greenstone wall rock screens mapped by Myers (1982). Myers obtained an unpublished U-Pb zircon age of ca. 160 Ma for the Harpster pluton (John Stacey, 1982, written commun.).

Mile 10.5: The unincorporated community of Harpster, Idaho.

Mile 11.2: Entering Nez Perce Indian Reservation (the 1863 boundary).

Mile 11.6: Outcrops on right show enclave-rich quartz diorite of the Harpster pluton.

Mile 12.7: Turn right on Sally Ann Creek Road and drive 0.5 miles.

Mile 13.2: Pullout on right on Sally Ann Creek Road.

STOP 2-7: Magma Mingling Relationships in the Harpster Pluton (46.0119 °N, 115.9532 °W)

The roadcuts for approximately the next 0.2 miles up Sally Ann Creek Road show beautiful mingling relationships between hornblende diorite-gabbro and host hornblende-biotite quartz diorite. Most of the quartz diorite contains diorite-gabbro enclaves. Hornblende phenocrysts in all lithologies here commonly contain relict clinopyroxene cores. A poorly outcropping rusty portion of the roadcut across from the pullout consists of highly altered epiclastic wall rock block or screen within the pluton. Columbia River basalt crops out just above these roadcuts.

Turn around and return to Highway 13.

Mile 13.7: Turn right onto Highway 13 northbound toward Kooskia. The highway from here to Stites passes back and forth across the very irregular Miocene unconformity. Some of this irregularity is due to faulting, but much can be ascribed to the old erosional surface having considerable relief that included deep canyons draining the paleo–Clearwater basin.
Mile 20.4: Entering Stites, Idaho.
Mile 22.0: Basalt dikes that were feeders to the Columbia River Basalt Group cut basalt flows in quarry on right.
Mile 23.7: Entering Kooskia, Idaho.
Mile 24.8: Crossing Middle Fork of the Clearwater River.
Mile 25.1: Junction with Highway 12. Turn left to go to Lewiston, Idaho, right to Lolo, Montana.

END OF ROAD LOG 2.

ROAD LOG 3—LEWISTON-OROFINO-LOWELL-LOLO (U.S. HIGHWAY 12)

This road log starts in Lewiston, Idaho, and continues eastward up the Clearwater and Lochsa rivers into Montana. Flood basalts of the Columbia River Basalt Group at Lewiston give way to underlying Mesozoic plutons and island-arc sedimentary (and volcanic?) rocks near Orofino. East of Kooskia, we cross the arc-continent boundary into Neoproterozoic metasedimentary rocks. At Lowell, we cross into the Coolwater culmination and back into rocks with arc affinity. A side trip up the Selway River provides a look at the Mesozoic metasedimentary and plutonic rocks within the culmination. Northeast of Lowell, we cross back into Proterozoic rocks and then into the Bitterroot lobe of the Idaho batholith, where we will review the results of the geochemical and geochronologic studies in the region. We will visit undeformed and deformed plutons, main-phase plutons and mafic dikes of the batholith, a hydrothermally altered zone, younger Tertiary plutons, and Mesoproterozoic country rocks of the batholith. Previous road logs along the eastern part of the route have been produced by Reid et al. (1979) and Hyndman (1989).

This U.S. Highway 12 road log has national significance because it covers much of the path that Meriwether Lewis and William Clark took through Idaho over two centuries ago. The distinctive geologic outcrops in this area have also played a prominent role in the oral tradition of the Nez Perce Indians. Brief comments on these are interspersed throughout the road log's narrative. For this information, we have relied primarily on an excellent historical and recreational guide by Hendrickson and Laughy (1999) for the area between Lewiston, Idaho, and Missoula, Montana.

Warning: Numerous stops are along busy highways. Watch for traffic! See Figure 21 for locations of field trip Stops 3-1 to 3-7.

Trip start: North Lewiston at intersection of U.S. Highway 12 and U.S. Highway 95 (46.4301 °N, 116.9914 °W, WGS84 datum).

Lewiston, Idaho, and Clarkston, Washington, are at the confluence of the Snake and Clearwater rivers in the Lewiston basin (Fig. 2). This is the lowest point in Idaho at 720 feet (220 m) elevation. Lewiston is Idaho's seaport, a result of slack water created by several dams and locks on the Columbia-Snake River system.

Lewiston Hill to the north of town is underlain by an east-west–trending asymmetrical anticline with a steeply dipping south limb that forms a dip slope of Columbia River basalt flows along much of the exposed hillsides (Garwood and Bush, 2001; Kauffman et al., 2009). The fold forms the northern edge of the Lewiston basin, and one or more north-dipping reverse faults occur at the bottom of the hill, where Columbia River basalt is thrust over late Pliocene (?) Clearwater gravels. The southern margin of the Lewiston basin is formed by the east-northeast–trending Waha front (Fig. 2), a complex monoclinal fold and fault system likely with multiple periods of movement, probably including pre-Miocene right-lateral strike-slip motion (Kauffman et al., 2009). The Snake River is entrenched in the basalt plateau to the south of this front, thereby forming Hells Canyon, which beautifully exposes rocks of the Wallowa terrane. Both the Lewiston Hill and Waha structures were active during Columbia River basalt emplacement and may have experienced Pliocene to Recent displacement (Kauffman et al., 2009; Alloway et al., 2013; Reidel et al., 2013).

Historical interest: Lewis and Clark reached the confluence of the Snake and Clearwater rivers in their dugout canoes on 10 October 1805. In his journal, Clark noted that the "water of the South fork [Snake] is a greenish blue, the north [Clearwater] as clear as cristial [*sic*]."

Travel 6.5 miles east up the Clearwater River to **MP-305.0.** Exit right to Highway 12 (Missoula). Highway 95 continues south to Grangeville and Boise.

All mileage now corresponds to milepost markers along Highway 12 from here east to Lolo Pass unless otherwise noted on side trips.

MP-11.0: Ant and Yellowjacket historical marker on right. Columbia River Basalt Group outcrops on left are part of Nez Perce Indian legend of Ant and Yellowjacket.

Historical interest: According to Nez Perce legend, a dispute developed between two mutually friendly colonies of ants and yellowjackets over the use of a certain sitting rock. The argument intensified to a point that Coyote, a heroic figure in many Nez Perce legends, intervened and ordered a halt to the fighting. When the head ant and head yellowjacket refused to cease fighting, Coyote drew upon his magical powers and turned them into stone. They are now preserved in motionless combat as basalt spires on the steep slope on the north side of the highway (Hendrickson and Laughy, 1999, p. 14).

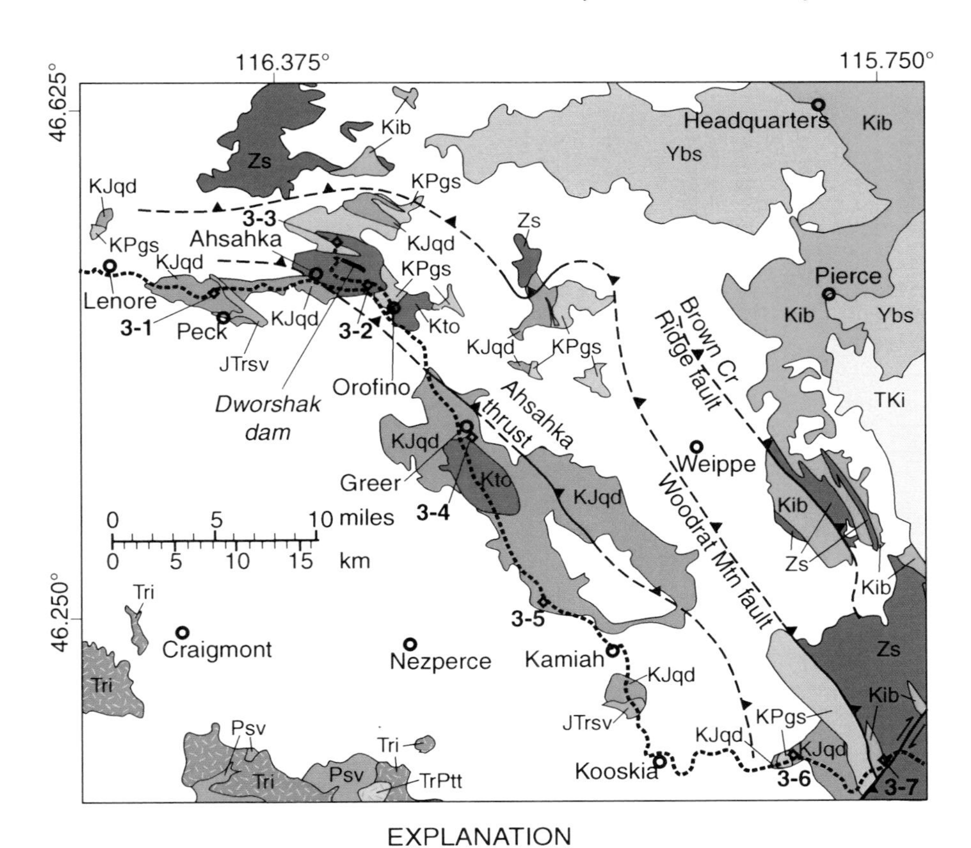

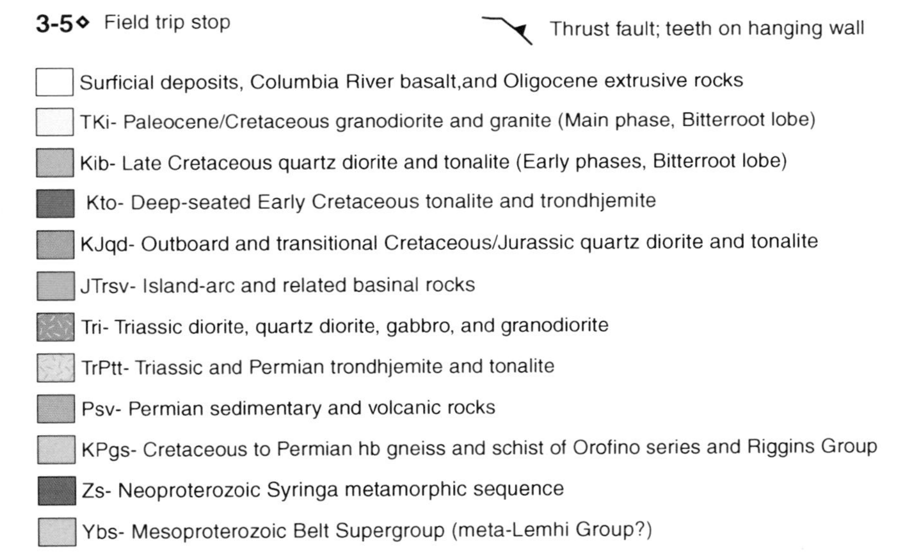

Figure 21. Map of Orofino area showing field trip Stops 3-1 to 3-7 and simplified geology modified from Lewis et al. (2005a) and Kauffman et al. (2009).

MP-14.9: State Highway 3 junction. Continue east on U.S. Highway 12.

MP-15.3: Opposite mouth of Potlatch Creek.

Historical interest: A canoe mishap in rapids near here on 8 October 1805 damaged one canoe and sent several members of the Lewis and Clark party into the river. Aid from a Nez Perce Indian camped on shore, along with that from other members of the expedition, prevented additional damage and loss of life. The expedition also camped near here on their way upstream on 5 May 1806 (Hendrickson and Laughy, 1999, p. 20).

MP-19.1: Gifford Road junction. Miocene interbeds within the Columbia River Basalt Group exposed in the roadcut dip 33° south-southeast on the south limb of an anticline with an axis parallel the river.

MP-23.1: Opposite mouth of Pine Creek. Orange material in the roadcuts above Clearwater River east of Pine Creek is an interbed between basalt flows.

MP-25.9: Opposite mouth of Bedrock Creek. Pre-Miocene basement rocks (impure quartzite, marble, granodiorite, tonalite, and quartz diorite) are exposed 1.5 miles north in the bottom of Bedrock Creek, as a result of downcutting of the creek below the Columbia River basalts. The quartzite and marble are presently assigned to the Orofino series (Lewis et al., 2005a). Foliation strikes east-west and dips to the north, and lineation plunges moderately to the northwest. Farther upstream from the metamorphosed sedimentary rocks, the tonalite is compositionally identical to the Payette River tonalite near McCall. A preliminary age of U-Pb on zircon of 91 Ma was reported by Byerly et al. (2013) on a sample from the Bedrock Creek exposure. Further, the pluton was sampled for paleomagnetic analyses, which are discussed below at Stop 3-2.

MP-27.6: Rattlesnake Creek Rest Area on left.

MP-28.3: Lenore Bridge.

Historical interest: Lewis and Clark camped upstream from the northeast end of the present bridge on their way west on 7 October 1805.

MP-29.0: Outcrops of columnar Columbia River basalt on opposite side of river. Exposures also show radiating columns, informally called "warbonnets."

MP-31.1: First exposures of pre-Miocene basement rocks along the Clearwater River (Jurassic or Cretaceous diorite and gabbro of the Wallowa terrane). They are overlain by Imnaha Basalt of the Columbia River Basalt Group.

MP-32.8: Cliffs on opposite side of river are Jurassic or Cretaceous diorite and gabbro.

MP-34.8: Junction of road to Peck.

MP-35.0: Pullout on left. Former site of Peck grain elevator.

STOP 3-1: Island-Arc Metasedimentary Rocks (46.4963 °N, 116.4332 °W)

The mixed unit contains biotite-quartz-plagioclase gneiss, biotite-hornblende gneiss, calc-silicate hornfels, muscovite-quartz schist, and lenses of gabbro and amphibolite (Kauffman et al., 2005b). Biotite-quartz-plagioclase gneiss is the predominant lithology at this outcrop. It is dark gray, fine grained, and contains 15–50 percent quartz and a similar range of plagioclase. The unit is more calcareous to the south, and calc-silicate hornfels is common in exposures in and near Peck. Minor marble is associated with the hornfels.

MP-36.1: Pullout on south side of highway. Largely undeformed quartz diorite pluton here yielded a 125.2 ± 2.5 Ma age (LA-ICPMS U-Pb zircon; Schmidt et al., 2009c).

MP-39.1: Pink House Hole Recreation Area.

MP-39.3: Crossing Ahsahka thrust. The Ahsahka thrust of Davidson (1990) and Snee et al. (2007) marks a major change in deformational style, metamorphic grade, and $^{40}Ar/^{39}Ar$ ages in the region. It is located southwest of the northwest-striking Orofino shear zone (Payne and McClelland, 2002; Payne, 2004; McClelland and Oldow, 2007) that is well exposed in the Brown Creek Ridge area, south of Pierce (Fig. 21; Lewis et al., 2007a), but less obvious in the Orofino area. The Ahsahka thrust is a northeast-dipping, ductile (gneissic) shear zone extending southeast from Orofino toward the area east of Kamiah (Fig. 21). Mesozoic rocks southwest of the shear zone lack pervasive gneissic foliation but are cut by numerous northeast-dipping ductile shear zones. Here the Ahsahka thrust is characterized by an ~250-m (820-ft)-wide zone of moderately to steeply northeast-dipping banded gneiss consisting of both mafic hornblende-rich bands and more felsic tonalite bands (Davidson, 1990). Rocks of the Ahsahka thrust are pervasively foliated to form mylonitic gneiss, much of which is relatively coarse grained and probably recrystallized (blastomylonite). Lineations are developed only locally in the Orofino area, but consistent offsets of late-stage pegmatite dikes provide evidence for northeast over southwest sense of shear (thrusting) in the shear zone (Davidson, 1990).

MP-40.0: Canoe Camp State Park. View of Dworshak dam to the northeast. Dworshak dam was constructed from 1966 to 1973. Log drives down the North Fork of the Clearwater River ended in 1971.

Historical interest: Lewis and Clark arrived here on foot and horseback in poor health on 26 September 1805, and spent 11 days making canoes and hunting. It was here they branded their horses and left them with the Nez Perce Indians for care over the winter. They proceeded west by canoe on 7 October (Hendrickson and Laughy, 1999, p. 31).

MP-44.0: Bridge to Orofino.

SIDE TRIP to Dworshak Dam Area

Reset odometer to zero. Turn left and cross bridge to Orofino.

Mile 0.1: Take immediate left on Highway 7 toward Ahsahka.

Mile 1.5: Outcrops on right are Cretaceous hornblende-biotite quartzite diorite and biotite tonalite. See Kauffman et al. (2005a) for mapping here.

Mile 1.7: MP-51 and Scoles Road (private drive—do not block).

STOP 3-2: Outboard (Low Sr_i) Cretaceous Plutons and Pegmatite (46.4953 °N, 116.2781 °W)

Strongly foliated hornblende-biotite quartz diorite here has an age of 115.5 ± 1.3 Ma (zircon LA-ICPMS method; Schmidt et al., 2009c), broadly similar to ages from the Hazard Creek complex in the McCall area (Manduca et al., 1993) and the Blacktail pluton east of Grangeville (McClelland and Oldow, 2007). The strong fabric here is probably related to the Ahsahka thrust (and possibly the northern extension of the western Idaho shear zone; Stetson-Lee et al., 2013). Undeformed pegmatite that cuts the fabric at this outcrop yielded a weighted mean age of 93.4 ± 1.9 Ma (zircon LA-ICPMS age; Schmidt et al., 2009c). These results indicate that motion here is older (> 94 Ma) than that along the Orofino shear zone that deforms 73 Ma rocks at Brown Creek Ridge southeast of Weippe (Fig. 21; Payne, 2004; McClelland and Oldow, 2007) and is more similar in age to the deformation in the south near McCall described by Giorgis et al. (2008).

Paleomagnetic and $^{40}Ar/^{39}Ar$ analyses were recently conducted using samples taken from this outcrop (note drill holes) and three localities to the west along the east-west portion of the arc-continent boundary (Byerly et al., 2013; Fig. 22). Paleomagnetic results from this outcrop were scattered, perhaps because this ca. 116 Ma quartz diorite is older and has a more complex history than the plutons to the west. Samples from three western localities yielded ca. 90 Ma U-Pb ages, similar to the late syndeformational Payette River tonalite (Manduca et al., 1993). The paleomagnetic directions from five western sites are similar and downward. The magnetic carrier is likely titanomagnetite, as determined by thermal demagnetization. The timing of the acquisition of the magnetization is constrained by thermal demagnetization temperatures to be between the ca. 90 Ma crystallization age and ca. 84 Ma $^{40}Ar/^{39}Ar$ age of hornblende. This is within the long normal chron (120–83 Ma), consistent with the downward direction, and close in age to the Cretaceous stillstand pole of Beck and Housen (2003). The observed

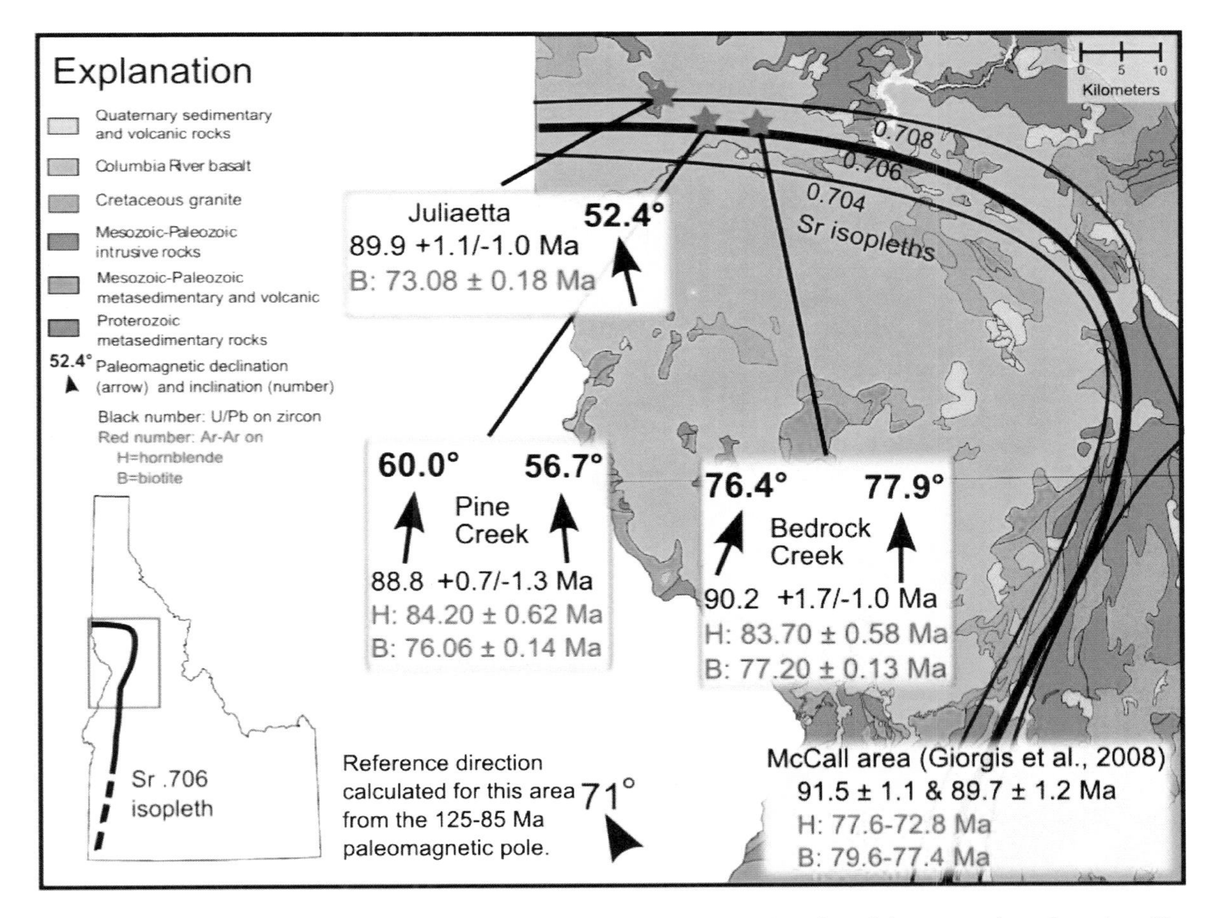

Figure 22. Map of area near Orofino showing Sr isopleths as a proxy for the location of the arc-continent boundary. The boundary was interpreted by Stetson-Lee et al. (2013) as the northern continuation of the western Idaho shear zone. Locations of samples from E-W section of the shear zone are indicated by stars. Arrows and bold numbers show the azimuth and inclination of the remanent magnetization (Byerly et al., 2013).

magnetization is ~30° clockwise of the direction expected for this location based on that pole. There are three straightforward interpretations of the work. First, this east-west portion of the North American boundary was originally oriented ~060°, parallel to transform faults associated with the late Precambrian rifting of North America to the northeast (e.g., Yates, 1968; Lund et al., 2003; Dickinson, 2004). Second, rotation occurred after emplacement of the Payette River tonalite and equivalent intrusive bodies (ca. 90 Ma). Third, the entire northern portion of this continental margin rotated ~30° clockwise since the end of deformation, ca. 85 Ma.

Mile 3.7: MP-53.
Mile 3.8: Road turnoff to Stop 3-3, just past Dworshak National Fish Hatchery. Turn right here, before the bridge that crosses the North Fork of the Clearwater River, and proceed toward Bruces Eddy launch area.
Mile 5.3: Road to Dworshak dam quarry turns off to the right. The quarry and surrounding area were studied in detail by Strayer et al. (1989), who described a 1.5-km (5000-ft)-thick, moderately northeast-dipping, northwest-striking mylonite zone with steeply pitching lineation developed in quartz diorite along and southwest of the 0.704/0.706 Sr isotopic boundary. Consistent reverse, top-to-southwest shear sense is given by offset of pegmatite dikes that intrude the quartz diorite. Strayer et al. (1989) estimated considerable thrust displacement (top-to-the-southwest) on this mylonite zone based on shear strain calculation using transposed mafic dikes. Quartz diorite from the quarry yielded a 116.7 ± 0.6 Ma U-Pb zircon age (single grain TIMS analysis; Snee et al., 2007), similar to the age of quartz diorite at the previous stop.
Mile 5.6: Bruces Eddy launch area, east end of Dworshak dam.

OPTIONAL STOP: Deformed Outboard (Low Sr_i) Cretaceous Plutons

The rock at this outcrop is foliated hornblende-biotite quartz diorite gneiss similar to that at the Dworshak dam quarry mentioned above. Structurally, it is located just south of the arc-continent boundary. It is a blastomylonite, indicating that deformation resulted in large, recrystallized grains. The unit contains 5–10 percent biotite, 5–15 percent hornblende, and epidote interpreted to be primary based on the presence of allanite cores. All sampled bodies have Sr_i < 0.704 (Criss and Fleck, 1987). The quartz diorite at the dam yielded $^{40}Ar/^{39}Ar$ plateau ages on hornblende (81.9 ± 0.4 Ma and 82.2 ± 0.4 Ma; Davidson, 1990; Snee et al., 2007). If the quartz diorite from the dam is the same age (ca. 117 Ma) as quartz diorite in the quarry, initial cooling was protracted or the hornblende was disturbed by later thermal events in the Late Cretaceous. Multiple pegmatite generations are visible here. Look for early ones stretched into foliation and dismembered, younger ones that are folded, and still younger ones that cross-cut the fabric. None of these have been dated, but late-stage pegmatite from the old Ahsahka grade 3 km (2 mi) west was dated by U-Pb (zircon) methods at 88.6 ± 1.4 Ma (Payne, 2004)

Return to Highway 7. Reset odometer to zero. Turn right onto Highway 7 and cross bridge to Ahsahka.
Mile 0.8: Road turnoff to Dworshak Dam Visitors Center and Big Eddy Marina. Turn right here.
Mile 4.1: Parking lot at Big Eddy Marina. Park here and walk to outcrops near small boat ramp on the opposite side of the building.

STOP 3-3: High Sr_i Pluton Intruded near Arc-Continent Boundary (46.5284 °N, 116.3053 °W)

Mylonitic granitic orthogneiss here has high Sr_i (>0.708; Criss and Fleck, 1987) and a U-Pb age on zircon of 96 ± 1 Ma (Unruh et al., 2008). The granitic gneiss contains potassium feldspar, which is uncommon in plutonic rocks in the Orofino area. Sr_i values decrease sharply to the northeast (Fig. 23), before climbing to 0.707–0.708 over a distance of ~3 km (2 mi). One explanation for this relationship is that this relatively young granitic gneiss was sourced from cratonal material that had been thrust under arc rocks and older plutons with low Sr_i ratios.

Good shear sense indicators can be seen on the outcrops near the swimming area. The rock displays reverse shear sense indicators on faces both perpendicular to foliation and parallel to lineation (Figs. 24A, 24B). However, it also contains *sinistral* shear sense indicators on faces perpendicular to foliation

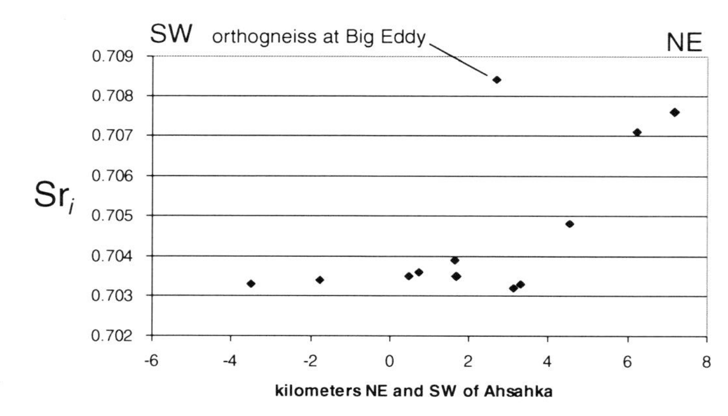

Figure 23. Initial $^{87}Sr/^{86}Sr$ values (Sr_i) plotted against distance northeast and southwest of Ahsahka (modified from Lewis et al., 2005a). Stop 3-3 is 2.3 km northeast of Ahsahka. Note data point ~2.5 km northeast of Ahsahka at Big Eddy Marina that shows anomalously high Sr_i relative to nearby samples. The 0.704/0.706 Sr isotopic "line" has generally been placed at this location, but Sr_i ~1 km (0.6 mi) farther northeast clearly decreases back to levels determined southwest of here, followed by a gradual increase in the ratio at distances >4 km (2.5 mi) northeast of Ahsahka.

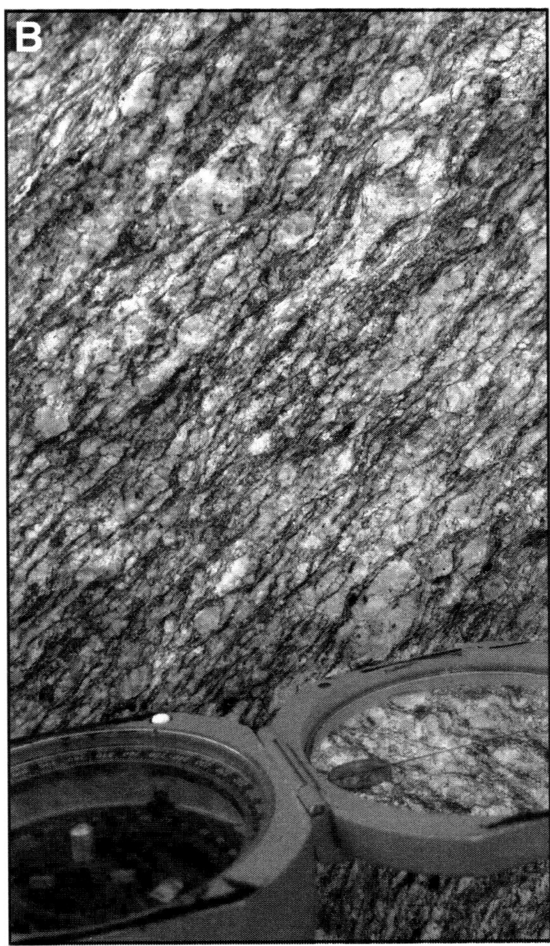

Figure 24. Field photograph of orthogneiss at Big Eddy Marina on Dworshak Reservoir (Stop 3-3). (A) Rock face is perpendicular to lineation (016, 47°) and foliation (297, 48°NE). Sigma porphyroclast indicates a left-lateral sense of shear—note the "tails" of the upper-left and lower-right sides of the porphyroclast below the 0.5 mm lead pencil. (B) Rock face is perpendicular to foliation (297, 48°NE) and parallel to lineation (016, 47°; compass oriented to point N). Weakly developed S-C fabrics and sigma porphyroclasts indicate reverse sense of shear.

and perpendicular to lineation. These senses of shear are typical of the fabric in the Dworshak reservoir area (Stetson-Lee et al., 2013), and are recorded by folded dikes, rotated porphyroclasts, and S-C fabrics in a few areas. This combination of shear sense indicators suggest the reverse and sinistral shear sense indicators formed together during triclinic deformation.

Stetson-Lee et al. (2013) and Byerly et al. (2013) have noted that the timing and style of deformation and exhumation in this area are nearly identical to that in the western Idaho shear zone near McCall (Giorgis et al., 2008). If correct, shear zone fabrics continue north from southern Idaho (Owyhee Mountains; Benford et al., 2010) and actually "turn the corner" in the Syringa embayment. McClelland and Oldow (2004, 2007) reasonably interpreted these fabrics as resulting from a different shear zone (e.g., Orofino shear zone), possibly because they contained kinematics (components of reverse and sinistral motion) not observed in the western Idaho shear zone near McCall. We argue, however, that reverse and sinistral motion are expected, controlled by the change in orientation of the North American margin. S. Giorgis (2013, personal commun.) has recently estimated a 50° angle of oblique convergence for the shear zone near McCall (estimated as an angle of 45–60° by Giorgis and Tikoff, 2004). If this convergence direction characterizes the overall movement of the accreted terranes with respect to North America, the same azimuth would resolve a *sinistral* shear sense along an EW-oriented boundary. There is also a slight difference in structural style around the Orofino "corner." Deformation appears strictly transpressional by McCall, but appears to be oblique-reverse near Orofino; these motions must be kinematically linked, a point made by McClelland and Oldow (2004).

Country rocks at this stop are part of the Orofino series, amphibolite-facies metasedimentary (and metavolcanic?) rocks first recognized near Orofino (Anderson, 1930; Hietanen, 1962). Lithologically varied at outcrop scale, these rocks are commonly sulfide-rich with iron-stained exteriors. The unit includes marble, some in large enough blocks to be mined to make cement, commonly associated with dark gray, fine-grained, garnet-diopside-hornblende gneiss. These metamorphic rocks appear to belong to the accreted terrane assemblage, but they straddle the Sr_i 0.704/0.706 line (Criss and Fleck, 1987) from Orofino to Kooskia. Two reconnaissance detrital zircon samples from garnet-hornblende-plagioclase quartzite near Orofino yielded ages no older than 280 Ma (Schmidt et al., 2003), indicating a detrital sedimentary source that did not include Laurentian rocks.

Return to Highway 12.

MP-44.0: Bridge to Orofino. Return to milepost mileage. Travel east.

MP-46.2: Outcrops on opposite side of canyon are marble of the Orofino series. See Lewis et al. (2005b) for detailed geology of this area.

MP-47.7: Opposite the mouth of Jim Ford Creek.

Historical interest: On 21 September 1805, explorer William Clark, five of his party, and a Nez Perce guide hiked down the ridge south of Jim Ford Creek to reach the Clearwater River. The guide located the nearby Nez Perce camp of Twisted Hair, who crossed the river to meet Clark. He and Clark smoked a pipe in the tradition of friendship. According to the Nez Perce account, Clark and his party had been spared from attack the previous night by the intercession of Wat-ku-ese, a Nez Perce woman who had previously been befriended by whites while a captive of another tribe far to the east (Hendrickson and Laughy, 1999, p. 39).

MP-48.0: Outcrops of Cretaceous or Jurassic quartz diorite from here to Greer.
MP-48.8: The road crosses back across the Ahsahka thrust and leaves the zone of pervasively foliated rocks.
MP-51.6: Bridge to Greer.

Historical interest: It was the discovery in 1860 of gold in the Pierce area northeast of here that increased encounters between Nez Perce tribal members and settlers of European descent. Settling here was a clear breach of an 1855 treaty, which had established a Nez Perce homeland. Conflict of ownership was resolved by another treaty offered by the U.S. Government in 1863, which dramatically decreased the size of the Nez Perce reservation. However, this treaty was signed only by the Nez Perce tribes that already lived on the new smaller reservation, led by a Nez Perce chief aptly named Lawyer. The 1863 treaty thus split the Nez Perce tribe into the "treaty" Nez Perce and the "non-treaty" Nez Perce, the latter of which never signed the treaty that gave away their land. Thus, the discovery of gold and the settlements associated with it led in a relatively direct—and largely inevitable—way to the Nez Perce war.

SIDE TRIP to Greer Area

Reset odometer to zero. Turn left and cross bridge to Greer.
Mile 0.5: Pullout on right side of road.

STOP 3-4: Relatively Undeformed Cretaceous Plutons with Low Sr_i (46.3851 °N, 116.1746 °W)

Good exposures of Cretaceous quartz diorite are present here on the lower part of Greer Grade. A sample yielded a ca. 89 Ma U-Pb zircon age (LA-ICPMS method, J. Vervoort, 2007, written commun.), similar to that of the Payette River tonalite. Unlike the Payette River tonalite, Sr_i values here are low (0.7033; Criss and Fleck, 1987). We are outboard (southwest) of the Ahsahka thrust.

Return to Highway 12. Return to milepost mileage. Travel east.
MP-53.4: Pullout on west side of highway. Pleistocene Glacial Lake Missoula outburst flood deposits in bank on west side of highway. Flood sediments (backwash) are present up the Clearwater River east as far as Kamiah.
MP-54.3: Fishing access site on east side of highway.
MP-55.1: Crossing into muscovite trondhjemite of the Sixmile Creek pluton.
MP-56.0: Pullout on left side of road. Outcrops are on opposite side of highway. *Note: This dangerous stop is not recommended for large groups.*

OPTIONAL STOP: Trondhjemite-Tonalite Suite Plutons

The Sixmile Creek pluton (Lee, 2004) is one of two epidote-bearing trondhjemite-tonalite plutons emplaced within the accreted terrane several kilometers west of the northern part of the arc-continent boundary. It is similar to other trondhjemite-tonalite-granite suite (TTG) intrusions in the region, including the trondhjemite near Riggins (Hamilton, 1963; Barker, 1979) and the Hazard Creek complex (Manduca et al., 1992, 1993). The pluton is zoned from a southern biotite-muscovite trondhjemite to a northern biotite tonalite. Pegmatites are locally abundant. Laser ablation U-Pb zircon ages by Lee (2004) are 119.1 ± 1.5 Ma for the tonalite and 115.7 ± 2.7 Ma for the trondhjemite. Crystallization, therefore, postdated crustal juxtaposition across the arc-continent boundary, but predated emplacement of the Idaho batholith. The pluton is foliated locally but predominantly massive. The rocks contain greater than 15 wt% Al_2O_3 and have high Na_2O/K_2O ratios (generally >5).

MP-61.9: Pullout on left side of road.

STOP 3-5: Jurassic Pluton of the Wallowa Terrane (46.2647 °N, 116.0913 °W)

Unit is massive to moderately foliated quartz diorite. Invariably hornblende-bearing; biotite subordinate (0–7 percent). Nearby sampling indicates that Sr_i values are low (0.7033–0.7037; Criss and Fleck, 1987). The U-Pb zircon age of 157.0 ± 0.5 Ma (McClelland and Oldow, 2007) from this outcrop and nearby hornblende $^{40}Ar/^{39}Ar$ plateau age of 157.9 ± 2.8 Ma (Criss and Fleck, 1987) indicates that at least some of the quartz diorite is Jurassic in age. The concordance of the zircon U-Pb and hornblende $^{40}Ar/^{39}Ar$ ages indicates that these rocks have been little affected by later thermal and deformational events in contrast to exposures nearer the arc-continent boundary.

MP-65.0: Flat area on opposite side of river now occupied by a lumber mill was the site of Lewis and Clark's "Long Camp."

Historical interest: Lewis and Clark occupied Long Camp from 14 May to 10 June 1806, waiting for the snow to melt so they could proceed east over the Lolo trail. On 10 June, against the Nez Perce advice, they started for Weippe and the Lolo trail. By 17 June, they were in snow 8–12 feet deep, and had to return to Weippe. Finally on 24 June, they set out again to cross the mountains. They reached Lolo Hot Springs on 29 June and enjoyed a good soak (Hendrickson and Laughy, 1999, p. 44).

MP-66.3: Kamiah. The valley broadens (Fig. 14), in part because of numerous landslides east and west of Kamiah in which large blocks of Columbia River basalt have moved downslope.
MP-68.5: Nez Perce National Historic Park site ("Heart of the Monster").

Historical interest: This site commemorates the creation of the Nee Me' Poo (or Ni Mii Puu), The People, today known as the Nez Perce. Cunning Coyote, magical creature of Nez Perce mythology, had been tearing down waterfalls along the Columbia River to the west to enable more salmon to travel upstream during their spawning migration. When Coyote heard of a monster devouring all the creatures of the Kamiah Valley, he armed himself and proceeded to the valley. He then tricked the monster into inhaling him, and, once inside, lit the fat surrounding the monster's heart. This enabled the previously ingested animals to escape in the smoke. Coyote then severed the monster's heart and made his own exit. Once free, he cut pieces from the dead monster's body and cast them to different places in the Pacific Northwest to create the other Indian tribes. Then Coyote used the blood from the enormous heart to create the Nee Me' Poo. Gradually the monster's heart turned to stone (basalt) and is still visible protruding from the earth at the center of the park (Hendrickson and Laughy, 1999, p. 49).

MP-72.1: Basalt cliffs (Columbia River Basalt Group). Bedrock is basalt from here to MP-81.3. All canyon exposures are Grande Ronde Basalt; uplands are Saddle Mountains Basalt.
MP-73.9: Junction of State Highway 13 that leads to Kooskia and the South Fork of the Clearwater River, and end of field trip 2.
MP-76.3: Tukatesp'e (Skipping stones) picnic area. Outcrops in river are Grande Ronde Basalt.
MP-81.3: Outcrops of foliated hornblende-biotite quartz diorite (Jurassic or Cretaceous) on north side of highway. Hornblende yielded an $^{40}Ar/^{39}Ar$ plateau age of 118.6 ± 0.8 Ma (Criss and Fleck, 1987). This location is less than a mile west of the $^{87}Sr/^{86}Sr$ 0.704/0.706 line. Samples have a low Sr_i ratio (0.7034; Criss and Fleck, 1987).
MP-82.3: Outcrops of strongly foliated hornblende-biotite quartz diorite (Jurassic or Cretaceous) with high Sr_i ratio (0.7091; Criss and Fleck, 1987).
MP-82.6: Intersection of Suttler Creek road to left: This is the approximate location of the 0.704/0.706 initial Sr_i isopleth identifying the arc-continent boundary (Armstrong et al., 1977; Fleck and Criss, 2007).

See Figure 25 for Stops 3-6 to 3-15.

MP-83.0: Pullout on right side of road.

STOP 3-6: Plutons Intruded into the Arc-Continent Boundary (46.1471 °N, 115.8397 °W)

This stop examines biotite tonalite east of 0.704/0.706 Sr_i isopleth. The western margin of this pluton is foliated, whereas the central part is massive to weakly foliated. No age has been obtained from here but it is mapped as Cretaceous. A sample collected from eastern part of pluton (MP-85.0) has anomalously low Sr_i (0.7037; Criss and Fleck, 1987). This result illustrates that in the Syringa embayment, there are irregular variations in Sr_i rather than the abrupt and simple boundary documented along the north-striking segment to the south.

MP-87.4: Swan Creek road. A hornblende-rich gneiss of the Mesozoic Orofino series is exposed along the lower part of Swan Creek road and in float back along the highway (Lewis et al., 2007a). The protolith was likely arc-derived sedimentary (and volcanic?) rocks, possibly of the Wallowa terrane. These rocks extend northeast to the Woodrat Mountain fault (about MP-88), a steeply dipping and locally folded structure that places Orofino series rocks on the west against continental rocks (Neoproterozoic quartzite) to the east. This marked difference in country rocks across the fault identifies the surface expression of the arc-continent boundary although the 0.704/0.706 Sr_i isopleth is mapped to the west at Suttler Creek (MP-82.6, near Stop 3-6).
MP-88.4: Outcrops of hornblende-biotite granodiorite orthogneiss occur ~70 m (230 ft) west of the Smith Creek road turnoff (MP-88.5). Potassium feldspar in this rock may be secondary because it is interstitial. The pluton was emplaced along and east of the Woodrat Mountain fault and underwent solid-state deformation. Samples at this locality yielded a K-Ar cooling age of 74.4 ± 0.5 Ma from hornblende and Sr_i of 0.7073–0.7075 (Criss and Fleck, 1987). The same rock is better exposed up the Swan Creek road, where anisotropy of magnetic susceptibility (AMS) is well defined with 350°, 73° strike and dip of flattening fabric and nearly down dip 72°, and 73° lineation trend and plunge (R.F. Burmester, 2013, written commun.). This AMS fabric is similar to that defined by deformed minerals immediately to the east at Stop 3-7.
MP-88.5: Smith Creek road turnoff. Turn left and proceed 0.3 mile up Smith Creek Road to first bold outcrop.

STOP 3-7: Syringa Quartzite (46.1376 °N, 115.7470 °W)

This outcrop consists predominantly of mylonitic quartzite. Thin calc-silicate gneiss intervals and thicker intervals of

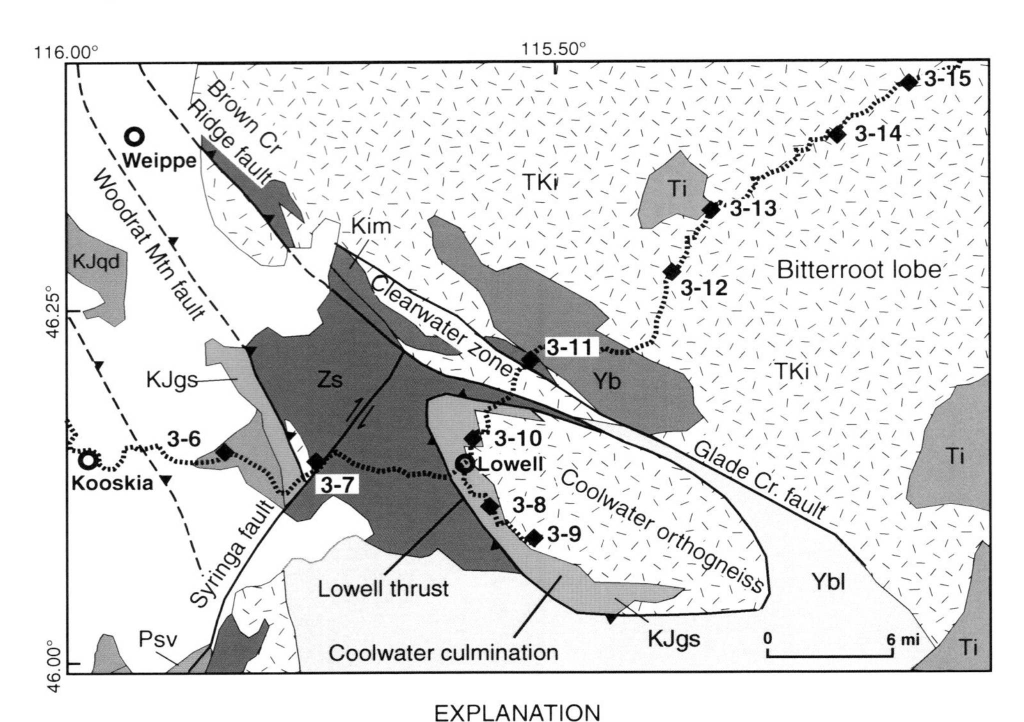

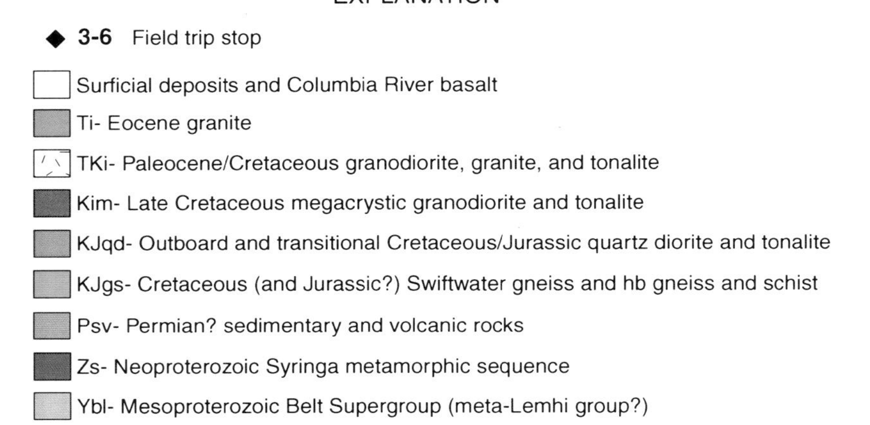

Figure 25. Map of Lowell area showing field trip Stops 3-6 to 3-15 and simplified geology modified from Lewis et al. (2012).

kyanite schist are exposed farther up the road. These rocks were once mapped as Mesoproterozoic Belt Supergroup (Greenwood and Morrison, 1973). However, an intimate association of pure quartzite and calc-silicate gneiss is a characteristic not found in the Belt Supergroup; furthermore, the thin carbonate intervals present throughout this succession are dissimilar to the thicker and more restricted carbonate intervals in the Belt Supergroup. Detrital zircon ages obtained from this outcrop are highly discordant, but one concordant grain dated at ca. 755 Ma and another at ca. 1130 Ma (Lewis et al., 2010) indicate these are younger than the Belt Supergroup (deposited ca. 1500–1400 Ma, Evans et al., 2000). This confirms the recognition by Lund et al. (2008) of a package of relatively young metasedimentary rocks from here east to Lowell. Similar rocks are mapped as the Neoproterozoic Umbrella Butte Formation (Lund et al., 2003), where it is exposed to the south in the Gospel-Hump Wilderness. Thus, the most likely correlation for this package is with the Neoproterozoic Windermere Supergroup or Neoproterozoic (?) Deer Trail Group, both Laurentian supracrustal units.

Pervasive linear-planar tectonite fabrics in this area have a range of attitudes (Fig. 26A) typical of high-strain structures to the south (e.g., Salmon River canyon region near Riggins; Blake et al., 2009; Gray, 2013). Here, metamorphic foliation dips moderately southeast and contains a steep mineral stretching lineation defined by elongate quartz (Figs. 26B, 26C). In thin section, mylonitized quartzite displays a strong shape-preferred orientation of white mica porphyroclasts, with the longest dimensions

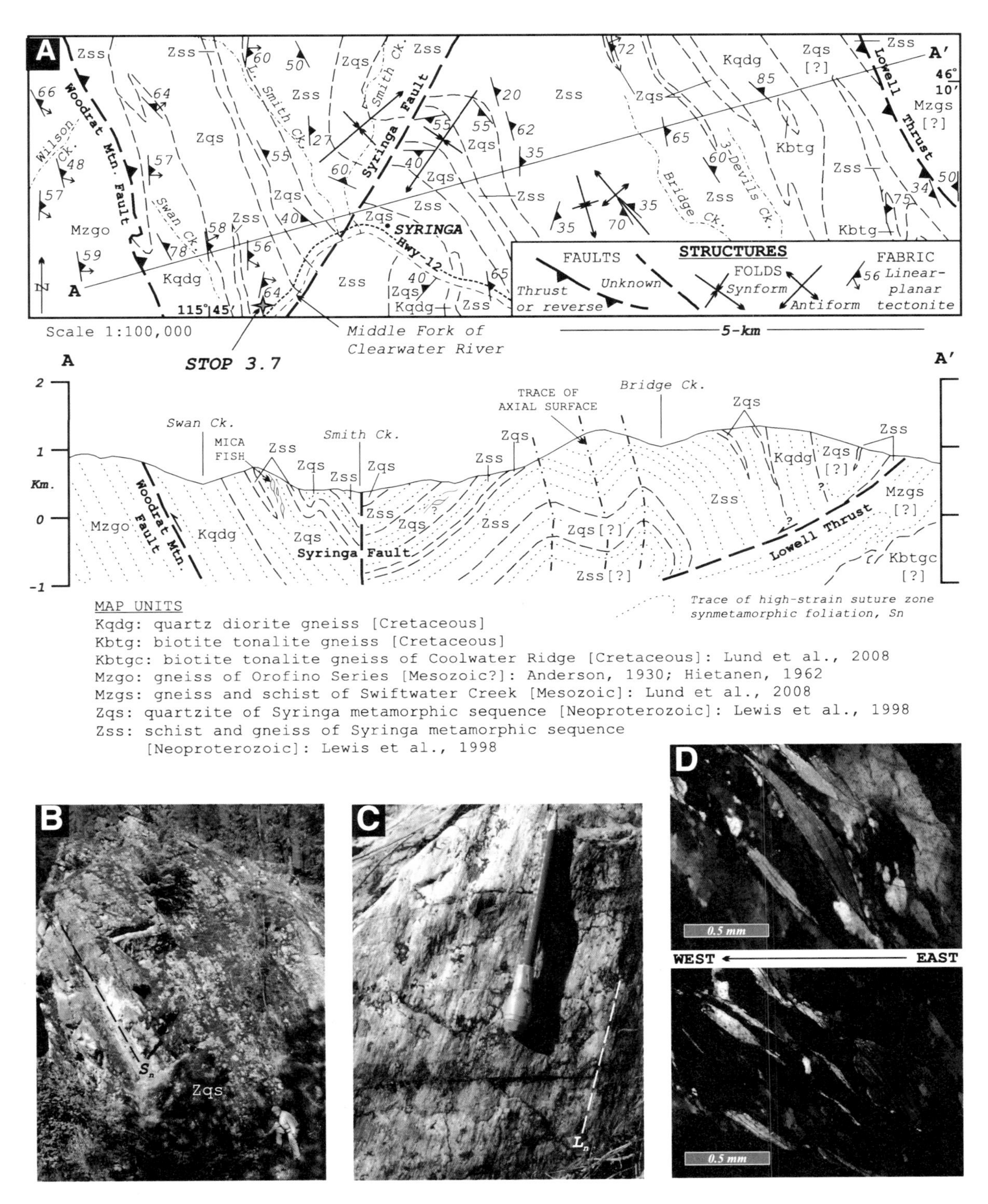

Figure 26. (A) Geologic map of the Syringa area, north-central Idaho; modified from Lewis et al. (2007a). Stop 3-7 indicated by star in lower left. Section A–A′ by K. Gray. Lowell thrust geometry (folded) acquired from Lund et al. (2008). (B) Field photograph of Neoproterozoic quartzite (Zqs unit) exposed at Stop 3-7. J. Watkinson shows dip of foliation (S_n = 012 64°SE). (C) In situ outcrop photograph of steeply plunging stretching lineation (L_n = 078 59°NE) on S_n. (D) Oriented thin section photomicrographs of mylonitic quartzite sampled along lower Smith Creek Road. View is to the north in each. The long axes of sigmoidal microstructures (mica fish) are gently inclined to the trace of foliation. Fish morphologies are largely lenticular and display high aspect ratios. Porphyroclast asymmetry and stacking patterns record top-to-the-west shear. Thin sections are cut parallel to lineation and sub-perpendicular to foliation strike. Crossed polars.

inclined at a small angle to foliation (Fig. 26D). Based on their asymmetrical, sigmoidal shape and stair-stepping of trails (cf. Lister and Snoke, 1984; ten Grotenhuis et al., 2003), mica fish locally record an east-over-west (reverse) sense of shear. When viewed on the lineation normal face (Y-Z plane of strain ellipsoid; Ramsay and Huber, 1983), textural asymmetry is not observed. As such, fabrics reveal no obvious field or microstructural evidence of transpressional strain (cf. Tikoff and Greene, 1997). These observations are consistent with arc-continent transitional structures reported in the Salmon River canyon east of Riggins (Gray et al., 2012).

Return to Highway 12. Continue east.

MP-90.0: Town of Syringa. The northeast-striking Syringa fault passes through Syringa and is probably responsible for the bend in the Middle Fork of the Clearwater River here. The regional northwest-striking foliation swings to the northeast near this fault and dikes of several ages intruded along it. The fault records both ductile and brittle deformation. Although access and exposures are limited south of the river, the Woodrat Mountain fault is truncated by, or merges with, the Syringa fault. As presently mapped (Lewis et al., 2007a), Syringa metamorphic sequence rocks are offset to the southwest on the southeast side of the fault in a right-lateral direction. However, some quartzite intervals north of Syringa appear to be offset left-laterally.

MP-93.0: Schist of the Syringa metamorphic sequence (Neoproterozoic). Unit is typically coarse-grained and feldspar-poor. Coarse-grained kyanite, accompanied by garnet and sillimanite, is relatively common and especially abundant to the northwest on Woodrat Mountain (Van Noy et al., 1970).

MP-94.3: Three Devils Picnic Area.

MP-95.3: Wild Goose campground. Park at pullout and proceed 0.2 mile up river to low roadcuts.

OPTIONAL STOP: Gneiss of Syringa

This stop includes exposures of interlayered quartzite and schist that are part of the same rock package as at Smith Creek (Stop 3-7).The gneiss of Syringa is predominately massive, rusty weathering, pyritiferous, garnet- and kyanite-bearing, muscovite-biotite-quartz-feldspar gneiss and schist. Relict arkosic or quartz-pebble grit zones, 200-m (660-ft)-thick quartzite units, and minor calc-silicate-bearing layers are present locally. This unit probably originated as immature feldspar-pebble graywacke interbedded with more mature quartz-rich sandstone. No primary sedimentary structures have been recognized. Thickness estimates are probably compromised by deformation but the unit may be ~1000 m (3300 ft) thick. A quartzite sample collected here contains detrital zircons having a range of 1500–775 Ma plus several Archean grains (Lund et al., 2008). Peaks between 900 and 775 Ma indicate the maximum age of the unit is Neoproterozoic and a source from the Rodinia rift belt. Peaks at ca. 1200–1100 indicate unidentified sources, and peak at 1450 Ma and the Archean grains indicate recycling from Mesoproterozoic strata. Thus, this unit had a Precambrian (Laurentian) provenance.

MP-96.7: Confluence of Lochsa and Selway Rivers, head of Middle Fork of Clearwater River. The road to the right leads up the Selway River.

SIDE TRIP up Selway River, Coolwater Culmination

The mileage restarts at the confluence of the Lochsa and Selway rivers.

MP-1.9: Straight stretch in road with wide shoulder but no large pullout.

OPTIONAL STOP: Gneiss of Swiftwater Creek

This stop exhibits a massive epidote-hornblende-biotite-quartz-feldspar gneiss containing lesser felsic conglomerate and feldspar-quartz gneiss. Other than local grit textures, primary sedimentary features are not preserved. Upper and lower contacts are structural or intrusive but as much as 500 m (1600 ft) thickness is preserved. The unit probably originated as volcanogenic, calcareous graywacke, and silty feldspathic sandstone (Fig. 27). Detrital zircons from this exposure are primarily 225 to 100 Ma grains plus a few 1800 to 1400 Ma grains (Lund et al., 2008). Detrital zircon data indicate that the unit was mostly sourced from oceanic terrane and plutonic rocks west of the arc-continent boundary but with minor contribution from Laurentia. Zircon

Figure 27. Gneiss of Swiftwater Creek along Selway River (MP-1.9), looking north.

data constrain depositional age to be between ca. 98 Ma (age of the youngest detrital zircon) and 86 Ma (age of the crosscutting Coolwater Ridge orthogneiss, see below). These rocks likely originated in an oceanic-floored, predominantly arc provenance, flysch basin between the arc and continent (Lund et al., 2008). Unit is exposed in the core of an antiform below the folded Lowell thrust fault (the "Coolwater culmination").

MP-2.8: Swiftwater bridge. Park at pullout by bridge and walk across bridge to upriver outcrop.

STOP 3-8: Ultramafite along Lowell Thrust Fault (46.1153 °N, 115.5727 °W)

A coarse-grained chlorite-amphibole ultramafite forms a massive outcrop. This exposure and several other small ultramafic and mafic (pyroxenite, dunite, garnet peridotite, and gabbro) bodies are strung along and near the western edge of the gneiss of Swiftwater (Anderson, 1931; Lewis et al., 2007a). The bodies are structural slivers, marking the Lowell thrust fault as a crustal-scale structure between the oceanic- and Laurentian-derived gneisses (Lund et al., 2008). Peridotite here contains abundant relicts of coarse-grained olivine. Minerals forming at the expense of olivine and possibly from orthopyroxene include talc, actinolite, chlorite, and carbonate. Bonnichsen and Godchaux (1994) described this exposure as well as numerous other ultramafic bodies that crop out along and west of the arc-continent boundary in western Idaho.

MP-4.6: Fenn Ranger Station.
MP-5.9: Pullout on right in bend in road.

STOP 3-9: Coolwater Ridge Orthogneiss (46.0927 °N, 115.5269 °W)

This is the pervasively deformed, migmatitic, fine-to medium-grained, biotite tonalite Coolwater Ridge orthogneiss (Morrison, 1968). Typically, quartz content is 20–30 percent and biotite content is 4–20 percent. Muscovite is present as less than or equal to 5 percent and is not conspicuous in hand specimen; garnet is present in trace amounts. Thin, lit-par-lit leucosomes are conspicuous and folded with tonalitic melanosome (Fig. 28). Planar fabric formed in solid state and was folded about shallow to steep eastward-plunging axes; down-dip plunging lineation was locally developed. This locality produced a complex set of U-Pb zircon ages with an interpreted emplacement age of 86.2 ± 0.8 Ma based on careful exclusion of leucosome material (Lund et al., 2008). Although the Coolwater Ridge orthogneiss is located 15 km (9 mi) east of the 0.708 Sr_i line of Criss and Fleck (1987), it has an Sr_i of 0.7066 and εNd of −2.91 (Lund et al., 2008). The anomalously low Sr_i value indicates a mixed origin from both oceanic and continental sources. Other conclusions from detailed SHRIMP (sensitive high-resolution ion microprobe) U-Pb data acquired for samples from this body include (1) xenocrystic zircon cores were recycled from the Blue Mountains province despite location of the Coolwater orthogneiss on the Laurentian side of the boundary, (2) zircon from leucosomes provided an age of 76.1 ± 1.2 Ma for the metamorphic (or melting?) event that resulted in migmatite, and (3) zircon overgrowths at 64.1 ± 1.2 Ma provided the age of a younger metamorphic event (Lund et al., 2008).

Turn around here and return to Lowell.

MP-97.3: Lowell.
MP-99.0: Mouth of Pete King Creek. Pullout on right.

STOP 3-10: Gneiss of Swiftwater (46.1660 °N, 115.5881 °W)

Massive weathering, biotite-quartz-plagioclase gneiss and schist at this exposure is interpreted as paragneiss in the window of Mesozoic island-arc rocks exposed well east of the arc-continent boundary, northwestern part of the Coolwater culmination (Lund et al., 2008). This and several other exposures contain trace to 2 percent, 5–10-mm-long black tourmaline crystals (Fig. 29). They are notable for high plagioclase content (15–65 percent) relative to other schist in the region and mapped by some workers as orthogneiss (e.g., Pitz, 1985). Other content includes quartz 15–50 percent, biotite 5–35 percent, muscovite 0–15 percent, garnet 0–5 percent, and potassium feldspar 0–3 percent. Because of high plagioclase content, the unit is enriched in Sr (>300 ppm), Na_2O (>2.9 percent), and CaO (>2.2 percent) relative to Neoproterozoic schist of the Syringa metamorphic sequence exposed west of Lowell. Zircons from this outcrop yield both Mesozoic and Proterozoic populations (U-Pb LA-ICPMS methods; Peter Oswald and Jeff Vervoort, 2009, written commun.), but Mesoproterozoic ages (many ca. 1380 Ma and thus similar to granite near

Figure 28. Coolwater orthogneiss in quarry along O'Hara Creek Road ~3 km southeast of Stop 3-9.

Elk City; Fig. 2) are predominant. Thus, these rocks likely formed from sediments with mixed arc and Laurentian sources.

MP-101.2: Northeast contact of Coolwater Ridge orthogneiss is located near here.
MP-101.6: Massive weathering schist and gneiss mapped as gneiss of Swiftwater Creek by Lewis et al. (2007a). εNd values here are about −1.1 (calculated at 90 Ma; Peter Oswald and Jeff Vervoort, 2009, written commun.), consistent with the field interpretation shown on earlier mapping (Lewis et al., 2007a).
MP-101.8: Likely crossing northeastern boundary of the Coolwater culmination (Lowell thrust fault) near here.
MP-102.2: Biotite quartzite and biotite-feldspar-quartz gneiss that are part of the Syringa metamorphic sequence or, less likely, Mesoproterozoic Belt Supergroup. εNd values here and in calc-silicate rocks 0.3 miles to the east are about –6.5 and –7.8, respectively (calculated at 680 Ma; Peter Oswald and Jeff Vervoort, 2009, written commun.), consistent with the age assignment and map interpretation as Syringa metamorphic sequence and not younger strata (Lewis et al., 2007a).
MP-103.0: Hornblende-biotite tonalite orthogneiss. Continental Sr_i value of 0.715 (Criss and Fleck, 1987). Rocks from here to MP-104.3 are typically mylonitic with near-vertical west-northwest–striking foliations. The fabric may be related to the Glade Creek fault discussed below.
MP-104.3: Apgar Campground.
MP-105.0: Glade Creek Campground. Park at campground and walk across highway.

Figure 29. Gneiss of Swiftwater Creek at mouth of Pete King Creek (Stop 3-10) with tourmaline (small black elongate grains) on schistose parting surface.

STOP 3-11: Augen Gneiss of Apgar Creek and Glade Creek Fault (46.2208 °N, 115.5266 °W)

The augen gneiss of Apgar Creek is megacrystic, coarse-grained, biotite- and hornblende-biotite granodiorite and subordinate tonalite orthogneiss (Fig. 30A). Unit contains potassium feldspar megacrysts as long as 5 cm that commonly form augen (Fig. 30B), some of which show rapikivi texture. Early U-Pb (TIMS) dating yielded an upper intercept age of 1751 ± 63 Ma,

Figure 30. Orthogneiss of Apgar Creek near Glade Creek campground (MP-105.0, Stop 3-11). (A) Steep foliation in outcrop. (B) Potassium feldspar augen.

resulting in an interpretation that augen gneiss of Apgar was part of the Paleoproterozoic basement to the Idaho batholith (Toth and Stacey, 1992). Lund et al. (2008) obtained a 94 ± 1 Ma age from zircon (SHRIMP U-Pb), Sr_i of 0.7083, and εNd of −5.22. Criss and Fleck (1987) report similar Sr_i as well as K-Ar cooling ages of 64.9 ± 2.0 Ma on hornblende and 53.0 ± 0.3 Ma on biotite. The cooling ages reflect either slow cooling from 94 to 53 Ma or, more likely, a reheating event. This unit and the similar one at MP-156.5 overlap in age and are petrographically and chemically similar to plutons of the early metaluminous suite and border zone suites more extensively exposed further south in the batholith (Gaschnig et al., 2010). The combination of outcrop occurrences and inherited zircons in younger phases of the batholith implies that ca. 100–90 Ma magmatism analogous to that in the Sierra Nevada occurred throughout the Idaho batholith.

The Glade Creek fault (Lewis et al., 1992b, 1998) is characterized by west-northwest (285°)–striking and vertical to steeply north-dipping foliation superposed on the plutonic rocks. Early down-dip ductile fabric overprinted by subhorizontal slickensides (Pitz, 1985) is interpreted to record initial southwest-directed reverse motion (postdating movement on the Lowell thrust fault) followed by possible left-lateral motion (Cretaceous?) and later right-lateral motion (Eocene?; Lewis et al., 2007a). Although shown as a single fault trace (Lewis et al., 1992b), the Glade Creek fault is one of a number of faults, including the Brown Creek Ridge shear zone (Lewis et al., 2007a), that are part of a broad zone of deformation at least several km in width and regional length collectively termed the "Clearwater zone" (Sims et al., 2005; Lund et al., 2008). Several major regional structures with other names have been interpreted in this general area, including the "trans-Idaho discontinuity" (Pitz, 1985, after a concept to explain abrupt changes in trends of Mesoproterozoic to Paleozoic rocks as presented by Yates, 1968, and in Mesozoic rocks by Armstrong et al., 1977) and "Orofino shear zone," which is interpreted as a sinistral offset of the arc-continent boundary and linking to thrust systems in the Rocky Mountain fold and thrust belt (McClelland and Oldow, 2004, 2007). Lund et al. (2008) suggested that this is a young manifestation of a long-lived fundamental crustal boundary that, in the context of Cretaceous tectonics, bounded the northeast side of the crustal wedge exposed in the Coolwater culmination. This old boundary may be the southwest limit of lower Belt Supergroup rocks as Prichard Formation and Ravalli Group rocks have not been recognized to the southwest.

MP-107.7: Major Fenn Campground. Calc-silicate rocks from here back west to MP-107 are probably metamorphosed Belt Supergroup rocks (Piegan Group).
MP-108.4: Knife Edge River access.
MP-110.7: Pullout on right.

OPTIONAL STOP: Western Edge of the Bitterroot Lobe

The outcrop is primarily biotite granodiorite and biotite tonalite. Country rock (quartzite and calc-silicate rocks) to the west is probably metamorphosed Belt Supergroup. Sr_i ratio of 0.7092 was obtained from granodiorite collected 1.6 km (1 mi) to the east near MP-111.7 (Criss and Fleck, 1987).

MP-113.4: Opposite mouth of Old Man Creek.
MP-116.0: Pullout on right.

STOP 3-12: Bitterroot Lobe of the Idaho Batholith (46.2869 °N, 115.3847 °W)

Muscovite-biotite granodiorite with 65.6 ± 3.6 Ma U-Pb zircon (LA-ICPMS) date (Gaschnig et al., 2010). A second U-Pb zircon analysis from a sample with subtle chemical and petrographic differences collected ~300 m (1000 ft) to the southwest along the highway yielded a notably younger 54.8 ± 1.6 Ma age (Gaschnig et al., 2010). Screens of country rock and migmatite are exposed to the southwest along the highway. The presence of muscovite in granodiorite may reflect minor assimilation of pelitic country rocks. Muscovite content typically increases in areas of the Bitterroot lobe with metamorphic screens.

MP-118.4: Pullout on right. Selway-Bitterroot Wilderness signs.
MP-120.1: Mouth of Fish Creek.
MP-120.2: Pullout on right with restrooms. Park here and walk back across Fish Creek bridge along highway.

STOP 3-13: Eocene Plutons of the Challis Magmatic Event (46.3325 °N, 115.3466 °W)

Closest exposures south of Fish Creek are biotite granodiorite. Deeply weathered rocks in the large roadcut to the south are coarse-grained biotite-hornblende syenite (Fig. 31). Dikes of the biotite granodiorite appear to cut the deeply weathered hornblende-rich rocks, but contact relations are not well established. The age of the biotite-hornblende syenite is 48.4 ± 3.5 Ma based on U-Pb zircon (LA-ICPMS) dating (Gaschnig et al., 2010). Age of the biotite granodiorite is less certain. LA-ICPMS dating identified a range of Eocene, Paleocene, and Cretaceous zircon components, and it is unclear whether this body is an Eocene pluton with Paleocene–Cretaceous inheritance or an older pluton that experienced Eocene metamorphic zircon growth during intrusion of the neighboring syenite. Basalt dikes, possibly related to the Columbia River Basalt Group, cut the biotite-hornblende granodiorite. From here east to No-see-um Creek (MP-124.8), contact and age relations are poorly understood. Two older U-Pb zircon dates have been obtained (54 Ma and 57 Ma; Toth and Stacey, 1992; Foster and Fanning, 1997), but small hornblende-bearing granitic intrusions similar to the 48.4 Ma pluton at Fish Creek are present locally. Several unmapped Eocene plutons are possibly present in this area, perhaps deep-level equivalents to the Cook Mountain intrusive complex exposed to the northeast that is more characteristic of typical hypabyssal Challis intrusions.

Figure 31. Eocene biotite-hornblende syenite south of Fish Creek (Stop 3-13, MP-120.2). Coarse grus like that in the cones at this outcrop is common to Eocene plutons in the region.

MP-121.6: Lochsa Historical Ranger Station.

MP-121.95: Pullout on south side of highway. Foster and Fanning (1997) obtained a U-Pb SHRIMP date of ca. 57 Ma near here.

MP-122.6: Turnoff to Wilderness Gateway Campground. From here back along the highway to MP-122 are several large andesitic bodies that have textures indicating mixing of a more mafic phase and a more felsic phase. We suspect they are Eocene (Challis) intrusive phases, both of which were intruded into Cretaceous granodiorite. At the bend in the highway (about MP-122.2) is a grus-rich exposure of likely Eocene hornblende-biotite granite or syenite, similar to that at Stop 3-13.

MP-124.6: U-Pb zircon date (54 Ma; Toth and Stacey, 1992). Numerous rock types are exposed here.

MP-124.8: Mouth of No-see-um Creek. Debris flow down No-see-um Creek in the mid-1990s closed the highway for an extended period. Faulted rock near headwaters may have contributed to slope instability.

MP-125.0: Pullout on south side of road.

OPTIONAL STOP: Mafic Paleocene (?) Dikes of the Batholith

This location is also Stop 4 of Hyndman (1989). Hyndman reported numerous synplutonic, or what we reinterpret as "late plutonic" or "early post-plutonic" mafic dikes, making up ~20 percent of the outcrop, and a modest range of magma mingling and mixing relationships in inhomogeneous granodiorite. Mylonitization of the dike rock is common, suggesting post-emplacement deformation as the dikes cooled.

MP-128.5: Bald Mountain Department of Transportation complex. Park at maintenance station and walk north along highway.

See Figure 32 for Stops 3-14 to 3-19.

STOP 3-14: Mafic Eocene Intrusive Complex (46.3850 °N, 115.2334 °W)

The outcrop consists of hornblende diorite, pyroxene-hornblende gabbro (Fig. 33) that may include cumulate phases, and biotite granodiorite. Hyndman (1989) described the outcrop as a synplutonic mafic complex related to the batholith, but a recently determined U-Pb age of 46.8 ± 1.0 Ma (Gaschnig et al., 2010) indicates that these rocks are part of the younger Challis intrusive episode. Overall, mafic rocks are sparse in the Bitterroot lobe and even more rare in the southern (Atlanta) lobe. Although Hyndman and Foster (1988) stressed the importance of mafic magmas in the generation of the batholith, physical evidence of multiple magmas being present simultaneously is rare at the current levels of exposure, and the mafic rocks and host granites have distinctly different isotopic compositions with no overlap. Mafic magmatic inclusions, so common in the Sierra Nevada batholith of California, are extremely rare throughout the widely exposed biotite granodiorite and granite of the Idaho batholith. Likewise, outcrops this mafic are atypical of the Bitterroot lobe. Fine-grained, presumably quenched contacts between mafic rocks and granodiorite are present as well as some hybridized contacts. It is uncertain which relationships are truly synplutonic and which relationships result from remobilization of small amounts of felsic rock by remelting along intrusive contacts. As noted above, textural evidence of magma mixing is more common in the Eocene (Challis) plutonic suite than in the Idaho batholith.

MP-131.3: Mouth of Holly Creek.

MP-132.5: Pullout on south side of road.

OPTIONAL STOP: Paleocene Main Phase Bitterroot Lobe

Porphyritic biotite granodiorite dated at 58.8 ± 2.9 Ma (U-Pb zircon; Gaschnig et al., 2010). Outcrop here has mylonitic foliation that dips 30° to the northeast and a subhorizontal lineation.

MP-133.7: Pullout on right.

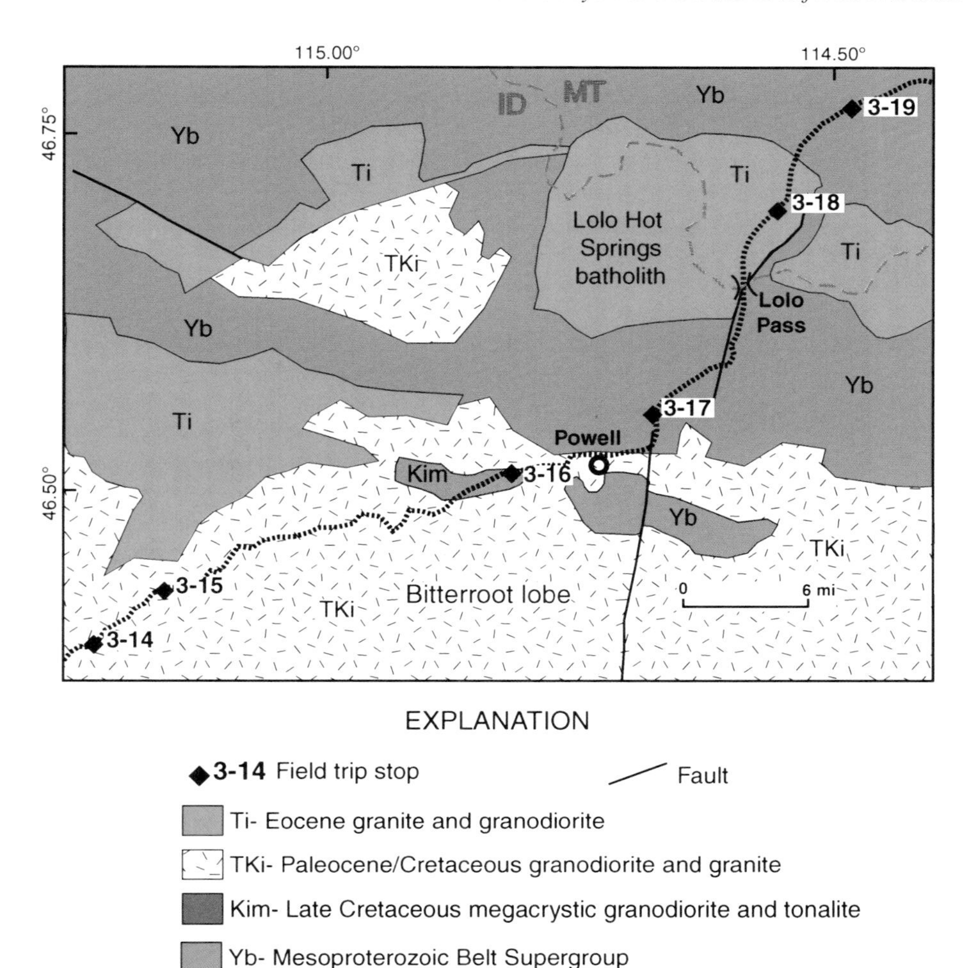

Figure 32. Map of Lolo Pass area showing field trip Stops 3-14 to 3-19 and simplified geology modified from Lewis et al. (2012). ID—Idaho; MT—Montana.

STOP 3-15: Mafic Paleocene Dikes of the Batholith (46.4234 °N, 115.1615 °W)

Outcrop of Paleocene biotite granodiorite with moderately dipping mafic dikes. The inclined dikes at this locality appear to feed into a flat-lying dike near the top of the outcrop. Dikes such as these are typically basaltic andesite and have Sr_i values between 0.7054 and 0.7066. While some are probably Eocene and associated with the Challis event, a $^{40}Ar/^{39}Ar$ date on hornblende from one of these dikes gave a Paleocene age of 56.5 ± 1.8 Ma (Foster and Fanning, 1997), indicating it was contemporaneous with Paleocene granitic magmatism of the Bitterroot lobe here.

MP-134.4: Mouth of Skookum Creek. Date of 58.7 ± 1.2 Ma obtained here (U-Pb zircon; Gaschnig et al., 2010).
MP-139.7: Saddle Camp Road.

OPTIONAL SIDE TRIP to Lolo Trail

Turn left on Saddle Camp Road and proceed north. Reset odometer to zero.

Mile 9.0: Saddle Camp and intersection with Lolo trail (Lolo motorway). Turn right and proceed east along Lolo motorway.

Historical interest: This is the trail that Lewis and Clark's "Corps of Discovery" followed on their epic journey in 1804–1806. Lewis and Clark had a particularly difficult time along the Lolo trail due to fallen timber, snow, and lack of game. Arguments continue regarding their exact path through Clearwater country, but it is likely that the road here closely follows their path. About 15 miles southwest, on their journey west in 1805, the explorers came to what is known as Willow Ridge. The 1500-ft-deep saddle to the west of the ridge (now given the name "Deep Saddle") looked like an improbable trail route, so they mistakenly veered south along Willow Ridge and down into the "Hungery" Creek drainage. Geology played a role in their difficulties, as the saddle was formed by erosion along the north- to northeast-trending Deep Saddle fault. They spent several days in the difficult country below the ridge before working their way west.

Mile 9.8: Small pullout on right. Large float blocks of intrusive rock.

Figure 33. Mafic Eocene intrusive rock near Bald Mountain Department of Transportation complex (MP-128.5).

OPTIONAL STOP: Eocene "Pink and White" Granite

This outcrop contains a type of Eocene granite originally mapped as the Horseshoe Lake stock (Lewis et al., 1992a, 1992b), but later mapped separately as one phase of the Cook Mountain intrusive complex (Lewis et al., 2007a, 2007b). This complex is considered part of the Eocene granodiorite and granite suite and not a part of the "pink granite" suite, an example of which is shown at Stop 3-18 for the Lolo Hot Springs batholith to the northeast. The granodiorite and granite suite contains both hornblende and biotite, and more plagioclase than is typical for the Eocene pink granites. The K-feldspar content is varied; because the crystals in this area are commonly pink (Fig. 34A), the rock is informally known as "fake pink granite." In addition to varied amounts of K-feldspar, the Eocene granodiorites and granites are distinguished from the Cretaceous biotite granodiorites by being finer grained and having distinctly zoned plagioclase.

The Eocene granodiorite and granite suite ranges from 65 to 74 percent SiO_2, has Rb/Sr ratios less than 1.2, Rb contents less than 180 ppm, and 180–550 ppm Sr. This contrasts with the Cretaceous biotite granodiorite, which has similar silica contents but contains less than 100 ppm Rb and more than 500 ppm Sr.

Mile 11.1: Devils Chair.

OPTIONAL STOP: Eocene Pink Granite

Typical exposure of Eocene granite that shows spire and tor weathering style and miarolitic cavities. Most Eocene granites were emplaced at epizonal levels and are characterized by perthitic alkali feldspar, pink color, high background radioactivity, and miarolitic cavities commonly filled with smoky quartz crystals (Bennett and Knowles, 1985; Motzer, 1985; Lewis and Kiilsgaard, 1991). Silica contents are above 74 percent; Rb/Sr ratios are greater than 1.2; Rb is 140–300 ppm; and Sr is less than 120 ppm. Rocks similar to these are exposed at Lolo Hot Springs northeast of Lolo Pass and in the Bungalow pluton along the North Fork of the Clearwater River. Nearly all of these Eocene plutons have less radiogenic $^{87}Sr/^{86}Sr$ and more radiogenic $^{143}Nd/^{144}Nd$ (and $^{176}Hf/^{177}Hf$) than the Paleocene Idaho batholith granites they intrude (Fig. 6), suggesting mantle input or perhaps a mafic lower crustal component unique to their genesis.

Return to Highway 12. Continue east.
MP-140.6: Small pullout at Ashpile Creek.

OPTIONAL STOP: Propylitic Alteration of the Batholith

Fractured and altered biotite granodiorite of probable Paleocene age. Alteration is typical of that found near Eocene plutons in the Atlanta lobe of the batholith. Rock contains abundant secondary chlorite and epidote along with pink feldspar suspected to be albite, which probably has replaced oligoclase (Fig. 34B). Feldspar from this outcrop has a $\delta^{18}O$ value of 0.75 per mil (VSMOW [Vienna standard mean ocean water]), indicating exchange with heated meteoric water (Peter Larson, 2007, written commun.). Rocks similar to these in the Atlanta lobe have $\delta^{18}O$ feldspar values as low as −8.2 per mil (Criss and Taylor, 1983; Lewis, 2001).

MP-141.9: Pullout on right.

OPTIONAL STOP: Paleocene Main Phase Bitterroot Lobe

Porphyritic biotite granodiorite dated at 61.9 ± 2.8 Ma by U-Pb zircon (LA-ICPMS) method (Gaschnig et al., 2010).

MP-144.8: Post Office Creek.
MP-148.0: Colgate Licks rest stop. Springs and associated mineral deposition that formed the licks are likely a result of hot water moving upward along a north-northeast–striking fault mapped here (Lewis and Stanford, 2002a).
MP-151.5: Warm Springs pack bridge.
MP-156.5: Pullout on right just east of curve.

STOP 3-16: Cretaceous Megacrystic Granodiorite (46.5042 °N, 114.8120 °W)

A complexly deformed early phase of the batholith. This location is the same as Stop 2 of Hyndman (1989), who noted

Figure 34. (A) Eocene hornblende-biotite monzogranite ("fake pink granite"). (B) Cretaceous biotite granodiorite at Ashpile Creek (MP-140.6) that has undergone propylitic alteration. Pink mineral is probably albite. (C) Calc-silicate gneiss of the Wallace Formation cut by a pegmatite dike at Stop 3-19. Dark area adjacent to dike is rich in hornblende that formed as a reaction rind in the otherwise green, diopside-rich gneiss. (D) Eocene granite of the Lolo Hot Springs batholith at Stop 3-18 north of Lolo Pass. Exposure is typical of epizonal Eocene granites of Idaho in containing widely spaced joints and weathering into spires.

complex veining by granitic dikes and K-feldspar megacrysts that average ~2.5 by 4 cm and have white plagioclase rims. Hyndman speculated that the megacrysts are neither typical phenocrysts (which would crystallize early from the magma) nor porphyroblasts (which crystallize in the solid rocks by replacement of preexisting grains). Rather, they are probably late magmatic grains that are large because of a high water content of the magma during late stages of crystallization. U-Pb zircon age of 95.1 ± 1.4 Ma has been obtained from a sample collected near here (J.N. Aleinikoff, 2013, written commun.).

MP-158.3: Wendover Campground.

Historical interest: Small parties of Nez Perce would sometimes leave the Lolo trail and fish in this area on their way to Montana to hunt buffalo. It is here that Lewis and Clark left the Lochsa on their route west in 1805 and climbed north to the Lolo trail (Hendrickson and Laughy, 1999, p. 92).

MP-159.3: Imnamatnoon Creek (formerly Papoose Creek). Pullout on right.

OPTIONAL STOP: Metamorphosed Wallace Formation

Metasedimentary rocks of the ca. 1454 Ma (Evans et al., 2000) Wallace Formation of the Belt Supergroup cut by aplite and pegmatite dikes. Calc-silicate minerals (actinolite and diopside) have formed in the layers that once contained carbonate. River gravels above the outcrop are part of an old terrace formed prior to the most recent down-cutting of the Lochsa River.

MP-161.8: Turnoff to Powell. Side road leads to Lochsa Lodge, Powell Campground, Powell Ranger Station, and the location of a Lewis and Clark Expedition campsite.
MP-163.4: Elk Summit Road.
MP-164.2: Calc-silicate gneiss of the Wallace Formation (Belt Supergroup).
MP-165.0: Bernard DeVoto Memorial Cedar Grove.

Historical interest: Bernard DeVoto was a historian and author, who frequently wrote in a very clear-sighted way about the American West. His books include *Across the Wide Missouri* (1947) and *The Course of Empire* (1952).

MP-165.8: Pullout on right.

STOP 3-17: Calc-Silicate Gneiss of the Wallace Formation (46.5506 °N, 114.6734 °W)

Calc-silicate gneiss of the Belt Supergroup (Wallace Formation). The green color is due to diopside and actinolite. Near pegmatite dikes the diopside has been replaced by black hornblende (Fig. 34C), a common feature of the calc-silicate gneiss in this area. Replacement by hornblende occurs at all scales, and in places with abundant granitic material the metasedimentary host rock lacks diopside and is hornblende-rich amphibolite.

MP-167.5: Pullout on right.

OPTIONAL STOP: Quartzite of the Ravalli Group

Feldspathic quartzite mapped as Ravalli Group of the Belt Supergroup (Lewis and Stanford, 2002b) and thus stratigraphically below the Wallace Formation.

MP-170.0 to MP-174: The highway follows a north-northeast–striking fault zone (Lewis and Stanford, 2002b). Crushed and altered rock and associated springs along this part of U.S. Highway 12 present a constant challenge to highway engineers. Lolo Pass owes its existence to this fault, which has pulverized the rock and allowed preferential erosion to produce a low gap in the Bitterroot Mountain Range.
MP-174.4: Lolo Pass and USFS Visitor Center.

Reset mileage to zero. Mileage continues to MP-32.6 at Lolo, Montana.
MP-5.0: Pullout on right.

STOP 3-18: Lolo Hot Springs Batholith (46.6957 °N, 114.5452 °W)

Massive biotite granite of the Eocene Lolo Hot Springs batholith is exposed as typical vertically jointed blocks on the west side of the road (Fig. 34D). Like other epizonal granites in Idaho and Montana, the Lolo Hot Springs batholith contains miarolitic cavities lined with feldspar and smoky quartz crystals.

MP-7.3: Lolo Hot Springs. The springs are along a northeast-striking fault system within the Eocene batholith (Lewis, 1998).

Historical interest: An ancient bathing spot used by Native Americans, the springs were also visited by the Lewis and Clark expedition in 1805. William Clark recorded in his journal: "…in this bath which had been prepared by the Indians by stopping the river with Stone and mud, I bathed and remained in 10 minits it was with dificuelty [*sic*] I could remain this long and it causd [*sic*] a prefuse swet [*sic*]" (Hendrickson and Laughy, 1999, p. 100).

MP-8.2: Hornfelsed Belt Supergroup mapped as Shepard Formation by Lewis (1998).
MP-12.2: Pullout on right.

STOP 3-19: Wallace Formation (46.7667 °N, 114.4709 °W)

Old quarry on north side of road exposes relatively low-grade metasedimentary rocks of the Wallace Formation (Belt Supergroup).

In contrast to the Wallace Formation at Stop 3-17, these rocks still contain dolomite and are stained orange where weathered.

MP-17.0: Lewis and Clark Campground.
MP-21.9: Fine-grained feldspathic quartzite, probably Ravalli Group (Belt Supergroup).
MP-28.0: Fort Fizzle.

Historical interest: It was here on 24 July 1877 that Captain Charles Rawn, four officers, 25 enlisted men, and a group of Flathead Indians attempted to block a band of non-treaty Nez Perce Indians who were fleeing General Howard's troops in Idaho after the start of the Nez Perce war. The troops were joined by civilian volunteers and together they built a log and brush barricade. Before dawn on 28 July, the Nez Perce broke camp and rode around the barricade on the north side without incident. Captain Rawn and his men were left guarding an arrangement of logs, the site of which later became known as "Fort Fizzle" (Hendrickson and Laughy, 1999, p. 104).

MP-31.6: Reddish talus on slopes north of the highway is from weathering of the Snowslip Formation (Belt Supergroup).
MP-32.5: Intersection with U.S. Highway 93 at Lolo, Montana.

END OF ROAD LOG 3.

ACKNOWLEDGMENTS

The authors would like to thank Thomas P. Frost, Robert J. Fleck, and Peter B. Larson for contributions to the 2005 Goldschmidt Conference road log that later formed the basis for the Lewiston to Lolo section (Road Log 3) presented here. We also thank John Watkinson for help with the kinematic interpretations from the Syringa quartzite presented at the Smith Creek stop (Road Log 3, Stop 3-7). Further thanks go to Paul Kelso and Ad Byerly for paleomagnetic analyses performed in Paul's lab at Lake Superior State University. Technical reviews by Russell F. Burmster and Stephen E. Box greatly improved the manuscript. An extra thank you is in order because these reviews were completed over the Christmas holiday. Significant portions of the research in the Dworshak dam area were funded by EarthScope grant #EAR-0844260. Geologic mapping efforts by the Idaho Geological Survey that form the basis for many of our interpretations were supported by the Statemap component of the U.S. Geological Survey National Cooperative Geologic Mapping Program.

FURTHER HISTORICAL READING

For those who are interested in the historical aspects of this trip, we recommend the following books:

Beal, M.D., 1963, I Will Fight No More Forever: Chief Joseph and the Nez Perce War: Seattle, Washington, University of Washington Press, 366 p.

Josephy, A.M., 1997, The Nez Perce Indians and the Opening of the Northwest: Boston, Houghton Mifflin Harcourt, 705 p.

McWhorter, L.V., 1983, Hear Me My Chiefs! Nez Perce Legend and History: Caldwell, Idaho, Caxton Press, 640 p.

Wilfong, C., 2006, Following the Nez Perce Trail: A Guide to the Nee-Me-Poo National Historic Trail with Eyewitness Accounts: Corvallis, Oregon, Oregon State University Press, 493 p.

REFERENCES CITED

Aliberti, E.A., 1988, A structural, petrographic, and isotopic study of the Rapid River area and selected mafic complexes in the northwestern United States: Implications for the evolution of an abrupt island arc–continent boundary [Ph.D. thesis]: Cambridge, Massachusetts, Harvard University, 194 p.

Alloway, M.R., Watkinson, A.J., and Reidel, S.P., 2013, A serial cross-section analysis of the Lewiston structure, Clarkston, Washington, and implications for the evolution of the Lewiston Basin, *in* Reidel, S.P., Camp, V.E., Ross, M.E., Wolff, J.A., Martin, B.S., Tolan, T.L., and Wells, R.E., eds., The Columbia River Flood Basalt Province: Geological Society of America Special Paper 497, p. 349–361.

Anderson, A.L., 1930, The Geology and Mineral Resources of the Region about Orofino, Idaho: Idaho Bureau of Mines and Geology Pamphlet 34, 63 p.

Anderson, A.L., 1931, Genesis of the anthophyllite deposits near Kamiah, Idaho: The Journal of Geology, v. 39, p. 68–81, doi:10.1086/623789.

Armstrong, R.L., Taubeneck, W.H., and Hales, P.O., 1977, Rb-Sr and K-Ar geochronometry of Mesozoic granitic rocks and their Sr isotopic composition, Oregon, Washington, and Idaho: Geological Society of America Bulletin, v. 88, p. 397–411, doi:10.1130/0016-7606(1977)88<397:RAKGOM>2.0.CO;2.

Ash, S.R., 1991, A new Jurassic flora from the Wallowa terrane in Hells Canyon, Oregon and Idaho: Oregon Geology, v. 53, p. 27–33.

Avé Lallemant, H.G., 1995, Pre-Cretaceous tectonic evolution of the Blue Mountains province, northeastern Oregon, *in* Vallier, T.L., and Brooks, H.C., eds., Geology of the Blue Mountains Region of Oregon, Idaho, and Washington: U.S. Geological Survey Professional Paper 1438, p. 271–304.

Avé Lallemant, H.G., Schmidt, W.J., and Kraft, J.L., 1985, Major Late-Triassic strike-slip displacement in the Seven Devils terrane, Oregon and Idaho: A result of left-oblique plate convergence: Tectonophysics, v. 119, p. 299–328, doi:10.1016/0040-1951(85)90044-7.

Barker, F., 1979, Trondhjemite: Definition, environment and hypotheses of origin, *in* Barker, F., ed., Trondhjemites, Dacites, and Related Rocks: New York, Elsevier, p. 1–12.

Beck, M.E., Jr., and Housen, B.A., 2003, Absolute velocity of North America during the Mesozoic from paleomagnetic data: Tectonophysics, v. 377, p. 33–54, doi:10.1016/j.tecto.2003.08.018.

Benford, B., Crowley, J., Schmitz, M., Northrup, C.J., and Tikoff, B., 2010, Mesozoic magmatism and deformation in the northern Owyhee Mountains, Idaho: Implications for along-zone variations for the western Idaho shear zone: Lithosphere, v. 2, no. 2, p. 93–118, doi:10.1130/L76.1.

Bennett, E.H., and Knowles, C.R., 1985, Tertiary plutons and related rocks in central Idaho, *in* McIntyre, D.H., ed., Symposium on the Geology and Mineral Deposits of the Challis 1° × 2° Quadrangle, Idaho: U.S. Geological Survey Bulletin 1658, p. 81–95.

Blake, D.E., Gray, K.D., Giorgis, S., and Tikoff, B., 2009, A tectonic transect through the Salmon River suture zone along the Salmon River Canyon in the Riggins region of west-central Idaho, *in* O'Connor, J.E., Dorsey, R.J., and Madin, I.P., eds., Volcanoes to Vineyards: Geologic Field Trips through the Dynamic Landscape of the Pacific Northwest: Geological Society of America Field Guide 15, p. 345–372.

Blake, M.C., Jr., Jayko, A.S., McLaughlin, R.J., and Underwood, M.B., 1988, Metamorphism and tectonic evolution of the Franciscan complex, northern California, *in* Ernst, W.G., ed., Metamorphism and Crustal Evolution of the Western United States: Englewood Cliffs, New Jersey, Prentice-Hall, p. 1035–1060.

Bonnichsen, B., and Godchaux, M.M., 1994, Geology of the Western Idaho Ultramafic Belt: Idaho Geological Survey Staff Report 94-3, 78 p.

Braudy, N., 2013, Deformation history of West Mountain, west-central Idaho: Implications for the western Idaho shear zone [M.S. thesis]: Madison, University of Wisconsin, 136 p.

Brooks, H.C., and Vallier, T.L., 1978, Mesozoic rocks and tectonic evolution of eastern Oregon and western Idaho, *in* Howell, D.G., and McDougall,

K.A., eds., Mesozoic Paleogeography of the Western United States: Society of Economic Paleontologists and Mineralogists, Pacific Section, Pacific Coast Paleogeography Symposium 2, p. 133–145.

Byerly, A., Kelso, P., Stetson-Lee, T., Gray, C., Roberts, L., Tikoff, B., Wilford, D., and Vervoort, J., 2013, Paleomagnetic studies along the western Idaho shear zone (WISZ): When did the 90 degree bend in the Sr 0.706 line occur?: Geological Society of America Abstracts with Programs, v. 45, no. 7, p. 361.

Camp, V.E., 1981, Geologic investigations of the Columbia Plateau, 2, Upper Miocene basalt distribution reflecting source locations, tectonism, and drainage history in the Clearwater embayment, Idaho: Geological Society of America Bulletin, v. 92, p. 669–678, doi:10.1130/0016-7606(1981)92<669:GSOTCP>2.0.CO;2.

Chase, R.B., Bickford, M.E., and Tripp, S.E., 1978, Rb-Sr and U-Pb isotopic studies of the northeastern Idaho batholith and its border zone: Geological Society of America Bulletin, v. 89, p. 1325–1334, doi:10.1130/0016-7606(1978)89<1325:RAUISO>2.0.CO;2.

Coffin, P., 1967, Geology of the Slate Creek quadrangle, Idaho County, Idaho [M.S. thesis]: Moscow, University of Idaho, 57 p.

Criss, R.E., and Fleck, R.J., 1987, Petrogenesis, geochronology, and hydrothermal systems of the northern Idaho batholith and adjacent areas based on $^{18}O/^{16}O$, D/H, $^{87}Sr/^{86}Sr$, K-Ar, and $^{40}Ar/^{39}Ar$ studies, *in* Vallier, T.L., and Brooks, H.C., eds., Geology of the Blue Mountains Regions of Oregon, Idaho, and Washington: The Idaho Batholith and Its Border Zone: U.S. Geological Survey Professional Paper 1436, p. 95–137.

Criss, R.E., and Taylor, H.P., Jr., 1983, An $^{18}O/^{16}O$ and D/H study of Tertiary hydrothermal systems in the southern half of the Idaho batholith: Geological Society of America Bulletin, v. 94, p. 640–663, doi:10.1130/0016-7606(1983)94<640:AOADSO>2.0.CO;2.

Davidson, G.F., 1990, Cretaceous tectonic history along the Salmon River suture zone near Orofino, Idaho: Metamorphic, structural and $^{40}Ar/^{39}Ar$ thermochronologic constraints [M.S. thesis]: Corvallis, Oregon State University, 143 p.

Davis, G.A., Monger, J.W.H., and Burchfiel, B.C., 1978, Mesozoic construction of the Cordilleran 'collage,' central British Columbia to central California, *in* Howell, D.G., and McDougall, K.A., eds., Mesozoic Paleogeography of the Western United States: Pacific Coast Paleogeography Symposium 2, Society of Economic Paleontology and Mineralogy, Los Angeles, California, p. 1–70.

Dickinson, W.R., 1979, Mesozoic forearc basin in central Oregon: Geology, v. 7, no. 4, p. 166–170, doi:10.1130/0091-7613(1979)7<166:MFBICO>2.0.CO;2.

Dickinson, W.R., 2004, Evolution of the North American Cordillera: Annual Review of Earth and Planetary Sciences, v. 32, p. 13–45, doi:10.1146/annurev.earth.32.101802.120257.

Dorsey, R.J., and LaMaskin, T.A., 2007, Stratigraphic record of Triassic-Jurassic collisional tectonics in the Blue Mountains Province, northeastern Oregon: American Journal of Science, v. 307, p. 1167–1193, doi:10.2475/10.2007.03.

Dorsey, R.J., and LaMaskin, T.A., 2008, Mesozoic collision and accretion of oceanic terranes in the Blue Mountains Province of northeastern Oregon: New insights from the stratigraphic record, *in* Spencer, J.E., and Titley, S.R., eds., Circum-Pacific Tectonics, Geologic Evolution, and Ore Deposits: Arizona Geological Society Digest 22, p. 325–332.

Doughty, P.T., and Sheriff, S.D., 1992, Paleomagnetic evidence for en echelon crustal extension and crustal rotations in western Montana and Idaho: Tectonics, v. 11, no. 3, p. 663–671, doi:10.1029/91TC02889.

Engebretson, D.C., Cox, A., and Gordon, R.G., 1985, Relative Motions between Oceanic and Continental Plates in the Pacific Basin: Geological Society of America Special Paper 206, 59 p.

Evans, K.V., Aleinikoff, J.N., Obradovich, J.D., and Fanning, C.M., 2000, SHRIMP U-Pb geochronology of volcanic rocks, Belt Supergroup, western Montana: Evidence for rapid deposition of sedimentary strata: Canadian Journal of Earth Sciences, v. 37, no. 9, p. 1287–1300, doi:10.1139/e00-036.

Fleck, R.J., 1990, Neodymium, strontium, and trace element evidence of crustal anatexis and magma mixing in the Idaho batholith, *in* Anderson, J.L., ed., The Nature and Origin of Cordilleran Magmatism: Geological Society of America Memoir 174, p. 359–373.

Fleck, R.J., and Criss, R.E., 1985, Strontium and oxygen isotopic variations in Mesozoic and Tertiary plutons of central Idaho: Contributions to Mineralogy and Petrology, v. 90, p. 291–308, doi:10.1007/BF00378269.

Fleck, R.J., and Criss, R.E., 2007, Location, age, and tectonic significance of the western Idaho suture zone, *in* Kuntz, M.A., and Snee, L.W., eds., Geological Studies of the Salmon River Suture Zone and Adjoining Areas, West-Central Idaho and Eastern Oregon: U.S. Geological Survey Professional Paper 1738, p. 15–50.

Follo, M.F., 1994, Sedimentology and stratigraphy of the Martin Bridge Limestone and Hurwal Formation (Upper Triassic to Lower Jurassic) from the Wallowa terrane, Oregon, *in* Vallier, T.L., and Brooks, H.C., eds., Geology of the Blue Mountains Region of Oregon, Idaho and Washington: Stratigraphy, Physiography, and Mineral Resources of the Blue Mountains Region: U.S. Geological Survey Professional Paper 1439, p. 1–27.

Foster, D.A., and Fanning, C.M., 1997, Geochronology of the northern Idaho batholith and the Bitterroot metamorphic core complex: Magmatism preceding and contemporaneous with extension: Geological Society of America Bulletin, v. 109, no. 4, p. 379–394, doi:10.1130/0016-7606(1997)109<0379:GOTNIB>2.3.CO;2.

Foster, D.A., Schafer, C., Fanning, C.M., and Hyndman, D.W., 2001, Relationships between crustal partial melting, plutonism, orogeny, and exhumation: Idaho-Bitterroot batholith: Tectonophysics, v. 342, no. 3–4, p. 313–350, doi:10.1016/S0040-1951(01)00169-X.

Foster, D.A., Doughty, P.T., Kalakay, T.J., Fanning, C.M., Conyer, S., Grice, W.C., and Vogl, J., 2007, Kinematics and timing of exhumation of metamorphic core complexes along the Lewis and Clark fault zone, northern Rocky Mountains, USA, *in* Till, A.B., Roeske, S.M., Sample, J.C., and Foster, D.A., eds., Exhumation Associated with Continental Strike-Slip Fault Systems: Geological Society of America Special Paper 434, p. 207–232.

Garwood, D.L., and Bush, J.H., 2001, Bedrock Geologic Map of the Lewiston Orchards North Quadrangle, Nez Perce County, Idaho: Idaho Geological Survey Technical Report 01-2, scale 1:24,000.

Garwood, D.L., Schmidt, K.L., Kauffman, J.D., Stewart, D.E., Lewis, R.S., Othberg, K.L., and Wampler, P.J., 2008, Geologic Map of the White Bird Quadrangle, Idaho County, Idaho: Idaho Geological Survey Digital Web Map 101, scale 1:24,000.

Gaschnig, R.M., Vervoort, J.D., Lewis, R.S., and McClelland, W.C., 2010, Migrating magmatism in the northern US Cordillera: In situ U-Pb geochronology of the Idaho batholith: Contributions to Mineralogy and Petrology, v. 159, p. 863–883, doi:10.1007/s00410-009-0459-5.

Gaschnig, R.M., Vervoort, J.D., Lewis, R.S., and Tikoff, B., 2011, Isotopic evolution of the Idaho batholith and Challis intrusive province, northern U.S. Cordillera: Journal of Petrology, v. 52, no. 12, p. 2397–2429, doi:10.1093/petrology/egr050.

Gaschnig, R.M., Vervoort, J.D., Lewis, R.S., and Tikoff, B., 2013, Probing for Proterozoic and Archean crust in the northern U.S. Cordillera with inherited zircons from the Idaho batholith: Geological Society of America Bulletin, v. 125, p. 73–88, doi:10.1130/B30583.1.

Getty, S.R., Selverstone, J., Wernicke, B.P., Jacobsen, S.B., Aliberti, E., and Lux, D.R., 1993, Sm-Nd dating of multiple garnet growth events in an arc-continent collision zone, northwestern U.S. Cordillera: Contributions to Mineralogy and Petrology, v. 115, p. 45–57, doi:10.1007/BF00712977.

Giorgis, S., and Tikoff, B., 2004, Constraints on kinematics and strain from feldspar porphyroclast populations, *in* Alsop, I., and Holdsworth R., eds., Transport and Flow Processes in Shear Zones: Geological Society of London Special Publication 224, p. 265–285.

Giorgis, S., Tikoff, B., Kelso, P., and Markley, M., 2006, The role of material anisotropy in the neotectonic extension of the western Idaho shear zone, McCall, Idaho: Geological Society of America Bulletin, v. 118, p. 259–273.

Giorgis, S., McClelland, W., Fayon, A., Singer, B.S., and Tikoff, B., 2008, Timing of deformation and exhumation in the western Idaho shear zone, McCall, Idaho: Geological Society of America Bulletin, v. 120, p. 1119–1133, doi:10.1130/B26291.1.

Goldstrand, P.M., 1994, The Mesozoic geologic evolution of the northern Wallowa terrane, northeastern Oregon and western Idaho, *in* Vallier, T.L., and Brooks, H.C., eds., Geology of the Blue Mountains Region of Oregon, Idaho, and Washington: Stratigraphy, Physiography, and Mineral Resources of the Blue Mountains Region: U.S. Geological Survey Professional Paper 1439, p. 29–53.

Gray, K.D., 2013, Structure of the Arc-Continent Transition in the Riggins Region of West-Central Idaho: Strip Maps and Structural Sections: Idaho Geological Survey Technical Report 13-1, 2 plates, scale 1:24,000.

Gray, K.D., and Oldow, J.S., 2005, Contrasting structural histories of the Salmon River belt and Wallowa terrane: Implications for terrane accretion in northeastern Oregon and west-central Idaho: Geological Society of America Bulletin, v. 117, no. 5, p. 687–706, doi:10.1130/B25411.1.

Gray, K.D., Watkinson, A.J., Gaschnig, R.M., and Isakson, V.H., 2012, Age and structure of the Crevice pluton: Overlapping orogens in west-central Idaho?: Canadian Journal of Earth Sciences, v. 49, no. 6, p. 709–731, doi:10.1139/e2012-016.

Greenwood, W.R., and Morrison, D.A., 1973, Reconnaissance Geology of the Selway-Bitterroot Wilderness Area: Idaho Bureau of Mines and Geology Pamphlet 154, 30 p.

Hamilton, W., 1963, Metamorphism in the Riggins Region, Western Idaho: U.S. Geological Survey Professional Paper 436, 95 p.

Hendrickson, B., and Laughy, L., 1999, Clearwater Country! The Traveler's Historical and Recreational Guide, Lewiston, Idaho-Missoula, Montana: Kooskia, Idaho, Mountain Meadow Press, 230 p.

Hietanen, A., 1962, Metasomatic Metamorphism in Western Clearwater County Idaho: U.S. Geological Survey Professional Paper 344-A, 113 p., scale 1:48,000.

Hillhouse, J.W., Gromme, C.S., and Vallier, T.L., 1982, Paleomagnetism and Mesozoic tectonics of the Seven Devils volcanic arc in northeastern Oregon: Journal of Geophysical Research, v. 87, p. 3777–3794, doi:10.1029/JB087iB05p03777.

Hooper, P.R., and Swanson, D.A., 1990, The Columbia River Basalt Group and associated volcanic rocks of the Blue Mountains province, *in* Walker, G.W., ed., Geology of the Blue Mountains Region of Oregon, Idaho, and Washington: Cenozoic Geology of the Blue Mountains Region: U.S. Geological Survey Professional Paper 1437, p. 63–99.

Hoover, A.L., 1986, Transect across the Salmon River suture, South Fork of the Clearwater River, western Idaho: Rare earth element geochemical, structural, and metamorphic study [M.S. thesis]: Corvallis, Oregon State University, 138 p.

House, M.A., Hodges, K.V., and Bowring, S.A., 1997, Petrological and geochronological constraints on regional metamorphism along the northern border of the Bitterroot batholith: Journal of Metamorphic Geology, v. 15, no. 6, p. 753–764, doi:10.1111/j.1525-1314.1997.00052.x.

House, M.A., Bowring, S.A., and Hodges, K.V., 2002, Implications of middle Eocene epizonal plutonism for the unroofing history of the Bitterroot metamorphic core complex, Idaho-Montana: Geological Society of America Bulletin, v. 114, no. 4, p. 448–461, doi:10.1130/0016-7606(2002)114<0448:IOMEEP>2.0.CO;2.

Hyndman, D.W., 1980, Bitterroot dome–Sapphire tectonic block, an example of a plutonic-core gneiss dome complex with its detached suprastructure, *in* Crittenden, M.D., Jr., Coney, P.J., and Davis, G.H., eds., Cordilleran Metamorphic Core Complexes: Geological Society of America Memoir 153, p. 427–444, doi:10.1130/MEM153-p427.

Hyndman, D.W., 1984, A petrographic and chemical section through the northern Idaho batholith: The Journal of Geology, v. 92, p. 83–102, doi:10.1086/628836.

Hyndman, D.W., 1989, Formation of the northern Idaho batholith and related mylonite of the western Idaho suture zone, *in* Chamberlain, V.E., Breckenridge, R.M., and Bonnichsen, B., eds., Guidebook to the Geology of Northern and Western Idaho and Surrounding Areas: Idaho Geological Survey Bulletin 28, p. 51–64.

Hyndman, D.W., and Foster, D.A., 1988, The role of tonalites and mafic dikes in the generation of the Idaho batholith: The Journal of Geology, v. 96, p. 31–46, doi:10.1086/629191.

Janecke, S.U., and Oaks, R.Q., Jr., 2011, Reinterpreted history of latest Pleistocene Lake Bonneville: Geologic setting of threshold failure, Bonneville flood, deltas of the Bear River, and outlets for two Provo shorelines, southeastern Idaho, USA, *in* Lee, J., and Evans, J.P., eds., Geologic Field Trips to the Basin and Range, Rocky Mountains, Snake River Plain, and Terranes of the U.S. Cordillera: Geological Society of America Field Guide 21, p. 195–222.

Jones, D.L., Silberling, N.J., and Hillhouse, J., 1977, Wrangellia—A displaced terrane in northwestern North America: Canadian Journal of Earth Sciences, v. 14, p. 2565–2577, doi:10.1139/e77-222.

Jones, R.W., 1982, Early Tertiary-age Kamiah volcanics, north-central Idaho, *in* Bonnichsen, B., and Breckenridge, R.M., eds., Cenozoic Geology of Idaho: Idaho Bureau of Mines and Geology Bulletin 26, p. 43–52.

Kauffman, J.D., Davidson, G.F., Lewis, R.S., and Burmester, R.F., 2005a, Geologic Map of the Orofino West Quadrangle, Nez Perce, Clearwater, and Lewis Counties, Idaho: Idaho Geological Survey Geologic Map 40, scale 1:24,000.

Kauffman, J.D., Davidson, G.F., Lewis, R.S., and Burmester, R.F., 2005b, Geologic Map of the Peck Quadrangle, Clearwater, Lewis, and Nez Perce Counties, Idaho: Idaho Geological Survey Geologic Map 38, scale 1:24,000.

Kauffman, J.D., Lewis, R.S., Schmidt, K.L., and Stewart, D.E., 2008, Geologic Map of the White Bird Hill Quadrangle, Idaho County, Idaho: Idaho Geological Survey Digital Web Map 96, scale 1:24,000.

Kauffman, J.D., Garwood, D.L., Schmidt, K.L., Lewis, R.S., Othberg, K.L., and Phillips, W.M., 2009, Geologic Map of the Idaho Parts of the Orofino and Clarkston 30 x 60 Minute Quadrangles, Idaho: Idaho Geological Survey Geologic Map 48, scale 1:100,000.

Kiilsgaard, T.H, Lewis, R.S., and Bennett, E.H., 2001, Plutonic and Hypabyssal Rocks of the Hailey 1° × 2° Quadrangle, Idaho: U.S. Geological Survey Bulletin 2064-U, 18 p.

Kirkham, V.R.D., and Johnson, M.M., 1929, Active faults near Whitebird, Idaho: The Journal of Geology, v. 37, p. 700–711, doi:10.1086/623653.

Kurz, G.A., 2001, Structure and geochemistry of the Cougar Creek complex, northeastern Oregon and west-central Idaho [M.S. thesis]: Boise, Idaho, Boise State University, 248 p.

Kurz, G.A., 2010, Geochemical, isotopic, and U-Pb geochronologic investigations of intrusive basement rocks from the Wallowa and Olds Ferry arc terranes, Blue Mountains Province, Oregon-Idaho [Ph.D. dissertation]: Boise, Idaho, Boise State University, 278 p.

Kurz, G.A., and Northrup, C.J., 2008, Structural analysis of mylonitic rocks in the Cougar Creek complex, Oregon–Idaho using the porphyroclast hyperbolic distribution method, and potential use of SC′-type extensional shear bands as quantitative vorticity indicators: Journal of Structural Geology, v. 30, p. 1005–1012, doi:10.1016/j.jsg.2008.04.003.

Kurz, G.A., Schmitz, M.D., Northrup, C.J., and Vallier, T.L., 2012, U-Pb geochronology and geochemistry of intrusive rocks from the Cougar Creek complex, Wallowa arc terrane, Blue Mountains Province, Oregon-Idaho: Geological Society of America Bulletin, v. 124, no. 3–4, p. 578–595, doi:10.1130/B30452.1.

LaMaskin, T.A., Dorsey, R.J., and Vervoort, J.V., 2008, Tectonic controls on mudrock geochemistry, Mesozoic rocks of eastern Oregon and western Idaho, USA: Implications for Cordilleran tectonics: Journal of Sedimentary Research, v. 78, p. 765–783, doi:10.2110/jsr.2008.087.

LaMaskin, T.A., Vervoort, J.D., and Dorsey, R.J., 2011, Early Mesozoic paleogeography and tectonic evolution of the western United States: Insights from detrital zircon U-Pb geochronology of the Blue Mountains Province, northeastern Oregon, USA: Geological Society of America Bulletin, v. 123, no. 9/10, p. 1939–1965, doi:10.1130/B30260.1.

Lee, R.G., 2004, The geochemistry, stable isotope composition, and U-Pb geochronology of tonalite trondhjemites within the accreted terrane, near Greer, north-central Idaho [M.S. thesis]: Pullman, Washington State University, 132 p.

Lewis, R.S., 1998, Geologic Map of the Montana Part of the Missoula West 30′ × 60′ Quadrangle: Montana Bureau of Mines and Geology Open-File Report MBMG 373, scale 1:100,000.

Lewis, R.S., 2001, Alteration and Mineralization in the Eastern Part of the Soldier Mountains, Camas County, Idaho: U.S. Geological Survey Bulletin 2064-V, 13 p.

Lewis, R.S., and Frost, T.P., 2005, Major oxide and trace element analyses for igneous and metamorphic rock samples from northern and central Idaho: Idaho Geological Survey Digital Analytical Data 2.

Lewis, R.S., and Kiilsgaard, T.H., 1991, Eocene plutonic rocks in south central Idaho: Journal of Geophysical Research, v. 96, no. B8, p. 13,295–13,311, doi:10.1029/91JB00601.

Lewis, R.S., and Stanford, L.R., 2002a, Geologic Map Compilation of the Hamilton 30 × 60 Minute Quadrangle, Idaho: Idaho Geological Survey Geologic Map 32, scale 1:100,000.

Lewis, R.S., and Stanford, L.R., 2002b, Geologic Map Compilation of the Missoula West 30 × 60 Minute Quadrangle, Idaho: Idaho Geological Survey Geologic Map 34, scale 1:100,000.

Lewis, R.S., Burmester, R.F., McFaddan, M.D., Eversmeyer, B.A., Wallace, C.A., and Bennett, E.H., 1992a, Geologic Map of the Upper North Fork of the Clearwater River Drainage, Northern Idaho: Idaho Geological Survey Geologic Map 20, scale 1:100,000.

Lewis, R.S., Burmester, R.F., Reynolds, R.W., Bennett, E.H., Myers, P.E., and Reid, R.R., 1992b, Geologic Map of the Lochsa River Area, Northern Idaho: Idaho Geological Survey Geologic Map 19, scale 1:100,000.

Lewis, R.S., Burmester, R.F., and Bennett, E.H., 1998, Metasedimentary rocks between the Bitterroot and Atlanta lobes of the Idaho batholith and their relationship to the Belt Supergroup, *in* Berg, R.B., ed., Belt Symposium III: Montana Bureau of Mines and Geology Special Publication 112, p. 130–144.

Lewis, R.S., Bush, J.H., Burmester, R.F., Kauffman, J.D., Garwood, D.L., Meyers, P.E., and Othberg, K.L., 2005a, Geologic Map of the Potlatch 30 × 60 Minute Quadrangle, Idaho: Idaho Geological Survey Geological Map 41, scale 1:100,000.

Lewis, R.S., Kauffman, J.D., Davidson, G.F., and Burmester, R.F., 2005b, Geologic Map of the Orofino East Quadrangle, Clearwater, and Lewis Counties, Idaho: Idaho Geological Survey Geologic Map 39, scale 1:24,000.

Lewis, R.S., Burmester, R.F., Kauffman, J.D., Breckenridge, R.M., Schmidt, K.L., McFaddan, M.D., and Myers, P.E., 2007a, Geologic Map of the Kooskia 30 × 60 Minute Quadrangle, Idaho: Idaho Geological Survey Digital Web Map 93, scale 1:100,000.

Lewis, R.S., Burmester, R.F., McFaddan, M.D., Kauffman, J.D., Doughty, P.T., Oakley, W.L., and Frost, T.P., 2007b, Geologic Map of the Headquarters 30 × 60 Minute Quadrangle, Idaho: Idaho Geological Survey Digital Web Map 92, scale 1:100,000.

Lewis, R.S., Vervoort, J.D., Burmester, R.F., McClelland, W.C., and Chang, Z., 2007c, Geochronological constraints on Mesoproterozoic and Neoproterozoic(?) high-grade metasedimentary rocks of north-central Idaho, USA, *in* Link, P.K., and Lewis, R.S., eds., Proterozoic Geology of Western North America and Siberia: Society for Sedimentary Geology (SEPM) Special Publication 86, p. 37–53.

Lewis, R.S., Vervoort, J.D., Burmester, R.F., and Oswald, P.J., 2010, Detrital zircon analysis of Mesoproterozoic and Neoproterozoic metasedimentary rocks of north-central Idaho: Implications for development of the Belt-Purcell basin: Canadian Journal of Earth Sciences, v. 47, p. 1383–1404, doi:10.1139/E10-049.

Lewis, R.S., Link, P.K., Stanford, L.R., and Long, S.P., 2012, Geologic Map of Idaho: Idaho Geological Survey Map 9, scale 1:750,000.

Lister, G.S., and Snoke, A.W., 1984, S-C mylonites: Journal of Structural Geology, v. 6, p. 617–638, doi:10.1016/0191-8141(84)90001-4.

Lund, K., 1995, Metamorphic and structural development of island-arc rocks in the Slate Creek–John Day Creek area, west-central Idaho, *in* Vallier, T.L., and Brooks, H.C., eds., Geology of the Blue Mountains Region of Oregon, Idaho, and Washington: Petrology and Tectonic Evolution of Pre-Tertiary Rocks of the Blue Mountains Region: U.S. Geological Survey Professional Paper 1438 p. 517–540.

Lund, K., 2004, Geology of the Payette National Forest, Valley, Idaho, Washington, and Adams Counties, West-Central Idaho: U.S. Geological Survey Professional Paper 1666, 89 p.

Lund, K., and Snee, L.W., 1988, Metamorphism, structural development, and age of the continent-island arc juncture in west-central Idaho, *in* Ernst, W.G., ed., Metamorphism and Crustal Evolution of the Western United States: Englewood Cliffs, New Jersey, Prentice-Hall, p. 296–331.

Lund, K., McCollough, W.F., and Price, E.H., 1993, Geologic Map of the Slate Creek–John Day Creek Area, Idaho County, Idaho: U.S. Geological Survey Miscellaneous Investigations Map I-2299, scale 1:50,000.

Lund, K., Aleinikoff, J.N., Evans, K.V., and Fanning, C.M., 2003, SHRIMP U-Pb geochronology of Neoproterozoic Windermere Supergroup, central Idaho: Implications for regional synchroneity of Sturtian glaciation and associated rifting: Geological Society of America Bulletin, v. 115, p. 349–372, doi:10.1130/0016-7606(2003)115<0349:SUPGON>2.0.CO;2.

Lund, K.I., Aleinikoff, J.N., Yacob, E.Y., Unruh, D.M., and Fanning, C.M., 2008, Coolwater culmination: Sensitive high-resolution ion microprobe (SHRIMP) U-Pb and isotopic evidence for continental delamination in the Syringa embayment, Salmon River suture, Idaho: Tectonics, v. 27, TC2009, doi:10.1029/2006TC002071.

Lund, K., Aleinikoff, J.N., Evans, K.V., duBray, E.A., Dewitt, E.H., and Unruh, D.M., 2010, SHRIMP U-Pb dating of recurrent Cryogenian and Late Cambrian–Early Ordovician alkalic magmatism in central Idaho: Implications for Rodinian rift tectonics: Geological Society of America Bulletin, v. 122, p. 430–453, doi:10.1130/B26565.1.

Manduca, C.A., Silver, L.T., and Taylor, H.P., 1992, $^{87}Sr/^{86}Sr$ and $^{18}O/^{16}O$ isotopic systematics and geochemistry of granitoid plutons across a steeply-dipping boundary between contrasting lithospheric blocks in western Idaho: Contributions to Mineralogy and Petrology, v. 109, p. 355–372, doi:10.1007/BF00283324.

Manduca, C.A., Kuntz, M.A., and Silver, L.T., 1993, Emplacement and deformation history of the western margin of the Idaho batholith near McCall, Idaho: Influence of a major terrane boundary: Geological Society of America Bulletin, v. 105, p. 749–765, doi:10.1130/0016-7606(1993)105<0749:EADHOT>2.3.CO;2.

McClelland, W.C., and Oldow, J.S., 2004, Displacement transfer between thick- and thin-skinned decollement systems in the central North American Cordillera, *in* Grocott, J., McCaffrey, K.J.W., Taylor, G., and Tikoff, B., eds., Vertical Coupling and Decoupling in the Lithosphere: Geological Society of London Special Publication 227, p. 177–195.

McClelland, W.C., and Oldow, J.S., 2007, Late Cretaceous truncation of the western Idaho shear zone in the central North American Cordillera: Geology, v. 35, no. 8, p. 723–726, doi:10.1130/G23623A.1.

McClelland, W.C., Tikoff, B., and Manduca, C.A., 2000, Two-phase evolution of accretionary margins: Examples from the North American Cordillera: Tectonophysics, v. 326, no. 1–2, p. 37–55.

McKay, M.P., Stowell, H.H., Schwartz, J.J., and Gray, K.D., 2011, P–T–t paths from the Salmon River suture zone, west-central Idaho: Continental growth by island arc accretion: Geological Society of America Abstracts with Programs, v. 33, no. 4, p. 76.

Mohl, G.B., and Thiessen, R.L., 1995, Gravity studies of an island-arc/continent suture zone in west-central Idaho and southeastern Washington, *in* Vallier, T.L., and Brooks, H.C., eds., Geology of the Blue Mountains Region of Oregon, Idaho, and Washington: Petrology and Tectonic Evolution of Pre-Tertiary Rocks of the Blue Mountains Region: U.S. Geological Survey Professional Paper 1438, p. 497–516.

Morrison, D.A., 1968, Reconnaissance geology of the Lochsa area, Idaho County, Idaho [Ph.D. dissertation]: Moscow, Idaho, University of Idaho, 126 p.

Morrison, R.K., 1963, Pre-Tertiary geology of the Snake River Canyon between Cache Creek and Dug Bar, Oregon-Idaho boundary [Ph.D. dissertation]: Eugene, Oregon, University of Oregon, 290 p.

Motzer, W.E., 1985, Tertiary epizonal plutonic rocks of the Selway-Bitterroot Wilderness, Idaho County, Idaho [Ph.D. dissertation] Moscow, Idaho, University of Idaho, 467 p.

Myers, P.E., 1982, Geology of the Harpster Area, Idaho County, Idaho: Idaho Bureau of Mines and Geology Bulletin 25, 46 p.

Nesheim, T.O., Vervoort, J.D., McClelland, W.C., Gilotti, J.A., and Lang, H.L., 2012, Mesoproterozoic syntectonic garnet within Belt Supergroup metamorphic tectonites: Evidence of Grenville-age metamorphism and deformation along northwest Laurentia: Lithos, v. 134, p. 91–107, doi:10.1016/j.lithos.2011.12.008.

Nolf, B., 1966, Geology and stratigraphy of part of the northern Wallowa Mountains, Oregon [Ph.D. dissertation]: Princeton, New Jersey, Princeton University, 138 p.

Northrup, C.J., Schmitz, M., Kurz, G., and Tumpane, K., 2011, Tectonometamorphic evolution of distinct arc terranes in the Blue Mountains Province, Oregon and Idaho, *in* Lee, J., and Evans, J.P., eds., Geologic Field Trips to the Basin and Range, Rocky Mountains, Snake River Plain, Terranes of the U.S. Cordillera: Geological Society of America Field Guide 21, p. 67–88.

O'Connor, J.E., 1993, Hydrology, Hydraulics, and Geomorphology of the Bonneville Flood: Geological Society of America Special Paper 274, 83 p.

Onasch, C.M., 1987, Temporal and spatial relations between folding, intrusion, metamorphism, and thrust faulting in the Riggins area, west-central Idaho, *in* Vallier, T.L., and Brooks, H.C., eds., Geology of the Blue Mountains Region of Oregon, Idaho, and Washington: The Idaho Batholith and Its Border Zone: U.S. Geological Survey Professional Paper 1436, p. 139–149.

Payne, J.D., 2004, Kinematic and geochronologic constraints for the truncation of the Salmon River suture zone [M.S. thesis]: Moscow, Idaho, University of Idaho, 43 p.

Payne, J.D., and McClelland, W.C., 2002, Kinematic and temporal constraints for truncation of the western Idaho shear zone: Geological Society of America Abstracts with Programs, v. 34, no. 5, p. 102.

Pitz, C.F., 1985, A remote sensing and structural analysis of the trans-Idaho discontinuity near Lowell, Idaho [M.S. thesis]: Pullman, Washington State University, 130 p.

Ramsay, J.G., and Huber, M.I., 1983, The Techniques of Modern Structural Geology, v. 1: London, UK, Academic Press, 391 p.

Reid, R.R., Bittner, E., Greenwood, W.R., Ludington, S., Lund, K., Motzer, W.E., and Toth, M., 1979, Geologic Section and Road Log across the Idaho batholith: Idaho Bureau of Mines and Geology Information Circular 34, 20 p.

Reidel, S.P., Hooper, P.R., Webster, G.D., and Camp, V.E., 1992, Geologic Map of Southeastern Asotin County, Washington: Washington Division of Geology and Earth Resources Geologic Map 40, scale 1:48,000.

Reidel, S.P., Camp, V.E., Tolan, T.L., Kauffman, J.D., and Garwood, D.L., 2013, Tectonic evolution of the Columbia River flood basalt province, *in* Reidel, S.P., Camp, V.E., Ross, M.E., Wolff, J.A., Martin, B.S., Tolan, T.L., and Wells, R.E., eds., The Columbia River Flood Basalt Province: Geological Society of America Special Paper 497, p. 293–324.

Schmidt, K.L., Burmester, R.F., Lewis, R.S., Link, P.K., and Fanning, C.M., 2003, New constraints on the western Idaho orocline: A primary feature in the Mesozoic collision zone or result of strike slip modification?: Geological Society of America Abstracts with Programs, v. 35, no. 6, p. 559.

Schmidt, K.L., Garwood, D.L., and Kauffman, J.D., 2005, Geologic Map of the Keuterville Quadrangle, Lewis and Idaho Counties, Idaho: Idaho Geological Survey Digital Web Map 38, scale 1:24,000.

Schmidt, K.L., Kauffman, J.D., Stewart, D.E., Othberg, K.L., and Lewis, R.S., 2007, Geologic Map of the Grangeville East Quadrangle, Idaho County, Idaho: Idaho Geological Survey Digital Web Map 86, scale 1:24,000.

Schmidt, K.L., Kauffman, J.D., Stewart, D.E., Garwood, D.L., Othberg, K.L., and Lewis, R.S., 2009a, Geologic Map of the Grave Point Quadrangle, Idaho County, Idaho, and Wallowa County, Oregon: Idaho Geological Survey Digital Web Map 111, scale 1:24,000.

Schmidt, K.L., Kauffman, J.D., Stewart, D.E., Garwood, D.L., Othberg, K.L., and Lewis, R.S., 2009b, Geologic Map of the Slate Creek Quadrangle, Idaho County, Idaho: Idaho Geological Survey Digital Web Map 110, scale 1:24,000.

Schmidt, K.L., Lewis, R.S., Gaschnig, R.M., and Vervoort, J.D., 2009c, Testing hypotheses on the origin of the Syringa embayment in the Salmon River suture zone, western Idaho, USA: Geological Society of America Abstracts with Programs, v. 41, no. 7, p. 223.

Schmidt, K.L., Schwartz, D.M., Lewis, R.S., Vervoort, J.D., LaMaskin, T.A., and Wilford, D.E., 2013, New detrital zircon ages constrain the origin and evolution of the Riggins Group assemblage along the Salmon River suture zone, western Idaho: Geological Society of America Abstracts with Programs, v. 45, no. 6, p. 66.

Schwartz, J.J., Snoke, A.W., Frost, C.D., Barnes, C.G., Gromet, L.P., and Johnson, K., 2010, Analysis of the Wallowa-Baker terrane boundary: Implications for tectonic accretion in the Blue Mountains province, northeastern Oregon: Geological Society of America Bulletin, v. 122, p. 517–536, doi:10.1130/B26493.1.

Schwartz, J.J., Snoke, A.W., Cordey, F., Johnson, K., Frost, C.D., Barnes, C.G., LaMaskin, T.A., and Wooden, J.L., 2011, Late Jurassic magmatism, metamorphism, and deformation in the Blue Mountains Province, northeast Oregon: Geological Society of America Bulletin, v. 123, no. 9/10, p. 2083–2111, doi:10.1130/B30327.1.

Selverstone, J., Wernicke, B.P., and Aliberti, E.A., 1992, Intracontinental subduction and hinged uplift along Salmon River suture zone in west-central Idaho: Tectonics, v. 11, p. 124–144, doi:10.1029/91TC02418.

Silberling, N.J., Jones, D.L., Blake, M.C., Jr., and Howell, D.G., 1987, Lithotectonic terrane map of the western conterminous United States, Pt. C., *in* Silberling, N.J., and Jones, D.L., eds., Lithotectonic Terrane Maps of the North American Cordillera: U.S. Geological Survey Miscellaneous Field Studies Map MF-1874-C, scale 1:250,000, 20 p.

Sims, P.K., Lund, K., and Anderson, E.D., 2005, Precambrian Crystalline Basement Map of Idaho—An Interpretation of Geomagnetic Data: U.S. Geological Survey Scientific Investigations Map 2829, scale 1:1,000,000.

Snee, L.W., Lund, K., Sutter, J.F., Balcer, D.E., and Evans, K.V., 1995, An ^{40}Ar/^{39}Ar chronicle of the tectonic development of the Salmon River suture zone, western Idaho, *in* Vallier, T.L., and Brooks, H.C., eds., Geology of the Blue Mountains Region of Oregon, Idaho, and Washington: Petrology and Tectonic Evolution of Pre-Tertiary Rocks of the Blue Mountains Region: U.S. Geological Survey Professional Paper 1438, p. 359–414.

Snee, L.W., Davidson, G.F., and Unruh, D.M., 2007, Geologic, geochemical, and ^{40}Ar/^{39}Ar and U-Pb thermochronologic constraints for the tectonic development of the Salmon River suture zone near Orofino, Idaho, *in* Kuntz, M.A., and Snee, L.W., eds., Geological Studies of the Salmon River Suture Zone and Adjoining Areas, West-Central Idaho and Eastern Oregon: U.S. Geological Survey Professional Paper 1738, p. 51–94.

Stanley, G.D., and Beauvais, L., 1990, Middle Jurassic corals from the Wallowa terrane, west-central Idaho: Journal of Paleontology, v. 64, p. 352–362.

Stanley, G.D., and Whalen, M.T., 1989, Triassic corals and spongiomorphs from Hells Canyon, Wallowa terrane, Oregon: Journal of Paleontology, v. 63, no. 6, p. 800–819.

Stetson-Lee, T., Tikoff, B., Byerly, A., Kelso, P., and Vervoort, J.D., 2013, The Idaho syntaxis: Western Idaho shear zone deformation that has gone round the Orofino bend: Geological Society of America Abstracts with Programs, v. 45, no. 7, p. 58.

Strayer, L.M., IV, Hyndman, D.W., Sears, J.W., and Myers, P.E., 1989, Direction and shear sense during suturing of the Seven Devils–Wallowa terrane against North America in western Idaho: Geology, v. 17, p. 1025–1028, doi:10.1130/0091-7613(1989)017<1025:DASSDS>2.3.CO;2.

Swanson, D.A., Anderson, J.E., Bentley, R.D., Byerly, G.R., Camp, V.E., Gardner, J.N., and Wright, T.L., 1979a, Reconnaissance Geologic Map of the Columbia River Basalt Group in Eastern Washington and Northern Idaho, Pullman 1° × 2° quadrangle: U.S. Geological Survey Open-File Report 79-1363, sheet 8 of 12, scale 1:250,000.

Swanson, D.A., Wright, T.L., Hooper, P.R., and Bentley, R.D., 1979b, Revisions in Stratigraphic Nomenclature of the Columbia River Basalt Group: U.S. Geological Survey Bulletin 1457-G, 59 p.

Taylor, H.P., Jr., and Margaritz, M., 1978, Oxygen and hydrogen isotope studies of the Cordilleran batholiths of western North America, *in* Robinson, B.W., ed., Stable Isotopes in the Earth Sciences: New Zealand Department of Scientific and Industrial Research Bulletin, v. 220, p. 151–173.

ten Grotenhuis, S.M., Trouw, R.A.J., and Passchier, C.W., 2003, Evolution of mica fish in mylonitic rocks: Tectonophysics, v. 372, p. 1–21, doi:10.1016/S0040-1951(03)00231-2.

Tikoff, B., and Greene, D., 1997, Stretching lineations in transpressional shear zones: An example from the Sierra Nevada batholith, California: Journal of Structural Geology, v. 19, p. 29–39, doi:10.1016/S0191-8141(96)00056-9.

Tikoff, B., Kelso, P., Manduca, C., Markely, M.J., and Gillaspy, J., 2001, Lithospheric and crustal reactivation of an ancient plate boundary: The assembly and disassembly of the Salmon River suture zone, Idaho, USA, *in* Holdsworth, R.E., Strachan, R.A., Magloughlin, J.F., and Knipe, R.J., eds., The Nature and Tectonic Significance of Fault Zone Weakening: Geological Society of London Special Publication 186, p. 213–231.

Toth, M.I., and Stacey, J.S., 1992, Constraints on the Formation of the Bitterroot Lobe of the Idaho Batholith, Idaho and Montana, from U-Pb Zircon Geochronology and Feldspar Pb Isotopic Data: U.S. Geological Survey Bulletin 2008, 14 p.

Tumpane, K.P., 2010, Age and isotopic investigations of the Olds Ferry terrane and its relations to other terranes of the Blue Mountains Province, Eastern Oregon and West-Central Idaho [M.S. thesis]: Boise, Idaho, Boise State University, 201 p.

Unruh, D.M., Lund, K., Kuntz, M.A., and Snee, L.W., 2008, Uranium–Lead Zircon Ages and Sr, Nd, and Pb Isotope Geochemistry of Selected Plutonic Rocks from Western Idaho: U.S. Geological Survey Open-File Report 2008-1142, 37 p.

Vallier, T.L., 1974, Preliminary Report on the Geology of Part of Snake River Canyon: Oregon Department of Geology and Mineral Industries, Geologic Map Series 6, scale 1:125,000, 15 p.

Vallier, T.L., 1977, The Permian and Triassic Seven Devils Group, Western Idaho and Northeastern Oregon: U.S. Geological Survey Bulletin 1437, 58 p.

Vallier, T.L., 1995, Petrology of pre-Tertiary igneous rocks in the Blue Mountains region of Oregon, Idaho, and Washington: Implications for the geologic evolution of a complex island arc, *in* Vallier, T.L., and Brooks, H.C., eds., Geology of the Blue Mountains Region of Oregon, Idaho, and Washington: Petrology and Tectonic Evolution of Pre-Tertiary Rocks of the Blue Mountains Region: U.S. Geological Survey Professional Paper 1438, p. 125–209.

Vallier, T.L., 1998, Islands and Rapids: A Geologic Story of Hells Canyon: Lewiston, Idaho, Confluence Press, 151 p.

Van Noy, R.M., Peterson, N.S., and Gray, J.J., 1970, Kyanite Resources in the Northwestern United States: U.S. Bureau of Mines Report of Investigations 7426, 81 p.

Walker, N.W., 1986, U/Pb geochronologic and petrologic studies in the Blue Mountains terrane, northeastern Oregon and westernmost-central Idaho: Implications for pre-Tertiary tectonic evolution [Ph.D. dissertation]: Santa Barbara, California, University of California Santa Barbara, 224 p.

White, D.L., and Vallier, T.L., 1994, Geologic evolution of the Pittsburg Landing area, Snake River Canyon, Oregon and Idaho, *in* Vallier, T.L., and Brooks, H.C., eds., Stratigraphy, Physiography, and Mineral Resources of the Blue Mountains Region: U.S. Geological Survey Professional Paper 1439, p. 55–74.

White, J.D.L., 1994, Intra-arc basin deposits within the Wallowa terrane, Pittsburg Landing area, Oregon and Idaho, *in* Vallier, T.L., and Brooks, H.C., eds., Stratigraphy, Physiography, and Mineral Resources of the Blue

Mountains Region: U.S. Geological Survey Professional Paper 1439, p. 75–90.
Wilford, D.E., 2012, Lu-Hf garnet geochronology of the Salmon River suture zone, west-central Idaho [M.S. thesis]: Pullman, Washington State University, 97 p.
Yates, R.G., 1968, The trans-Idaho discontinuity, *in* Proceedings of the 23rd International Geological Congress: Prague, Czechoslovakia, 1968, v. 1, p. 117–123.
Zen, E-an, 1988, Tectonic significance of high-pressure plutonic rocks in the western Cordillera of North America, *in* Ernst, W.G., ed., Metamorphism and Crustal Evolution of the Western United States: Englewood Cliffs, New Jersey, Prentice-Hall, p. 41–67.
Zen, E-an, and Hammarstrom, J.M., 1984, Magmatic epidote and its petrologic significance: Geology, v. 12, p. 515–518.
Zirakparvar, N.A., Vervoort, J.D., McClelland, W.C., and Lewis, R.S., 2010, Insights into the metamorphic evolution of the Belt-Purcell Basin: Evidence from Lu-Hf garnet geochronology: Canadian Journal of Earth Sciences, v. 47, p. 161–179, doi:10.1139/E10-001.

Manuscript Accepted by the Society 20 February 2014

Printed in the USA

The Geological Society of America
Field Guide 37
2014

Sedimentary record of glacial Lake Missoula along the Clark Fork River from deep to shallow positions in the former lakes: St. Regis to near Drummond, Montana

Larry N. Smith*
Department of Geological Engineering, Montana Tech of The University of Montana, Butte, Montana 59701, USA

Michelle A. Hanson*
Saskatchewan Geological Survey, Ministry of the Economy, 200-2101 Scarth Street, Regina, Saskatchewan S4P 2H9, Canada

ABSTRACT

Glacial Lake Missoula was repeatedly dammed by the Purcell Trench Lobe of the Cordilleran ice sheet during the last glaciation to maximum altitudes near 4200 ft (1280 m). Studies from outside of the lake basin suggest that the lake filled and drained multiple times in the late Pleistocene. Deposits and landforms within the former glacial lake basin provide evidence for a complex lake-level history that is not well understood for this famous impoundment. At least two general lake phases are evident in the stratigraphy: an earlier phase of catastrophic drainage that was responsible for large-scale dramatic erosional and depositional features, and a later, less-catastrophic, phase responsible for the preservation of fine-grained glaciolacustrine sediments. Features of the earlier lake phase include giant gravel dunes and openwork gravel with anomalously large clasts (erratics). Deposits from the later phase are mostly low-energy glaciolacustrine sediments that record a history of lake-bottom sedimentation and repeated lake-floor exposure.

A focus of this field trip is to look at evidence for the two lake phases as well as evaluate the record of exposure surfaces, and therefore lake-level lowerings, during the second phase at multiple locations in the lake basin. One of the second phase sites is close to a highstand, full basin position in the lake (near Garden Gulch), representing a maximum water depth at this site of ~100 m, whereas others (Rail line and Nine-mile) are at lower altitudes in regions that may have been under as much as 300 m of water.

Fine-grained glaciolacustrine sediments are rippled very fine sandy silt and fining-upward sequences of laminated silt and clayey silt of glaciolacustrine origin. Periglacial features, contorted bedding, desiccation, and paleosols in outcrop provide clear evidence of multiple exposure surfaces; each represent a lake-lowering event.

*E-mails: lsmith@mtech.edu; Michelle.Hanson@gov.sk.ca.

Smith, L.N., and Hanson, M.A., 2014, Sedimentary record of glacial Lake Missoula along the Clark Fork River from deep to shallow positions in the former lakes: St. Regis to near Drummond, Montana, *in* Shaw, C.A., and Tikoff, B., eds., Exploring the Northern Rocky Mountains: Geological Society of America Field Guide 37, p. 51–63, doi:10.1130/2014.0037(02). For permission to copy, contact editing@geosociety.org.

Optically stimulated luminescence (OSL or "optical dating") ages on quartz from the three sections (Ninemile, Rail line, and Garden Gulch) allow for preliminary correlations that suggest approximately the same phase of glacial Lake Missoula sedimentation. The exposure surfaces suggest that the glacial-lake level rose and fell at least 8–12 times to elevations above and below the sections (936–1180 m), filling to within 100 m of full pool (1280 m). Optical dating shows that this occurred after 20 ka and the last inundation of the lake before 13.5 ka. Correlation of specific exposure surfaces throughout the basin will be required to develop a lake-level history.

INTRODUCTION

During the last glaciation, glacial Lake Missoula was formed by repeated blocking of the Clark Fork River by the Purcell Trench Lobe of the Cordilleran ice sheet near the present state boundary between Idaho and Montana (Fig. 1; Pardee, 1910; Breckenridge, 1989). The lake volume has been estimated to have been between 2200 and 2600 km^3, using the highest shorelines, at ~4200 ft (1280 m) near Missoula, Montana, and the present topography (Pardee, 1910; Smith, 2006). Multiple failures of a succession of ice dams repeatedly released water, which flowed into glacial Lake Columbia in northern Washington (Atwater, 1986), and across the Channeled Scablands in central Washington, where they carved and deposited landforms of the now-famous scabland topography, including coulees and dry falls, and giant ripples and gravel bars (Bretz, 1969; Baker, 1973). Like

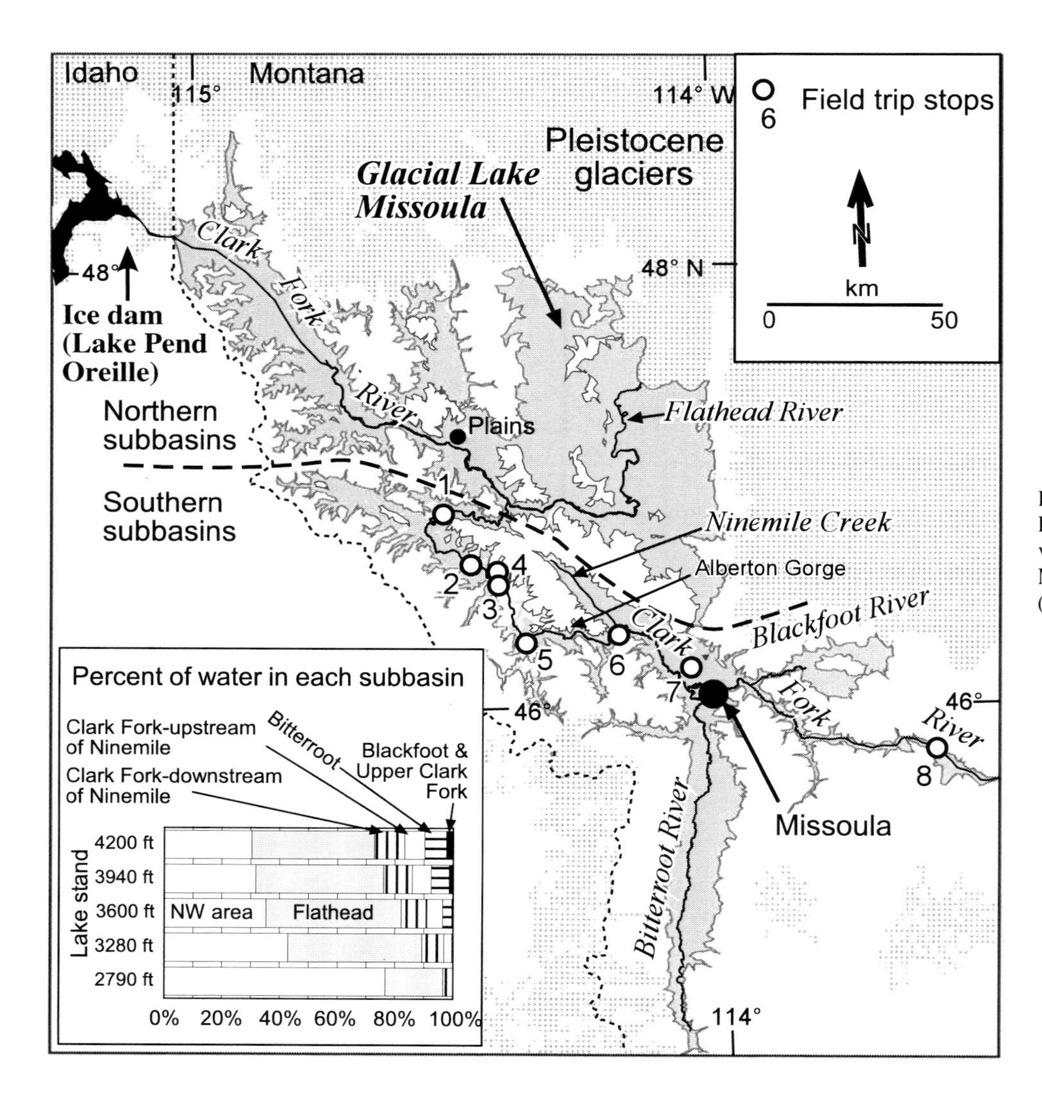

Figure 1. Location map, route for glacial Lake Missoula field trip, and relative volumes of subbasins of glacial Lake Missoula at different lake-level stands (modified from Smith, 2006).

any mountainous reservoir, GLM had many subbasins defined by drainage basins that are tributary to the Clark Fork River (Fig. 1). The northern subbasins, the Flathead River and the lower Clark Fork River areas, held more than 75 percent of GLM water, in comparison to the southern subbasins, the Bitterroot Valley, the Missoula Valley, and the Clark Fork River above Missoula, where this field trip takes place (Fig. 1; Smith, 2006).

Langton (1935) called deposits of glaciolacustrine silt and clay deposited in lake-bottom positions throughout the glacial Lake Missoula basin the "Lake Missoula beds." The original studies of the stratigraphy and geomorphic features formed by rapid lake drainage focused in the northern subbasins, in the areas east and west of Plains, Montana (Pardee, 1910, 1942). Most subsequent work on the lake deposits and landforms has concentrated on the area downstream of the Clark Fork River's confluence with the Flathead River (Baker, 1973; Breckenridge, 1989; Alho et al., 2010). In contrast, this field trip focuses on sedimentary deposits and landforms preserved along the Clark Fork River upstream of its confluence with the Flathead River (Fig. 1).

The purpose of this field trip is to view and discuss sedimentary sections and geomorphic features along the Clark Fork River including the Ninemile area, Alberton Gorge, Superior, St. Regis, Missoula Valley, and upstream reaches of the Clark Fork. We will look at two groupings of deposits that represent two phases of lake filling and draining—an early phase of high-energy flood discharges and a later phase of lower energy discharges. We will look at examples of erosional features and depositional features such as gravelly alluvium that show that discharges from the earlier GLM stands were significantly greater than the final lake phase (Stops 1, 2, and 3; Pardee, 1942; Chambers, 1971; Smith, 2006). Deposits of the later phase are represented by lake-bottom glaciolacustrine sediments; these deposits contain stratigraphic and some chronologic evidence for correlation to flood deposits outside of the lake basin. At Stops 5, 6, 7, and 8 we will look at exposures of the Lake Missoula beds that record a long history of lake-bottom sedimentation (2000–3000 years according to varve counts) and repeated lake-floor exposure. This repeated lake-floor exposure does not necessarily indicate that discharges were high-energy or that the lakes drained completely during lake-level lowering events.

The area where Ninemile Creek enters the Clark Fork River upstream of Alberton Gorge is the location of an exceptionally well-exposed and accessible section of glaciolacustrine sand, silt, and clay. This section has been studied repeatedly and referenced in many works on the glacial Lake Missoula flooding (Alt and Chambers, 1970; Chambers, 1971, 1984; Waitt, 1980, 1985; Fritz and Smith, 1993; Shaw et al., 1999, 2000; Atwater et al., 2000; Booth et al., 2004; Hanson et al., 2012). Based on his study of flood slackwater sediments in central Washington, Waitt (1980, 1985) interpreted the Ninemile section to represent more than 40 catastrophic glacial Lake Missoula drainings. Similarly, Atwater (1986) interpreted sections of relatively low-energy lake-bottom sediment in glacial Lake Columbia to contain coarser sediment beds resulting from 89 influxes from glacial Lake Missoula.

LATE PLEISTOCENE GLACIAL LAKE MISSOULA CHRONOLOGY

There are many ages associated with the existence of glacial Lake Missoula during the late Pleistocene. The onset of flooding from the lake has been constrained to at, or after, 23.0–19.0 cal. ka B.P.[1] by radiocarbon ages associated with: (1) the advance of the Cordilleran ice sheet into eastern Washington; (2) glacial Lake Missoula flood sediments in Oregon; (3) low salinity anomalies inferred from freshwater diatom abundances off the southern Oregon coast that have been attributed to glacial Lake Missoula flooding, and; (4) glacial Lake Missoula sediment in the North Pacific (Clague et al., 1980; Benito and O'Connor, 2003; Lopes and Mix, 2009; Gombiner et al., 2010). Similarly, optical ages of 21.0 and 20.4–17.0 ka[2] predate fine-grained, lake-bottom sediment (Baker et al., 2009).

Based on varve counts, optical dating of the clay portion of the varves, and stratigraphic interpretation of glacial Lake Missoula sediments in Mission Valley, Montana, Levish (1997) concluded that the lake existed continuously from ca.19.2–16.0 ka. In glacial Lake Columbia, Atwater (1986) described glaciolacustrine sediment that is interbedded with 89 sand and silt beds that he inferred were deposited by glacial Lake Missoula floods. Based on a radiocarbon age and varve counts, Atwater (1986) estimated that glacial Lake Columbia existed over a period of 2000–3000 years between 19.4 and 14.7 cal. ka B.P. and that glacial Lake Missoula flooded into it during most of this time. Studying multiple sequences of sand and silt slackwater sediments in central Washington that Waitt (1980, 1985) inferred recorded multiple catastrophic drainings of glacial Lake Missoula, Clague et al. (2003) used tephra (Mount St. Helens set S) and paleomagnetic secular variation records to constrain flooding to either between 21.0 and 16.1 or between 18.5 and 13.1 cal. ka B.P. Two age ranges were produced based on two different possible ages for the tephra and based on different estimates of the timing between flooding events.

Radiocarbon ages on glacial Lake Missoula turbidites in the North Pacific constrain the end of flooding, to between 13.0 and 12.6 cal. ka B.P. (Zuffa et al., 2000). The last flood from glacial Lake Missoula, however, must have occurred before ca. 13.7–13.4 ka (Kuehn et al., 2009), when Glacier Peak tephras G and B blanketed parts of the lake basin (Levish, 1997; Sperazza et al., 2002). Lastly, optical ages and estimated varve counts between glacial lake draining events at the Ninemile Creek (Stop 6) and Rail line (Stop 7) sites by Hanson et al. (2012) place these two exposures between 15.1 and 12.4 ka—toward the end of the existence of glacial Lake Missoula and within the second phase of lake filling and draining—consistent with the stratigraphy that we will see on this trip.

[1]All radiocarbon ages are reported as cal. ka B.P. and were calibrated with OxCal 4.1 using the IntCal09 calibration curve (Bronk Ramsey, 2009; Reimer et al., 2009).

[2]All optical ages are reported in ka.

Route and Road Log

The field trip will officially start from St. Regis, Montana, and travel southeast on I-90, through Missoula, to the Drummond area. Attendees leaving Bozeman, Montana, will travel northwestward through much of the area on the way to the starting point, thus some pertinent features are pointed out on a brief road log from east to west. Altitudes are given for reference to the highstand of glacial Lake Missoula at ~4200 ft (1280 m); locations use the WGS84 datum.

Mileage	*Directions*
0.0	Route to the starting point from the 7th Street and Interstate-90 (I-90) intersection in Bozeman, Montana.
139	The community of Gold Creek is at 4200 ft (1280 m) altitude, which is about equivalent to the highest recognized shoreline of GLM without considering unrecognized isostatic rebound adjustments in the lake basin.
152	The lowest altitude areas around Drummond are ~3950–4100 ft (1204–1250 m), suggesting the area was beneath ~100–250 ft (30–76 m) of water at the maximum lake level. Glacial outwash sediment entered the lake from the south along the Flint Creek valley.
158	Six miles (9.7 km) west of Drummond is a well-exposed section of lake-bottom facies strata on the north side of the highway. Note the continuity of bedding, along with subtle scour features. The stratigraphic, sedimentologic, and periglacial features at this site will be the subject of Stop 8, the last stop on the trip, as we return to Bozeman.
195	Near the community of East Missoula, I-90 crosses the Clark Fork River twice. Look for boulder-sized glacial erratics in the river and gravel and glaciolacustrine silts and clay in the freeway roadcuts (on the north). Gravel, deposited by significant currents, and interbedded silt and clay occur in exposures north of I-90 in East Missoula (Berg, 2006). The stratigraphy of this site has not yet been described in detail, but it likely represents multiple lake stands and drainage events.
199	Water draining from the upper Clark Fork and Blackfoot rivers flowed through Hellgate Canyon during drainage events, causing erosion of large clasts of Belt Supergroup rocks from the canyon walls, which were deposited near the University of Montana-Missoula campus (Alt, 2000, 2001).
207	The Missoula International Airport sits atop glacial Lake Missoula silt and clay (at an altitude of 3200 ft, 975 m); we will visit these deposits at the Rail line section of Hanson et al. (2012) at Stop 7.
223	The Ninemile section of lake-bottom deposits is best exposed along I-90, 0.25–0.6 mi (0.4–1 km) west of the Ninemile Road intersection. This outcrop is the most complete and well-studied section of lake-bottom deposits in the basin, and is the site of Stop 6. From the freeway, note the cyclic and overall fining-upward character of the section. West of Ninemile is the upper Alberton Gorge—a bedrock gorge with local scabland topography and a few lake deposits preserved.
234	At Cyr, where the gorge widens, eddy bars and streamlined bars record catastrophic flooding; overlying glacial Lake Missoula silt and clay record later lake-filling events (Stop 5).
242	I-90 cuts through a gravel bar formed on the inside bend in the flood route; we will visit an exposure here (Stop 3). Note the gravels on the northern freeway cut. Downstream of the bar, lake-bottom deposits of silt and clay overlie the gravels and the upper surface displays a characteristic rolling topography that will be discussed at Stop 4.
271	Take Exit 33 at St. Regis, turn northeast on Highway MT-135, travel ~1 mi (1.6 km) to the second street, and turn right on Lobo Loop Road to the Montana Visitor Information Center.

DESCRIPTION OF FIELD STOPS

Meeting Point—St. Regis, Montana Visitor Information Center, 230 Lobo Loop, St. Regis, Montana; phone: (406) 649–2290 (272 mi [438 km] west of Bozeman, Montana, ~4 h travel time).

Depart from Visitor Information parking lot and return to MT-135. Turn right and proceed northwest on MT-135 for 4.2 mi (6.7 km). Exit to the right onto a small dirt road. About 1 mi (1.6 km) south of the turn, MT-135 enters a narrow canyon and turns east. Slow down; the turnoff will be the first road on the right, just NE of the end of a guardrail. Take the dirt road ~0.25 mi (320 m) down to an unofficial crossing of the rail line.

STOP 1: Imbricated Boulders at Toole Siding
Location: 47°19′37″N, 115°02′02″W; 2630 ft (802 m)

The rail line is active, so be aware of the possibility of trains coming from either direction! This stop is at the top of a canyon reach of the Clark Fork River, downstream from St. Regis. Bedrock in the area is Revett Formation quartzite of the Belt Supergroup (Purcell Supergroup in Canada). The geology and landforms of the area are shown in Figure 2. The large accumulation of boulders with diameters of 2–6 m is unusual and represents a basal channel deposit of flood gravel; in places, it is a lag gravel where smaller grain sizes have been swept away. Boulders are well exposed and some are in original depositional position along the north side of the rail cut. These large clasts were deposited for an unknown distance above the rail grade and near the active channel of the Clark Fork River. Rough calculations of the fluid shear stresses required to transport the clasts shows they could not have been moved by normal-sized floods. Their position on the bank and within the channel suggests they originated from erosion of the upstream portion of the bedrock knob and accumulated where the channel widened slightly in this area. Their position very near or at stream grade (2600–2630 ft; 792–802 m) here shows that the Clark Fork River has downcut very little, or not at all, at this location since the late Pleistocene.

Return to MT-135 by the same route, turn left, and return to eastbound I-90. Travel ~13.5 mi (22 km) east to Exit 47, turn right and immediately left (SE) onto Diamond Road and then 6 mi (9.7 km) to the gravel pit near Mountain West Bark Products plant.

STOP 2: Flood Gravels near the Mouth of Trout Creek
Location: 47°08′24″N, 114°50′54″W; 2800 ft (853 m)

In road cuts and the gravel quarry above this site, there are gravelly deposits with large-scale cross-stratification. This eddy deposit is similar to others near confluences of the Clark Fork River and its tributaries. Stratification in many eddy deposits is not always evident; some exposures are composed of massive,

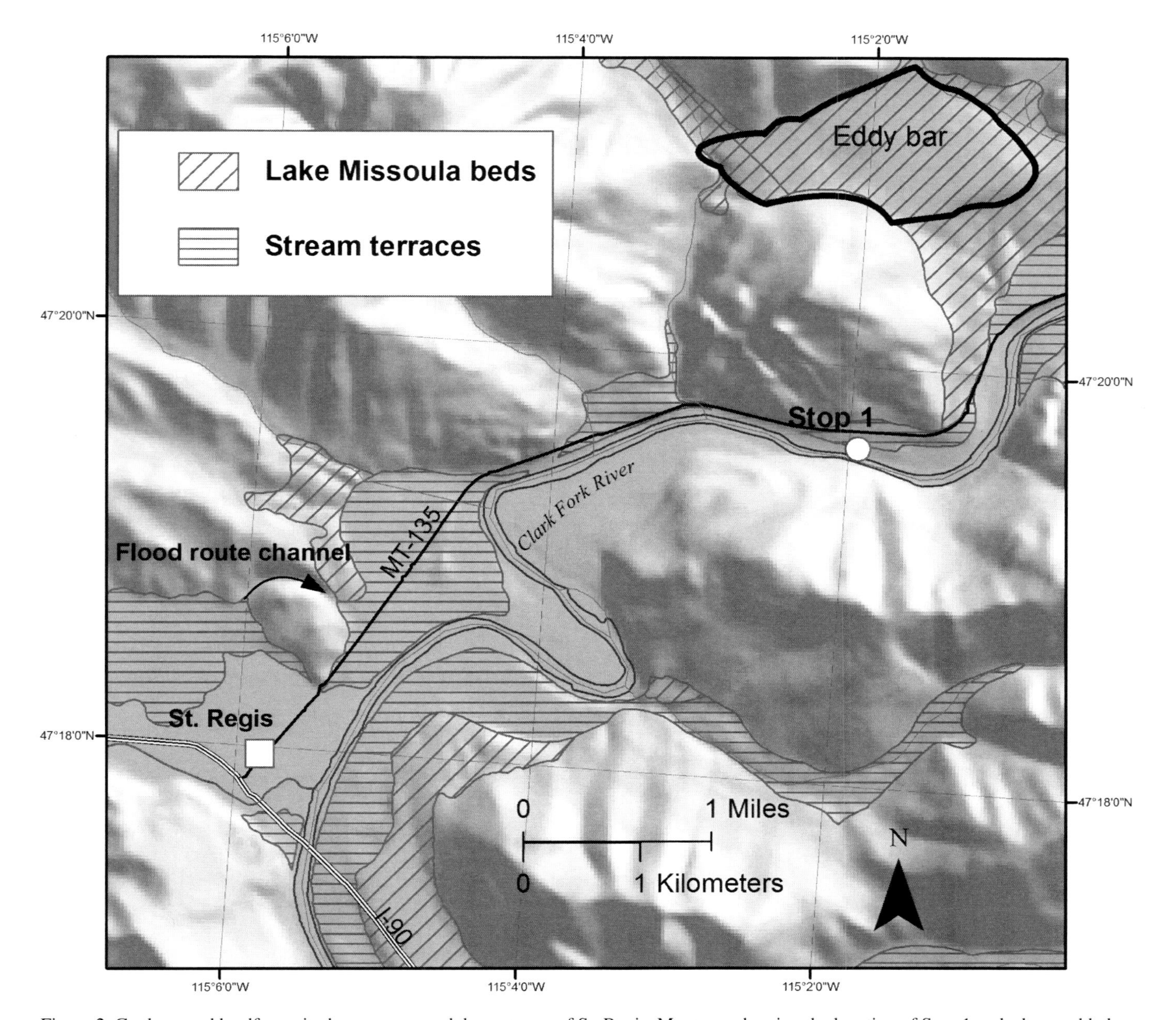

Figure 2. Geology and landforms in the area near and downstream of St. Regis, Montana, showing the location of Stop 1 and a large eddy bar downstream of a canyon reach (modified from Lonn and McFaddan, 1999).

very poorly sorted gravel. The sediment in this exposure, however, was deposited in a protected embayment by floodwaters flowing up Trout Creek, away from the main flood route, thus allowing for preservation of unidirectional stratification. Quarrying of the area above the road cut has made interpretation of the entire area difficult. Holocene colluvial deposits that may contain Mazama tephra, which has an age of 7627 ± 150 cal. ka B.P. (Zdanowicz et al., 1999), overlie gravel higher in the section.

Return to I-90 at Superior and travel east. Note the well-developed flat and rolling topography on top of the Lake Missoula beds. The rolling topography is especially evident east of the freeway in the half mile before Tarkio. After traveling 16 mi (26 km) from Superior, take Exit 61 toward Tarkio. Pass under the freeway toward the Tarkio Fishing Access Site, turn left at ~0.8 mi (1.3 km), and continue on the old rail grade. Travel ~0.3 mi (0.5 km) to a wide spot near a driveway on the right.

STOP 3: Flood Gravels and Lake Missoula Beds at Tarkio Rail Cut Exposure
Location: 47°00′36″N, 114°44′11″W; 2930 ft (893 m)

The cut along the rail grade near Tarkio exposes large-scale cross-stratification developed in granule- to boulder-sized gravel capped by fine-grained Lake Missoula beds (Fig. 3). Openwork gravels with large-scale cross-stratification were deposited during catastrophic drainage and development of the huge gravel bar. Some of the gravel is cemented by calcium carbonate. No age control is available for the gravel deposits.

A slump occurred in the gravel, causing minor movement and soft-sediment deformation in the fine-grained glaciolacustrine beds. The contact at the base of the Lake Missoula beds must represent a transgressive sequence where alluvial deposits are overlain by lakebed deposits. Irregularities at this contact and within the gravel suggest that slumping of the gravel deposit occurred during transgression of the lake after one catastrophic flooding event.

Return the same way, go under the freeway, travel northeast ~0.2 mi (0.3 km), turn to the north on Nemote Creek Road 454, and travel ~0.5 mi (0.8 km) to the crest of the low hills. Pull off on the side of the road.

Figure 3. Gravelly alluvium with large-scale cross-stratification (indicated by white arrows) overlain by Lake Missoula silts at Stop 2; black arrows show contact between gravels and silts; 5 ft measuring staff at black and white arrow (from Smith, 2006).

STOP 4: Rolling Topography on Lake Missoula Beds
Location: 47°01′57″N, 114°44′05″W; 3120 ft (951 m)

Along roads traversing the upper surface of the Lake Missoula beds, note the topography. At this location (A in Fig. 4), the east-west trending crests and troughs have been interpreted to have been dunes that formed during high-energy, north-flowing drainage events (Alt, 2001, p. 76–77). Aerial-photo interpretation of the hills and valleys shows dendritic patterns to the rolling topography, suggesting that they are not depositional landforms but networks of dry paleovalleys incised into the Lake Missoula beds (Fig. 4; Smith, 2006).

Key features to observe about the drainage basins and stream networks developed on the Lake Missoula beds are that most of the basins do not have active streams; if they do, the streams are underfit in much larger valleys. Where these small tributary drainage networks end at terraces along the Clark Fork River, alluvial fans are either not present, or they contain a much smaller amount of sediment than that expected to have been eroded from the stream network (at location B in Fig. 4). Thus, the side-valley systems appear to grade to terrace levels below the bench we are standing on (the upper surface of the Lake Missoula beds) and the alluvial sediments on these terraces were eroded before the terraces were abandoned. The lack of fans suggests that the drainage basins were mostly cut into the Lake Missoula beds during the final flood. If this is a typical record of water draining from fine-grained glaciolacustrine sediment, and the Lake Missoula beds represent multiple lake stands, we should expect to find gullies in the stratigraphic record. Another interpretation would be that the drainage basins are mostly Holocene gully systems and the sediments were carried off the terraces by unrecognized (filled-in?) channels.

Return to I-90 and travel east ~5 mi (8 km) to Exit 66, turn left on Fish Creek Road and then right on Old Hwy 10 to Cyr, cross the Clark Fork River bridge (to I-90 east), go under the freeway and park on the right before the on-ramp.

STOP 5: Gravel Bars and Lake Missoula Beds—Cyr Bridge over the Clark Fork River
Location: 47°0′13″N, 114°34′36″W; 2978 ft (908 m)

The widening of the Alberton Gorge into this valley led to the decreased flow velocities during the early phase of catastrophic lake drainages. The wooded ridge due south of the valley is a 300 ft-high (91 m) gravelly eddy bar deposited by currents flowing up the Sawmill Gulch tributary to the Clark Fork River (Fig. 5). In the exposure below the roadside along the east-bound on-ramp are Lake Missoula beds that are distorted by slumping and possibly periglacial action between lake stands. The series of stream terraces cut into glaciolacustrine deposits on the south side of the river may have formed during, or after, the last drainage of the lake.

Continue east on I-90 for ~12 mi (19 km) to Exit 82 (Ninemile Road). The Alberton Gorge has few preserved glacial sediments, except for the upper reach between the community of Alberton and Ninemile Creek. Scabland erosion of bedrock surfaces is common. Turn south to Bighorn Road, park and we will walk to the north-facing exposure along the freeway.

STOP 6: Ninemile Section of Lake Missoula Beds
Location: 47°01′13″N, 114°22′43″W; 3190 ft (972 m)

The Ninemile section of Lake Missoula beds is exposed on both the north-facing and the south-facing road cuts (Fig. 6). The striking cyclic pattern of bedding is evident at a distance on both sides; however, the higher moisture on the north-facing side allows for better exposure of sedimentologic details. The south-facing outcrops are hardened by desiccation of the fine-grained sediments. These exposures have long been recognized as a type example of fine-grained glaciolacustrine sedimentation in glacial Lake Missoula. Published studies began with Dave Alt and his MSc student Richard Chambers from the University of Montana (Alt and Chambers, 1970; Chambers, 1971, 1984).

The rhythmic nature of the units is so striking that Waitt (1980, 1985) theorized that a bed-for-bed correlation could be done with the flood slackwater sediments in central Washington,

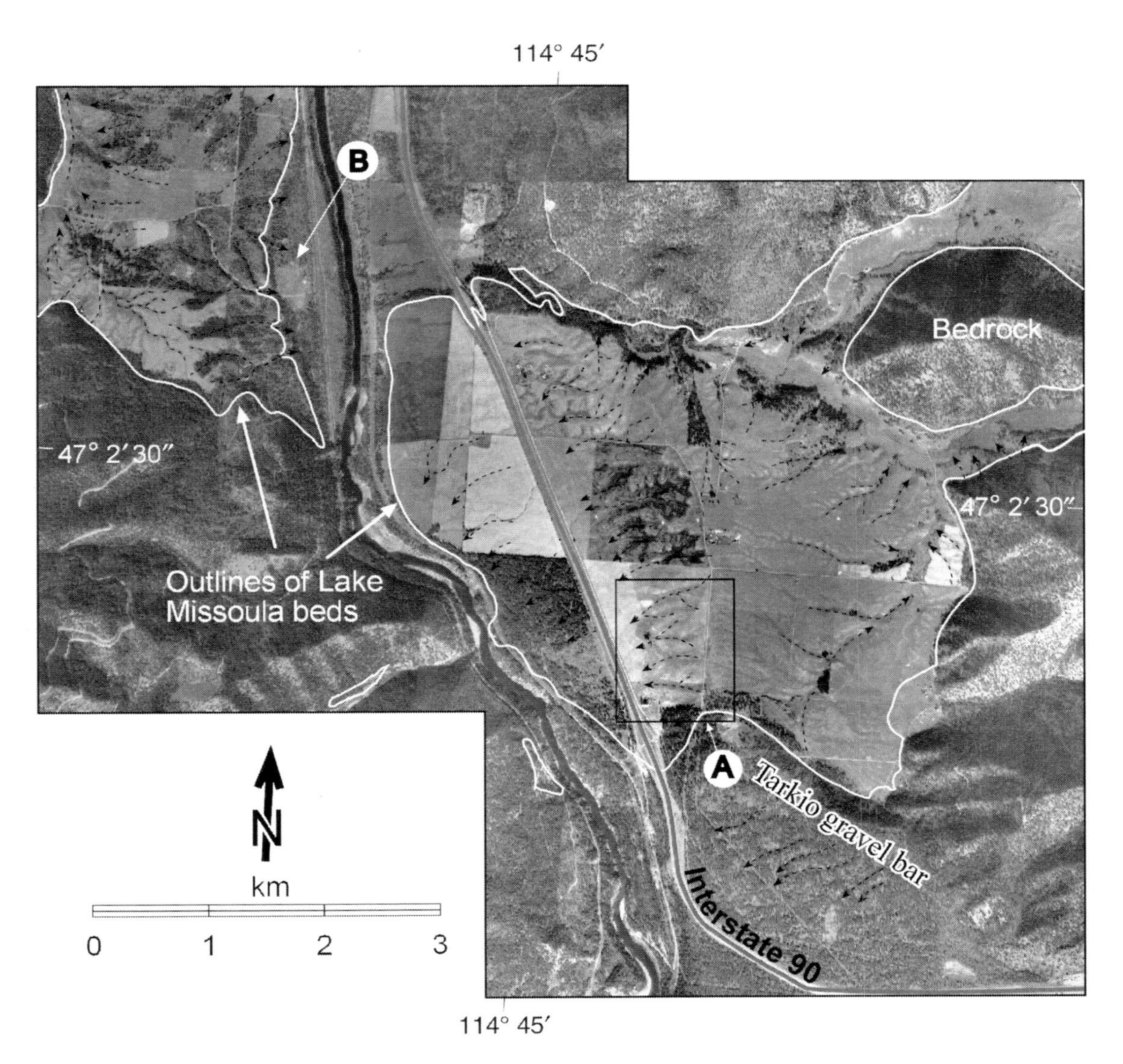

Figure 4. Digital orthophotos show the drainage networks that developed on the Lake Missoula beds in the Tarkio and Quartz flats areas and some of the drainage basins. Site A is rolling topography previously interpreted as a down-valley train of giant dunes. Site B is the location where drainage basins end in a depositional environment at a stream terrace remnant. The alluvial fans in this area are very small compared to the amount of sediment eroded from the drainage basin; dashed lines with arrows show some of the channels in lake-bed silts (from Smith, 2006).

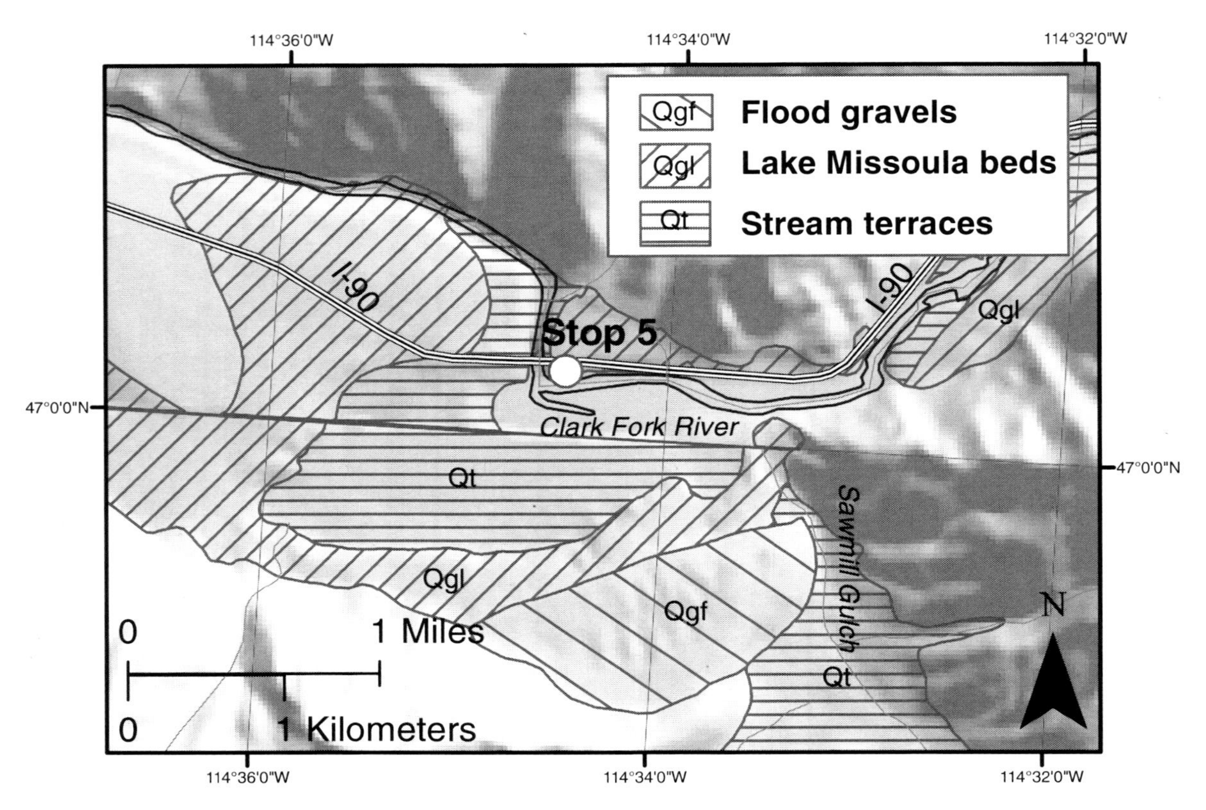

Figure 5. Geologic map of the area near Cyr, Montana (modified from Lonn et al., 2007).

and Atwater (1986) attempted such a correlation to the flood deposits in glacial Lake Columbia based on varve counts. Additionally, if the rhythmic character of the Lake Missoula beds is due to filling and draining cycles of the lake, these beds should be correlative to other deposits within the lake basin. Indeed, these sediments likely overlap with sediments at the Rail line site (Stop 7) based on sedimentary characteristics and optical dating (Hanson et al., 2012).

This exposure consists of 34 units of rhythmically stratified, graded beds of fine sand, silt, and clay. These rhythmic beds overly coarse- to fine-grained sand and pebble-cobble gravel at the base of the exposure—exposed only toward the western end. The main sediments of this site can be divided into three lithofacies: a gravel-sand facies, a sand-silt facies, and a silt-clay facies. The gravel-sand facies comprises thin beds of coarse sand and granule to medium pebble gravel and lies at the base of nine of the 34 units. This facies was likely deposited by non-channelized fluvial flow in shallow water on the lake floor before the lake began to fill (Chambers, 1971; Hanson et

Figure 6. (A) North-facing exposure at the Ninemile section along I-90, looking southeast (Stop 6); (B) climbing ripples in the sand-silt facies; (C) rhythmically bedded sand, silt, and clay unit—one bed of the sand-silt facies is overlain by the silt-clay facies (unit is 320 cm thick) (B and C from Hanson et al., 2012).

al., 2012). The sand-silt facies is present in every unit and comprises rippled or laminated very fine sand or silt. This facies was likely deposited by turbidity currents at relatively shallow depths as the lake began to fill (Chambers, 1971, 1984; Hanson et al., 2012). Alternate suggestions for the deposition of this facies include: (1) shallow-water, high-energy currents generated by the complete drainage of glacial Lake Missoula (Alt and Chambers, 1970; Curry, 1977); or, (2) turbidity currents deep within the lake (Shaw et al., 1999). The silt-clay facies is also present in each of the 34 rhythmically stratified units and comprises sets of silt and clay couplets that decrease in thickness up-unit. These couplets and similar ones elsewhere in the lake basin have been interpreted as varves (Alt and Chambers, 1970; Chambers, 1971, 1984; Waitt, 1980; Fritz and Smith, 1993; Levish, 1997; Hanson et al., 2012).

Evidence of exposure of the lake floor between rhythmic units (periglacial features or desiccation of the deposits) has lead several researchers to conclude that there were multiple complete or partial drainages of the lake (Fig. 7; Chambers, 1971, 1984; Hanson et al., 2012). It is important to note that the base of this section is ~300 m above the base of the ice dam, ~200 km to the west. Thus, it is difficult to conclude from the stratigraphy and sedimentology of this exposure if: (1) the lake that drained was at full pool; and (2) the draining event was complete and catastrophic or if it only drained to some level below the base of the exposure. It is likely that these sediments would only be preserved during smaller, less energetic lake draining events (Alho, et al., 2010; Hanson et al., 2012), which is consistent with the fact that the optical age at the base of the exposure (15.1 ka) places it toward the end of the existence of the lake. The amount of time between each drainage event is difficult to estimate because of an unknown amount of erosion of the lake floor during drainage. Minimum estimates can be based on existing varve counts, which vary from two to 44 for individual units at this site (Hanson et al., 2012).

Return to Exit 82 and continue I-90 to Exit 101 (N. Reserve Road – Hwy 93). Notice along the way the rolling topography on Lake Missoula beds.

STOP 7. Rail Line Exposure of Lake Missoula Beds
Location: 46°53′44″N, 114°5′45″W; 3170–3200 ft (966–975 m)

Travel ~2 mi south on Highway 93 (N. Reserve Road) and turn west on Mullan Road for ~1.7 mi (2.7 km). Turn right on Hiawatha Road. Follow the road until it bends. At the bend, you will see a power station. Beyond the station is the exposure. Park near the edge of the road. The Rail line site is an exposure along the abandoned Chicago, Milwaukee, St. Paul, and Pacific Railroad. Note, there are no known restrictions for accessing the exposure (it is on Missoula Airport Authority land), but you should ask permission from landowners to park your vehicle on Hiawatha Road as it is private property. The exposure continues for ~1.4 km but is best exposed closest to Hiawatha Road.

This exposure contains 29 upward-thinning units of rhythmically bedded fine sand, silt, and clay. Three lithofacies are present: a sand facies, a silt facies, and a silt-clay facies (Fig. 8). The sand facies, which resembles the gravel-sand facies at the Ninemile site, occurs at the base of 10 units and comprises well-sorted, very fine- to fine-grained sand that is laminated or rippled. This facies was likely deposited as non-channelized fluvial sediments in shallow water on the floor of the lake before the lake refilled (Hanson et al., 2012). The silt facies, which resembles the sand-silt facies at the Ninemile site, consists of thinly laminated, coarse to fine silt with type-B ripple-drift cross-lamination. This facies was likely deposited by turbidity currents at relatively shallow depths as the lake began to fill (Hanson et al., 2012). The silt-clay facies is the dominant facies at this site (Fig. 9) and it consists of silt-clay couplets similar to those at the Ninemile site, which are interpreted to be varves (Hanson et al., 2012). Boundaries between all units are unconformable and show evidence of erosion and lake-bottom exposure in the form of dessication and periglacial features (where ice-wedge cracks are filled with sediment from the walls; Fig. 9). Sedimentology and two optical ages (14.8 and 12.6 ka; Hanson et al., 2012) potentially place this exposure stratigraphically overlapping and above the Ninemile Creek exposure.

Return to I-90 and travel east ~36 mi (58 km) to Exit 138 (Bearmouth), turn north, crossing the Clark Fork River, turning east onto the I-90 frontage road, and travel 11 mi (18 km). As the road reenters the floodplain and is close to a river, turn south through a ranch gate onto private property. We have permission to enter the land, driving ~0.4 mi (640 m) to near the outcrop.

STOP 8: Garden Gulch Section of Lake Missoula Beds
Location 46°42′30″N, 113°15′28″W; 3850–3890 ft (1173–1186 m)

The Garden Gulch section was measured on a 0.4 km-long, nearly vertical exposure of bedrock, sand and gravel, and silty lake sediments near the Clark Fork River. The exposure was formed ca. 1890 when a meander loop of the Clark Fork River was cut off during the construction of the Northern Pacific railroad. The rail and subsequent highway routes were constructed parallel to the modified channel, which cuts into bedrock of greenschist-grade, metamorphosed mudstone and fine-grained sandstone of the Blackleaf Formation. During construction of the artificial channel, unconsolidated fluvial gravel and lacustrine sands, silts, and clays were left near the channel; minor amounts of spoils from the excavation were placed on top of the sediment. The silty sediments are periodically undercut during high-discharge events on the river, which has helped to maintain a nearly vertical exposure.

As described here, the Garden Gulch section is but one composite section on this large outcrop. It was measured from east to west up toward the major gulley that bisects the outcrop. The deposits include gravelly fluvial sediments of a paleo–Clark Fork River (2 m) that is overlain by ~7 m of fining-upward

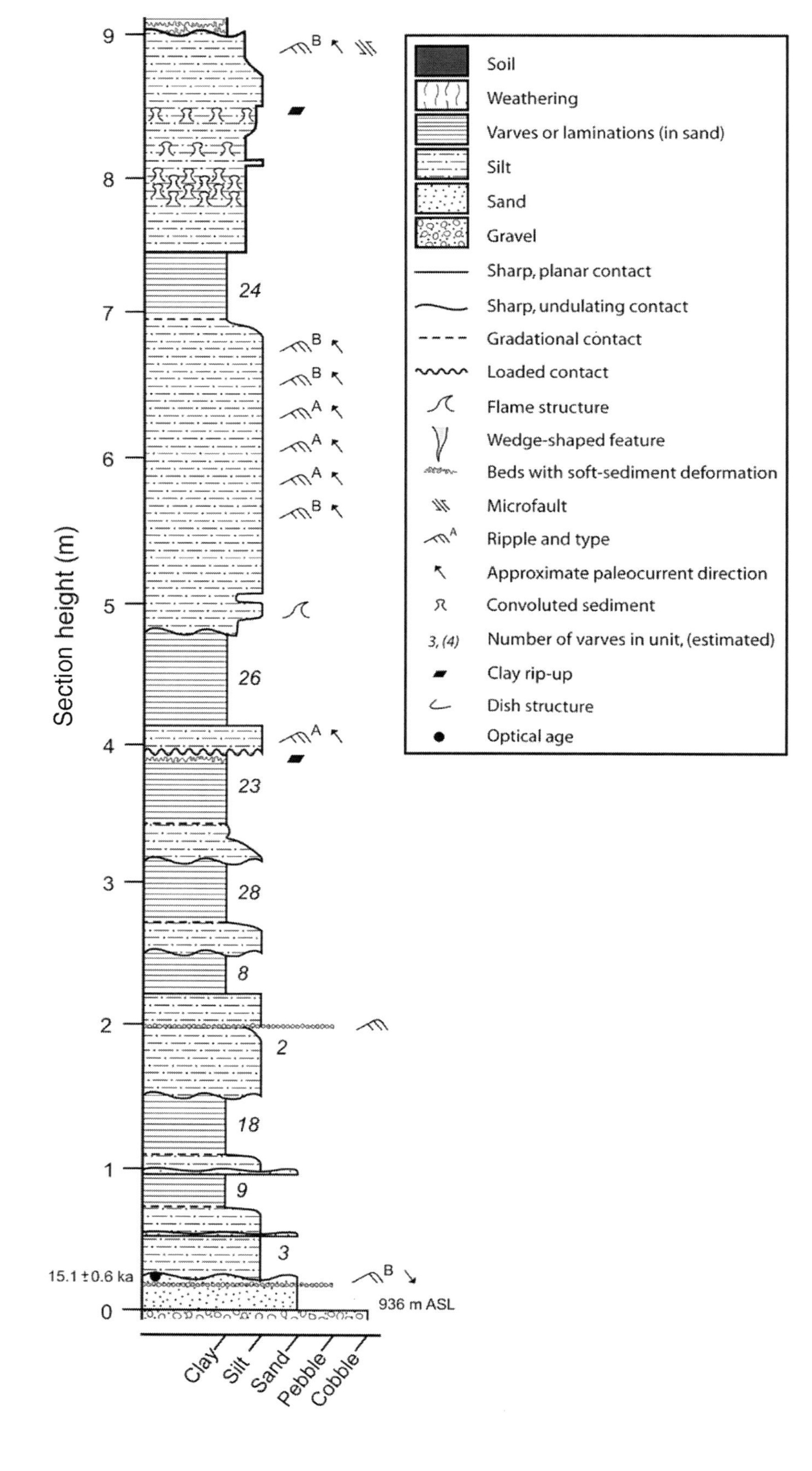

Figure 7. Measured section and location of an optical age for the lower 9 m of Lake Missoula beds exposed at Stop 6 (from Hanson et al., 2012).

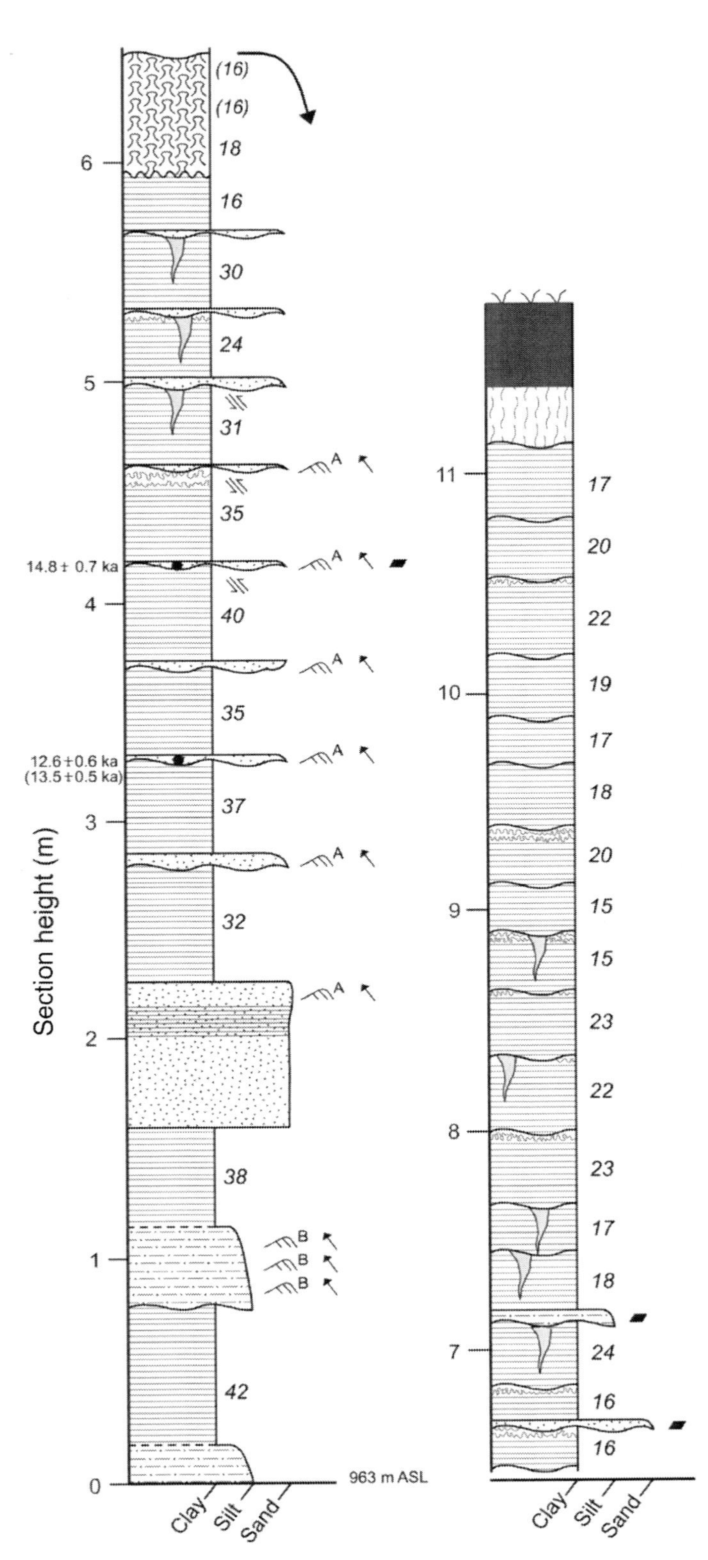

Figure 8. Measured section and location of an optical age for the lower 9 m of Lake Missoula beds exposed at Stop 7 (see Fig. 7 for key to symbols; from Hanson et al., 2012).

Figure 9. (A) A thinning-upward sequence of silt-clay varves, overlain and underlain by the rippled sand-silt facies at the Rail line section (tape is 40 cm long); (B) downward-tapering periglacial feature, filled with brecciated sediment from the walls of the feature, in Lake Missoula beds at the Rail line section (tape is 43 cm long).

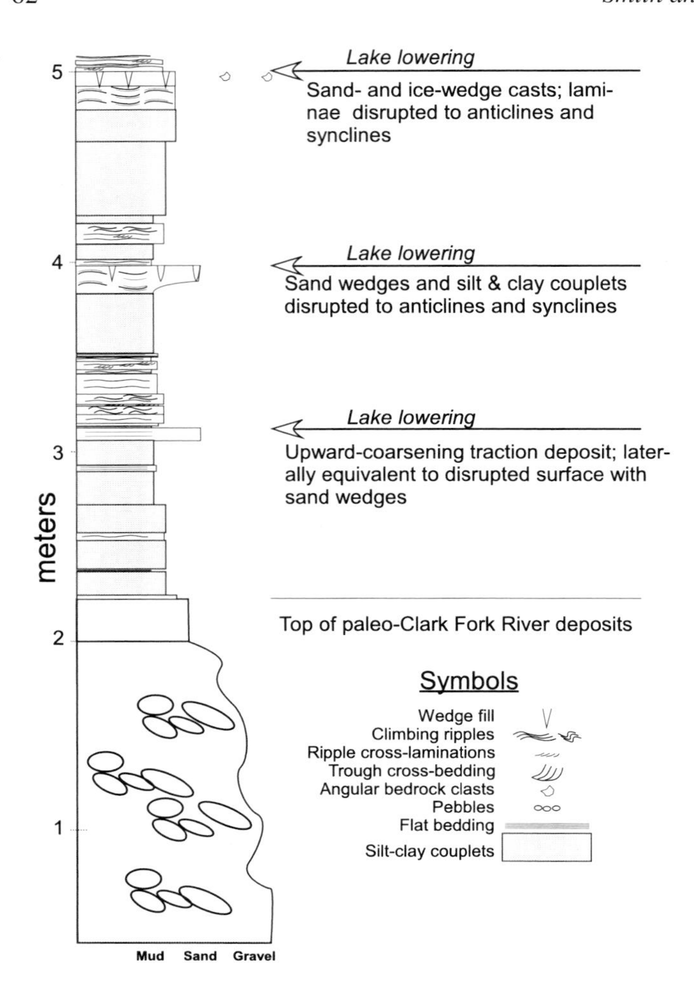

Figure 10. Lower 5 m of the 9 m of exposure at the Garden Gulch section.

Figure 11. Panorama of the lower 6.5 m of the 9 m of the Garden Gulch section. Inset photograph of a 70-cm exposure of contorted glaciolacustrine deposits cut by an ice-wedge cast (between measuring tape and pipe); tape and trowel for scale.

sequences of laminated silt and clayey silt with beds of rippled very fine sandy silt of mostly lacustrine origin (Figs. 10, 11). Two quartz sand samples were dated by optical dating, one in traction deposits between the top of the paleo–Clark Fork River deposits and the lake deposits (east of Figs. 10 and 11), and one from massive, very fine grained, silty sand near the top of the outcrop. Preliminary results give an age of 20.4–17 ka for the lower sample and 10.5 ka for the upper sample (O.B. Lian, 14 July 2011, personal commun.). The lower sample is interpreted to be fluvial or shallow-water lacustrine (beach or delta) sediment deposited during initial transgression of GLM. The uppermost unit, a massive very fine-grained sandy silt with multiple soil horizons, may be loess and colluvium. Sand wedges, contorted bedding, and a paleosol provide evidence of 8–12 exposure surfaces in the outcrop, each representing a lake-lowering event. This record of lowering (which could be either partial or full lake drainings) from full, or nearly full-pool stands of GLM after 20.4–17 ka likely correlate to intervals at the Rail line and Ninemile sections.

Return to the frontage road, turn right, and travel east, pass under the freeway, keep left, travel through Drummond, and reenter the freeway on the east side of town.

Stay on I-90 through Butte to Bozeman, to Exit 306.

ACKNOWLEDGMENTS

The Montana Bureau of Mines and Geology Groundwater Assessment Program supported most of the initial fieldwork for LNS, with the purpose of understanding the distribution of aquifers in the area. Montana Tech seed grant funded the OSL dating at the Garden Gulch section. Helpful reviews by Vic Baker, Phyllis Hargrave, and Colin Shaw improved the paper's clarity. J.J. Clague kindly provided Figure 6A.

REFERENCES CITED

Alho, P., Baker, V., and Smith, L.N., 2010, Paleohydraulic reconstruction of the largest Glacial Lake Missoula draining(s): Quaternary Science Reviews, v. 29, p. 3067–3078, doi:10.1016/j.quascirev.2010.07.015.

Alt, D., 2000, The catastrophic drainage of Glacial Lake Missoula, *in* Roberts, S., and Winston, D., eds., Geologic Field Trips, Western Montana and Adjacent Areas: Rocky Mountain Section of the Geological Society of America, Missoula, University of Montana, p. 31–39.

Alt, D., 2001, Glacial Lake Missoula and Its Humongous Floods: Missoula, Mountain Press, 208 p.

Alt, D., and Chambers, R.L., 1970, Repetition of the Spokane flood: American Quaternary Association Meeting 1, Yellowstone Park and Bozeman, Montana, Abstracts, Montana State University, Bozeman, p. 1.

Atwater, B.F., 1986, Pleistocene Glacial-Lake Deposits of the Sanpoil River Valley, Northeastern Washington: U.S. Geological Survey Bulletin 1661, 39 p.

Atwater, B.F., Smith, G.A., and Waitt, R.B., 2000, The Channeled Scabland: back to Bretz?: Geology, v. 28, p. 574–575, doi:10.1130/0091-7613(2000)28<576:TCSBTB>2.0.CO;2.

Baker, V.R., 1973, Paleohydrology and Sedimentology of Lake Missoula Flooding in Eastern Washington: Geological Society of America Special Paper 144, 73 p., doi:10.1130/SPE144-p1.

Baker, V.R., Bjornstad, B.N., Greenbaum, N., Porat, N., Smith, L.N., and Zreda, M.G., 2009, Possible revised chronology of late Pleistocene megaflooding, northwestern US: Geological Society of America Abstracts with Programs, v. 41, no. 7, p. 168.

Benito, G., and O'Connor, J.E., 2003, Number and size of last-glacial Missoula floods in the Columbia River valley between the Pasco Basin, Washington, and Portland, Oregon: Geological Society of America Bulletin, v. 115, p. 624–638, doi:10.1130/0016-7606(2003)115<0624:NASOLM>2.0.CO;2.

Berg, R.B., 2006, Geologic Map of the Upper Clark Fork Valley between Bearmouth and Missoula, Southwestern Montana: Montana Bureau of Mines and Geology Open-File Report MBMG 535, 18 p., 1 sheet, scale 1:24,000.

Booth, D.B., Troost, K.G., Clague, J.J., and Waitt, R.B., 2004, The Cordilleran ice sheet, *in* Gillespie, A.R, Porter, S.C., and Atwater, B.F., eds., The Quaternary Period in the United States: Developments in Quaternary Science 1: Amsterdam, Elsevier, p. 17–43.

Breckenridge, R.M., 1989, Lower glacial lakes Missoula and Clark Fork ice dams, *in* Breckenridge, R.M., ed., Glacial Lake Missoula and the Channeled Scabland: Missoula, Montana to Portland, Oregon: American Geophysical Union, 28th International Geological Congress, Field Trip Guidebook T310, p. 13–21.

Bretz, J.H., 1969, The Lake Missoula floods and the Channeled Scabland: The Journal of Geology, v. 77, p. 505–543, doi:10.1086/627452.

Bronk Ramsey, C., 2009, Bayesian analysis of radiocarbon dates: Radiocarbon, v. 51, p. 337–360.

Chambers, R.L., 1971, Sedimentation in glacial Lake Missoula [M.S. thesis]: Missoula, The University of Montana, 100 p.

Chambers, R.L., 1984, Sedimentary evidence for multiple glacial lakes Missoula, *in* McBane, J.D., and Garrison, P.B., eds., Northwest Montana and Adjacent Canada: Billings, Montana Geological Society, p. 189–199.

Clague, J.J., Armstrong, J.E., and Mathews, W.H., 1980, Advance of the late Wisconsinan Cordilleran ice sheet in southern British Columbia since 22,000 yr BP: Quaternary Research, v. 13, p. 322–326, doi:10.1016/0033-5894(80)90060-5.

Clague, J.J., Barendregt, R., Enkin, R.J., and Foit, F.F., Jr., 2003, Paleomagnetic and tephra evidence for tens of Missoula floods in southern Washington: Geology, v. 31, p. 247–250, doi:10.1130/0091-7613(2003)031<0247:PATEFT>2.0.CO;2.

Curry, R.R., 1977, Glacial geology of Flathead Valley and catastrophic drainage of Glacial Lake Missoula: Missoula, University of Montana, Geological Society of America, Rocky Mountain Section Field Guide, v. 4, p. 14–38.

Fritz, W.J., and Smith, G.A., 1993, Revisiting the Ninemile section: Problems with relating glacial Lake Missoula stratigraphy to the Scabland-floods stratigraphy: Eos (Transactions American Geophysical Union), v. 74, no. 43 (supplement), p. 302.

Gombiner, J., Hendy, I.L., Hemming, S.R., Fleisher, M.Q., Pierce, E., Mesko, G., and Dale, C.L., 2010, Spatial and temporal variation of the last ice age mega-floods in the Pacific Northwest: Sediment provenance using single-aliquot K/Ar dating: American Geophysical Union 2010 Fall Meeting, abstract PP21A-1663.

Hanson, M.A., Lian, O.B., and Clague, J.J., 2012, The sequence and timing of large late Pleistocene floods from glacial Lake Missoula: Quaternary Science Reviews, v. 31, p. 67–81, doi:10.1016/j.quascirev.2011.11.009.

Kuehn, S.C., Froese, D.G., Carrara, P.E., Foit, F.F., Jr., Pearce, N.J.G., and Rotheisler, P., 2009, Major- and trace-element characterization, expanded distribution, and a new chronology for the latest Pleistocene Glacier Peak tephras in western North America: Quaternary Research, v. 71, p. 201–216, doi:10.1016/j.yqres.2008.11.003.

Langton, C.M., 1935, Geology of the northeastern part of the Idaho Batholith and adjacent region in Montana: The Journal of Geology, v. 43, p. 27–60, doi:10.1086/624269.

Levish, D.R., 1997, Late Pleistocene sedimentation in glacial Lake Missoula and revised glacial history of the Flathead Lobe of the Cordilleran Ice Sheet, Mission valley, Montana [Ph.D. dissertation]: Boulder, University of Colorado, 191 p.

Lonn, J.D., and McFaddan, M.D., 1999, Geologic Map of the Wallace 30′ × 60′ Quadrangle, Montana Bureau of Mines and Geology Open-File Report MBMG 388, 15 p., 2 sheets, scale 1:100,000.

Lonn, J.D., Smith, L.N., and McCulloch, R.B., 2007, Geologic Map of the Plains 30′ × 60′ Quadrangle, Western Montana, Montana Bureau of Mines and Geology Open-File Report MBMG 554, 43 p., 1 sheet, scale 1:100,000.

Lopes, C., and Mix, A.C., 2009, Pleistocene megafloods in the northeast Pacific: Geology, v. 37, p. 79–82, doi:10.1130/G25025A.1.

Pardee, J.T., 1910, The glacial Lake Missoula, Montana: The Journal of Geology, v. 18, p. 376–386, doi:10.1086/621747.

Pardee, J.T., 1942, Unusual currents in glacial Lake Missoula, Montana: Geological Society of America Bulletin, v. 53, p. 1569–1599.

Reimer, P.J., Baillie, M.G.L., Bard, E., Bayliss, A., Beck, J.W., Blackwell, P.G., Bronk Ramsey, C., Buck, C.E., Burr, G.S., Edwards, R.L., Friedrich, M., Grootes, P.M., Guilderson, T.P., Haidas, I., Heaton, T.J., Hogg, A.G., Hughen, K.A., Kaiser, K.F., Kromer, B., McCormac, F.G., Manning, S.W., Reimer, R.W., Richards, D.A., Southon, J.R., Talamo, S., Turney, C.S.M., van der Plicht, J., and Weyhenmeyer, C.E., 2009, IntCal09 and Marine09 radiocarbon age calibration curves, 0–50,000 years cal BP: Radiocarbon, v. 51, p. 1111–1150.

Shaw, J., Munro-Stasiuk, M., Sawyer, B., Beaney, C., Lesemann, J.-E., Musacchio, A., Rains, B., and Young, R.R., 1999, The Channeled Scabland: Back to Bretz?: Geology, v. 27, p. 605–608, doi:10.1130/0091-7613(1999)027<0605:TCSBTB>2.3.CO;2.

Shaw, J., Munro-Stasiuk, M., Sawyer, B., Beaney, C., Lesemann, J.-E., Musacchio, A., Rains, B., and Young, R.R., 2000, The Channeled Scabland: Back to Bretz?: Geology, v. 28, p. 576, doi:10.1130/0091-7613(2000)28<577:TCSBTB>2.0.CO;2.

Smith, L.N., 2006, Stratigraphic evidence for multiple drainings of glacial Lake Missoula along the Clark Fork River, Montana, USA: Quaternary Research, v. 66, p. 311–322, doi:10.1016/j.yqres.2006.05.009.

Sperazza, M., Thomas, G., Hendrix, M.S., and Moore, J.N., 2002, Record of late Pleistocene through Holocene climate change in a regional lake system: Flathead Lake basin, northwestern Montana: Eos (Transactions, American Geophysical Union), v. 83, p. F941.

Waitt, R.B., 1980, About forty last-glacial Lake Missoula jökulhlaups through southern Washington: The Journal of Geology, v. 88, p. 653–679, doi:10.1086/628553.

Waitt, R.B., 1985, Case for periodic, colossal jökulhlaups from Pleistocene glacial Lake Missoula: Geological Society of America Bulletin, v. 96, p. 1271–1286, doi:10.1130/0016-7606(1985)96<1271:CFPCJF>2.0.CO;2.

Zdanowicz, C.M., Zielinski, G.A., and Germaini, M.S., 1999, Mount Mazama eruption: Calendrical age verified and atmospheric impact assessed: Geology, v. 27, p. 621–624, doi:10.1130/0091-7613(1999)027<0621:MMECAV>2.3.CO;2.

Zuffa, G.G., Normark, W.R., Serra, F., and Brunner, C.A., 2000, Turbidite megabeds in an oceanic rift valley recording jökulhlaups of late Pleistocene glacial floods of the western United States: The Journal of Geology, v. 108, p. 253–274, doi:10.1086/314404.

Manuscript Accepted by the Society 21 February 2014

Printed in the USA

The Geological Society of America
Field Guide 37
2014

Neotectonics and geomorphic evolution of the northwestern arm of the Yellowstone Tectonic Parabola: Controls on intra-cratonic extensional regimes, southwest Montana

Chester A. Ruleman
U.S. Geological Survey, Geosciences and Environmental Change, Science Center, P.O. Box 25046, Denver Federal Center, MS 980, Denver, Colorado 80225, USA

Mort Larsen
Wyoming State Geological Survey, P.O. Box 1347, Laramie, Wyoming 82073, USA

Michael C. Stickney
Earthquake Studies Office, Montana Bureau of Mines and Geology, Montana Tech of the University of Montana, 1300 W. Park Street, Butte, Montana 59701, USA

ABSTRACT

The catastrophic Hebgen Lake earthquake of 18 August 1959 (M_W 7.3) led many geoscientists to develop new methods to better understand active tectonics in extensional tectonic regimes that address seismic hazards. The Madison Range fault system and adjacent Hebgen Lake–Red Canyon fault system provide an intermountain-active tectonic analog for regional analyses of extensional crustal deformation. The Madison Range fault system comprises fault zones (~100 km in length) that have multiple salients and embayments marked by preexisting structures exposed in the footwall. Quaternary tectonic activity rates differ along the length of the fault system, with less displacement to the north. Within the Hebgen Lake basin, the 1959 earthquake is the latest slip event in the Hebgen Lake–Red Canyon fault system and southern Madison Range fault system. Geomorphic and paleoseismic investigations indicate previous faulting events on both fault systems. Surficial geologic mapping and historic seismicity support a coseismic structural linkage between the Madison Range and Hebgen Lake–Red Canyon fault systems.

On this trip, we will look at Quaternary surface ruptures that characterize prehistoric earthquake magnitudes. The one-day field trip begins and ends in Bozeman, and includes an overview of the active tectonics within the Madison Valley and Hebgen Lake basin, southwestern Montana. We will also review geologic evidence, which includes new geologic maps and geomorphic analyses that demonstrate preexisting structural controls on surface rupture patterns along the Madison Range and Hebgen Lake–Red Canyon fault systems.

Ruleman, C.A., Larsen, M., and Stickney, M.C., 2014, Neotectonics and geomorphic evolution of the northwestern arm of the Yellowstone Tectonic Parabola: Controls on intra-cratonic extensional regimes, southwest Montana, *in* Shaw, C.A., and Tikoff, B., eds., Exploring the Northern Rocky Mountains: Geological Society of America Field Guide 37, p. 65–87, doi:10.1130/2014.0037(03). For permission to copy, contact editing@geosociety.org.

INTRODUCTION

The Madison Range and Madison Valley, located in southwestern Montana, have experienced multiple stages of tectonic activity ranging in age from Early Proterozoic (Erslev, 1981, 1982, 1993) to Quaternary (Pardee, 1950; Shelden, 1960; Myers and Hamilton, 1964; Witkind, 1975; Mathieson, 1983; Schneider, 1985; Lundstrom, 1986; Ruleman, 2001, 2002a, 2002b, 2009). Currently, the region is under tectonic influences of both the northern Basin and Range province and Yellowstone caldera (Anders et al., 1989; Pierce and Morgan, 1992). We will examine the influence of preexisting structures on Neogene tectonic activity.

The north-northwest-trending Madison Valley is ~100 km in length and lies between the Madison and Gravelly Ranges, both of which are cored with Archean rocks (Figs. 1 and 2). Tertiary extension and deposition of sediments have created thicknesses of basin fill greater than 2000 m (Peterson and Witkind, 1975; Gary, 1980; Schofield, 1981). Previous work has revealed how the Yellowstone caldera and associated hotspot track have influenced tectonic activity (Anders et al., 1989; Pierce and Morgan, 1992; McCalpin and Warren, 1993). Surficial geologic mapping and morpho-metric analyses (Ruleman, 2002b) show that Quaternary displacements decrease to the north away from the caldera, suggestive of proximal thermo-tectonic influences from the caldera complex.

More than 100 fault scarp profiles and other morpho-metric analyses (e.g., mountain-front sinuosity, valley-floor width to valley height ratios, and a reconnaissance basal faceted spur analysis) have been used to characterize the Madison Range fault system. Prehistoric most recent offset along the fault system has previously been determined to be 500–6000 yr B.P. by soil morphology on offset deposits (Schneider, 1985) and <4000 yr B.P. by diffusion-equation modeling of fault scarps (Lundstrom, 1986). Soils on the youngest offset deposits are generally very weak consisting of 5–10 cm of A horizon with little to no B horizon on top of parent material or C horizon (i.e., coarse gravel and sand). Our work along the Madison Range and Hebgen Lake–Red Canyon fault systems indicates slip rates of 0.1–0.5 mm/yr and maximum credible earthquakes ranging from M_w 6.5–7.2, consistent with the 1959 Hebgen Lake earthquake M_w 7.3.

Within the footwall of the Madison Range fault system, Laramide and older structures coincide with embayments and salients along the range front, as well as truncated seismogenic surface ruptures. This field trip begins at the northern extent of the fault system and follows its trace southward in the direction of increasing Pleistocene offset. We will also examine the paleoseismic and historical structural interrelationships of the Madison Range and Hebgen Lake–Red Canyon fault systems compared to preexisting structures.

Normal faults in the Hebgen Lake–Red Canyon fault system (Hebgen and Red Canyon segments) overlap Laramide, basement-involved thrust faults and folds at the southeast end of the ancestral Madison-Gravelly arch (Kellogg et al., 1995; Tysdal, 1986). The Hebgen normal fault scarp overlaps the trace of closely spaced imbricate faults of the Beaver Creek thrust system, whereas the Red Canyon normal fault scarp closely follows the overturned limb of a Laramide fold and the Divide thrust fault (Fig. 1). Normal faults of the fault system are likely in the early stages of extension in this tectonic setting (Lageson, 2009).

STOP 1—Cedar Creek Alluvial Fan
(UTM 451754E; 5016540N; elev. 5880 ft)

Drive south of Ennis, Montana, on U.S. Hwy 287 ~4.1 mi and turn left on gravel ranch road up the Cedar Creek fan. Drive ~3.8 mi and park at the fork in the road.

At this stop, we will look at middle to late Pleistocene displacement along the northern Madison Range fault system. At Cedar Creek, deposits of two different ages are offset ~2.4 and 4.9 m (Fig. 3). To the north, at Jourdain Creek at the northern margin of the fault system, late Pleistocene fan deposits are offset ~1.3 m, most likely reflecting a tapering of displacement at the northern tip of this segment (Fig. 4). Based on contrasting surficial morphology, morphostratigraphic position, and soil development, deposits are differentiated as late middle Pleistocene (Bull Lake age) and late Pleistocene (Pinedale age). Based on consistent middle to late Pleistocene single-event displacements profiled along the entire length of the fault system, characteristic displacements range between 2 and 3 m (Fig. 5). Single-event displacements fewer than 2 m only occur at section boundaries.

STOP 2—Shell Creek Alluvial Fan and Preexisting Structural Controls in the Paleozoic Sedimentary Section on Quaternary Surface Rupture Patterns
(UTM 448988E; 5012297N; elev. 5340 ft)

From Stop 1, return to U.S. Hwy 287 and turn south. Drive ~2.4 mi to County Rd. 212 (Airport Road) and turn left. Drive ~2.4 mi east and park at the turn off for the County Big Sky Airstrip.

At this location, surficial displacement patterns and bedrock mapping of the footwall of the Madison Range fault system (Kellogg and Williams, 2000; Ruleman, 2002b) show a preexisting structural control on late Quaternary faulting patterns. Here, preexisting Laramide thrust faults and lateral ramps in the Mississippian and Pennsylvanian sections preserved in the footwall caused Pleistocene faulting to step laterally (Figs. 6A and 6B). Cross sections show the cross-cutting relationships between Laramide structures and Neogene normal faulting (Fig. 7).

STOP 3—Mill Creek to Bear Creek Showing Preexisting Structural Controls in the Cretaceous Sedimentary Section on Quaternary Surface Rupture Patterns and an Example of a Perched Basement Wedge
(UTM 453154E; 5001432N; elev. 5850 ft)

From Stop 2, return to U.S. Hwy 287 and turn left. Proceed ~4.4 mi south and turn left to head east on Bear Creek Loop Road (toward the Bear Creek Ranger Station and trailhead). Stop after crossing Bear Creek.

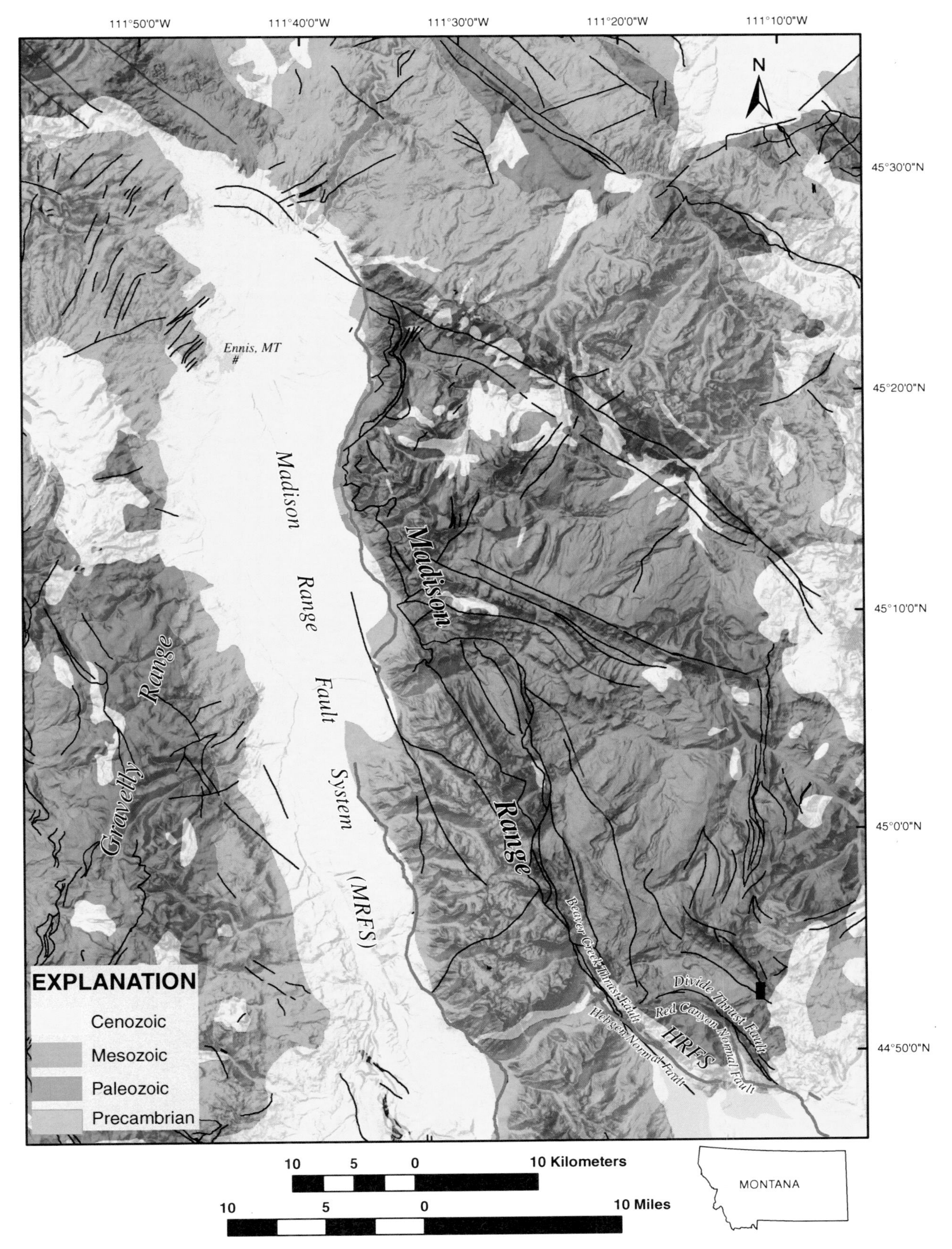

Figure 1. Generalized geologic map showing distribution of Precambrian and Phanerozoic rocks. Black lines show preexisting Laramide and older faults. Red lines show traces of Pleistocene surface ruptures. The Red Canyon and Hebgen normal faults make up the Hebgen Lake–Red Canyon fault system (HRFS) from Montana Bureau of Mines and Geology (2013).

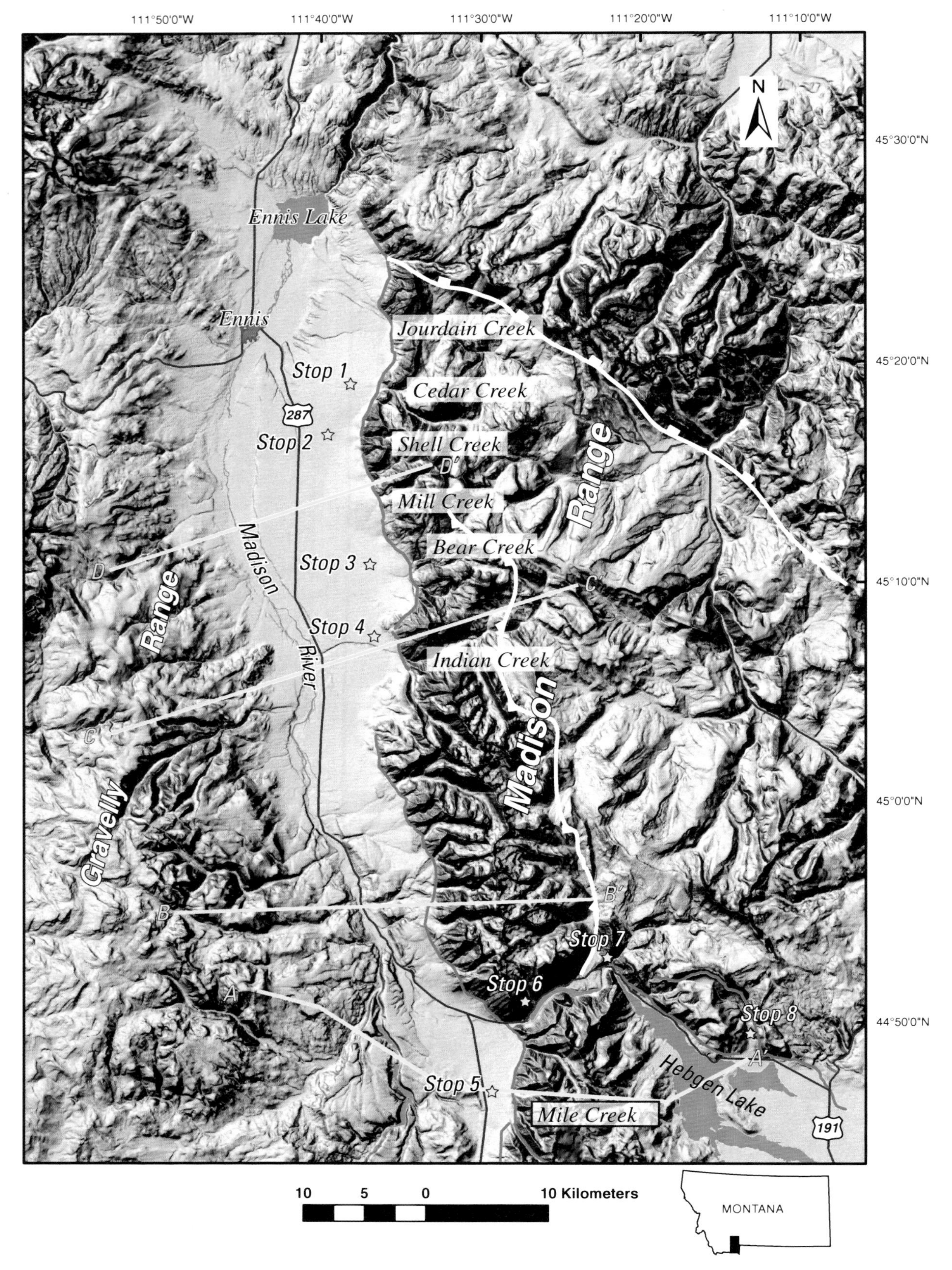

Figure 2. Physiographic map of the study region showing field trip stops, cross sections presented herein, location of Pleistocene surface ruptures (red), and major preexisting faults (white, teeth on hanging wall of thrust faults and rectangles on hanging wall of high-angle reverse fault).

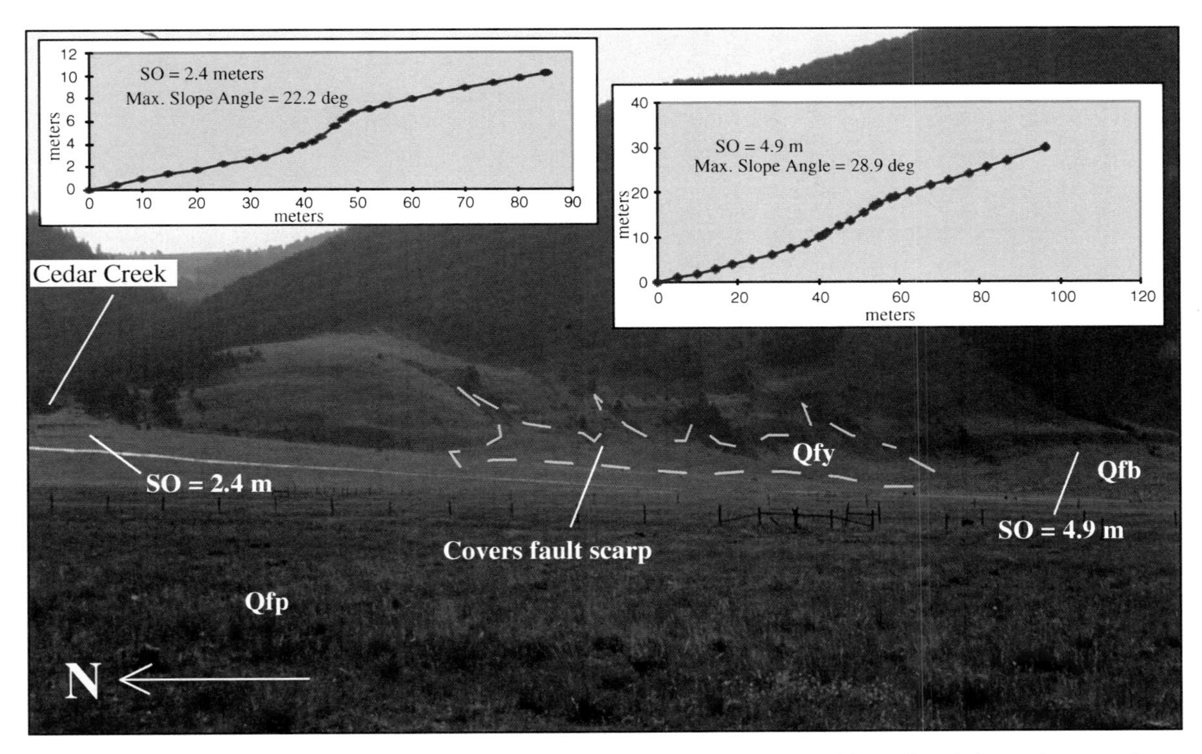

Figure 3. View looking east at Cedar Creek. Foreground is a Pinedale outwash fan (Qfp), offset 2.4 m at the range front. To the south, Bull Lake fan deposits (Qfb) are displaced 4.9 m. Qfy deposits are unfaulted Holocene fan deposits covering the trace of surface ruptures. Profiles show amounts of surface offset and maximum slope angles. Both horizontal and vertical axes of the profiles are in meters and blue dots show locations of profile measurements. SO is the surface offset and Max. Slope Angle is the highest angle measured on the fault scarp.

Figure 4. Pinedale age fan deposits (Qfp) offset 1.3 m at Jourdain Creek. This is the northernmost surface rupture along the Madison Range fault system. SO is the surface offset.

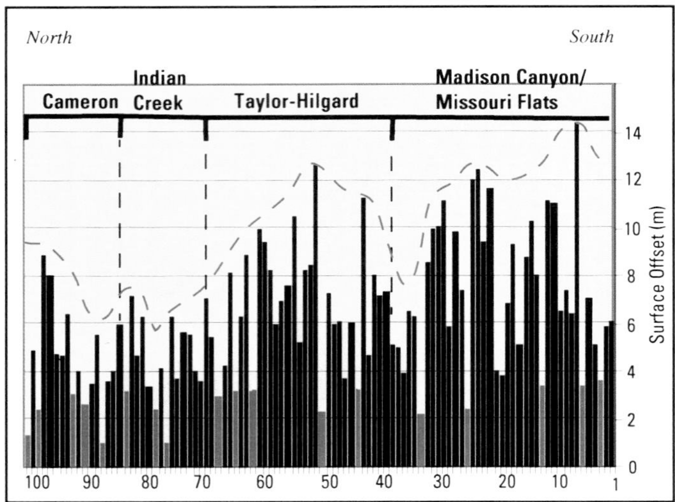

Figure 5. Surface offset along the Madison Range fault system. Sections and boundaries listed at the top. Numbers along the X-axis are the number of profiles from north to south along the fault system. The central regions of all the sections, except Indian Creek, appear to have the greatest amount of offset. Along the Indian Creek section, displacement increases to the north and south away from the central portion. This further supports the idea of seismic "spillover" from the north and south rather than an independent seismogenic section of the fault system. Dashed red line shows the southward increase in magnitude of surface displacement. Proposed single-event fault scarps are shown in red.

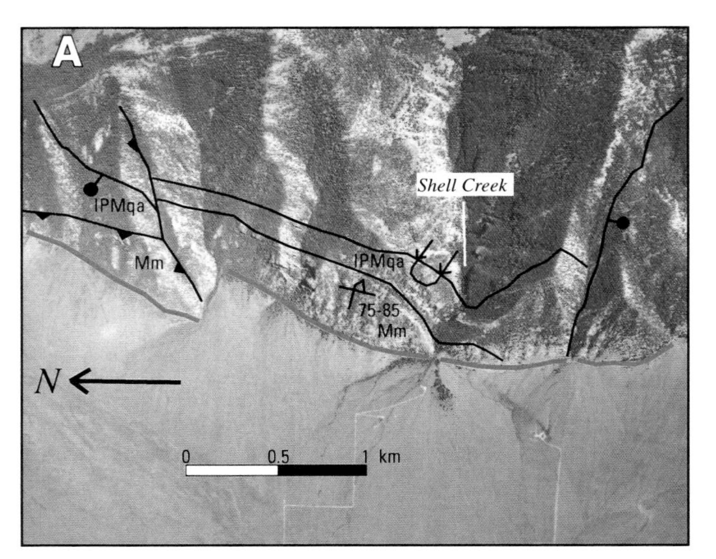

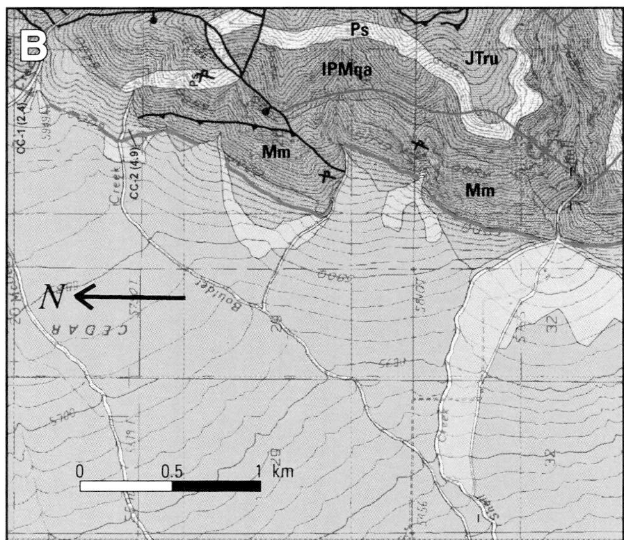

Figure 6. Thrusts and overturned syncline with normal faults at Shell Creek. (A) Aerial photograph of the Shell Creek area. Red line shows trace of late Pleistocene surface rupture. Mm—Mississippian Madison Formation; IPMqa—Pennsylvanian-Mississippian Quadrant Formation; Ps—Permian Shedhorn Sandstone; and JTRU—Morrison Formation, Ellis Group, Woodside Siltstone and Dinwoody Formation undivided. (B) Geologic map showing relationships in A.

At this stop, late Pleistocene surficial ruptures follow the trend of Laramide thrust faults preserved in the footwall of the Madison Range fault system. Varying amounts of displacement along multiple strands of this fault zone suggest coseismic rupture of multiple fault strands during a single late Pleistocene seismogenic event (Fig. 8). At the northern end of the geologic map (Fig. 8), a small remnant of Precambrian rock (Aqf) has been thrust over Cretaceous rocks.

From this location, the example of a "perched basement wedge" (Lageson and Zim, 1985) can be seen. Toward the east, within the footwall of the fault system, Archean metamorphic rocks have been thrust over Paleocene conglomerates during Laramide contraction. Neogene extension beheaded this thrust fault with high-angle normal faults, leaving the preexisting Archean hanging wall and Paleozoic footwall syncline "perched" in the current physiographic configuration (Fig. 9A and 9B).

STOP 4—Indian Creek Section of the Madison Fault Zone: Coseismic Spillover from Northern or Southern Sections or Independent Seismogenic Segment? *(UTM 453395E; 4995681N; elev. 5915 ft)*

Continue south on Bear Creek Loop Road ~3.7 mi. Stop before crossing Indian Creek.

The Indian Creek section of the Madison Range fault system is ~4 km in length. Varying amounts of displacement may indicate coseismic "spillover" from either the southern or the northern fault system (Fig. 10). Minimum displacement is ~1.0 m and maximum displacement 7.1 m. The 1.0 m offset occurring on the youngest deposits and not at a section boundary is likely explained by seismogenic "spillover" from events either occurring from the north or the south. Paleoseismic analyses have yet to determine the structural linkages of the Indian Creek section with the northern and southern Madison Range fault system.

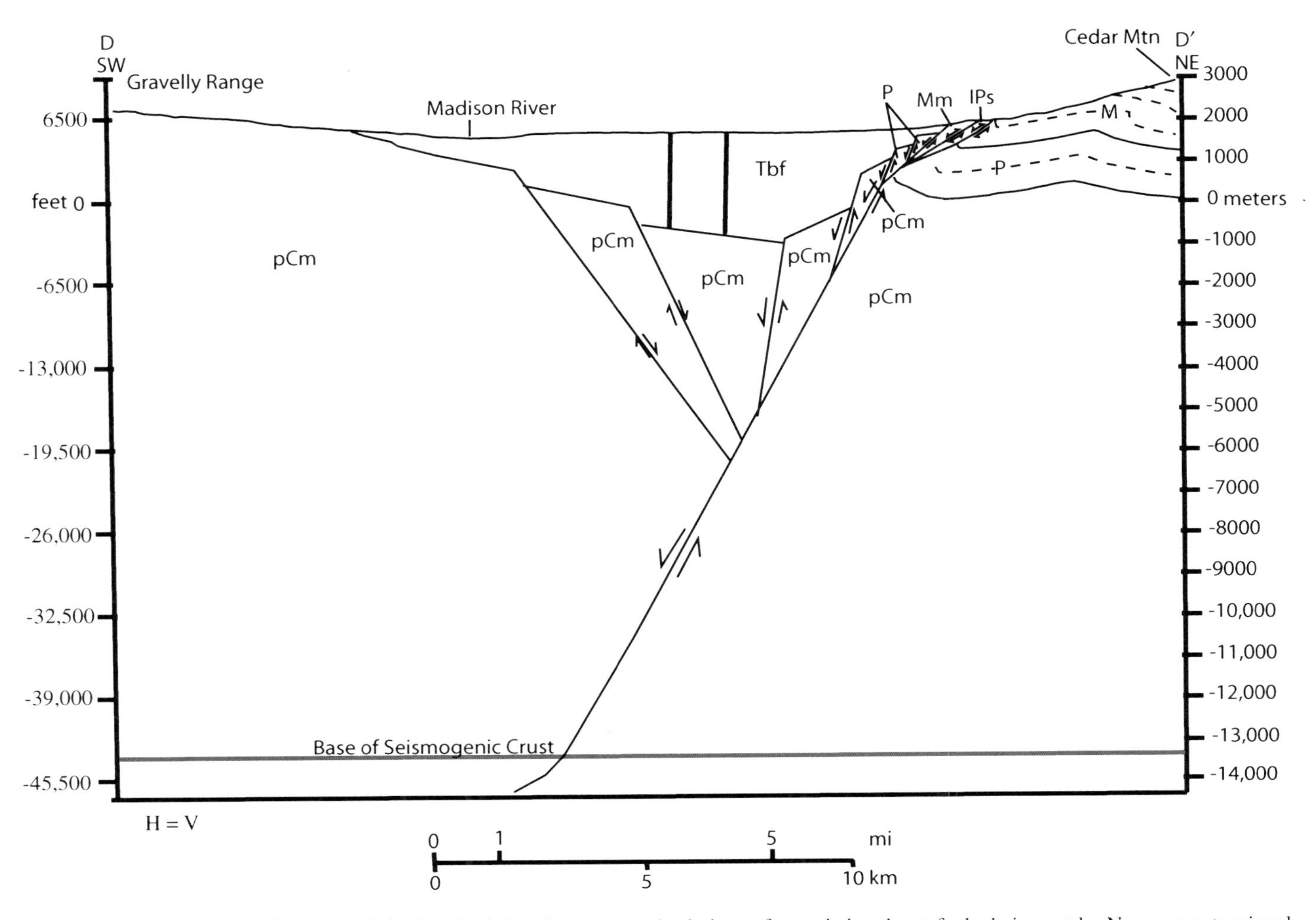

Figure 7. Cross section D–D′ (see Fig. 2 for location) showing structural relations of preexisting thrust faults being cut by Neogene extensional normal faults. Tbf—Tertiary basin fill; M—Mesozoic rocks; P—Permian rocks; IPs—Pennsylvanian rocks; Mm—Mississippian Madison Limestone; and pCm—Precambrian metamorphic rocks. Basin depths from Kellogg et al. (1995).

Cross sections reveal the cross-cutting relationships between preexisting Laramide structures and Neogene extensional structures (Figs. 11A and 11B).

STOP 5—Little Mile Creek and Mile Creek: Southern Section of the Madison Fault Zone and Coseismic Rupture during the 1959 Hebgen Lake Earthquake *(UTM 463092E; 4956793N; elev. 6690 ft)*

Continue south across Indian Creek and then west on Bear Creek Loop Road to U.S. Hwy 287. Proceed south on U.S. Hwy 287 ~15.0 mi and turn south on State Hwy 87 and continue south ~4.9 mi. Turn left at the Mile Creek Road. Stop just off the highway or drive up to the trailhead at the range front.

At this stop, we see the relationship between the Madison Range and Hebgen Lake fault systems. Historic seismicity and cross sections demonstrate coseismic behavior between the two fault systems. At Little Mile Creek, displacement along the Madison Range fault system increased by varying amounts during the 1959 Hebgen Lake earthquake (Myers and Hamilton, 1961, 1964; Ruleman, 2002b). The maximum displacement was 14.4 m (Figs. 12A and 12B). This is ~2 m greater than maximum offset measured to the north, indicating reactivation of that amount during the 1959 Hebgen earthquake, consistent with displacement along the Hebgen Lake–Red Canyon fault system during that event. Figure 13 shows the surficial geologic relationships with rupture patterns and amounts of displacement.

Figure 14 shows the structural relationships between the Madison Range and Hebgen Lake–Red Canyon fault systems (locations of cross sections on Fig. 2). Doser (1985) demonstrated focal mechanisms from the 1959 event on the Hebgen Lake–Red Canyon fault system that support the structural interpretation of the southern Madison Range fault system and potential coseismic activation of multiple faults during a single earthquake event.

To conclude, morpho-metric analyses support the subdivision of the Madison Range fault system into four and

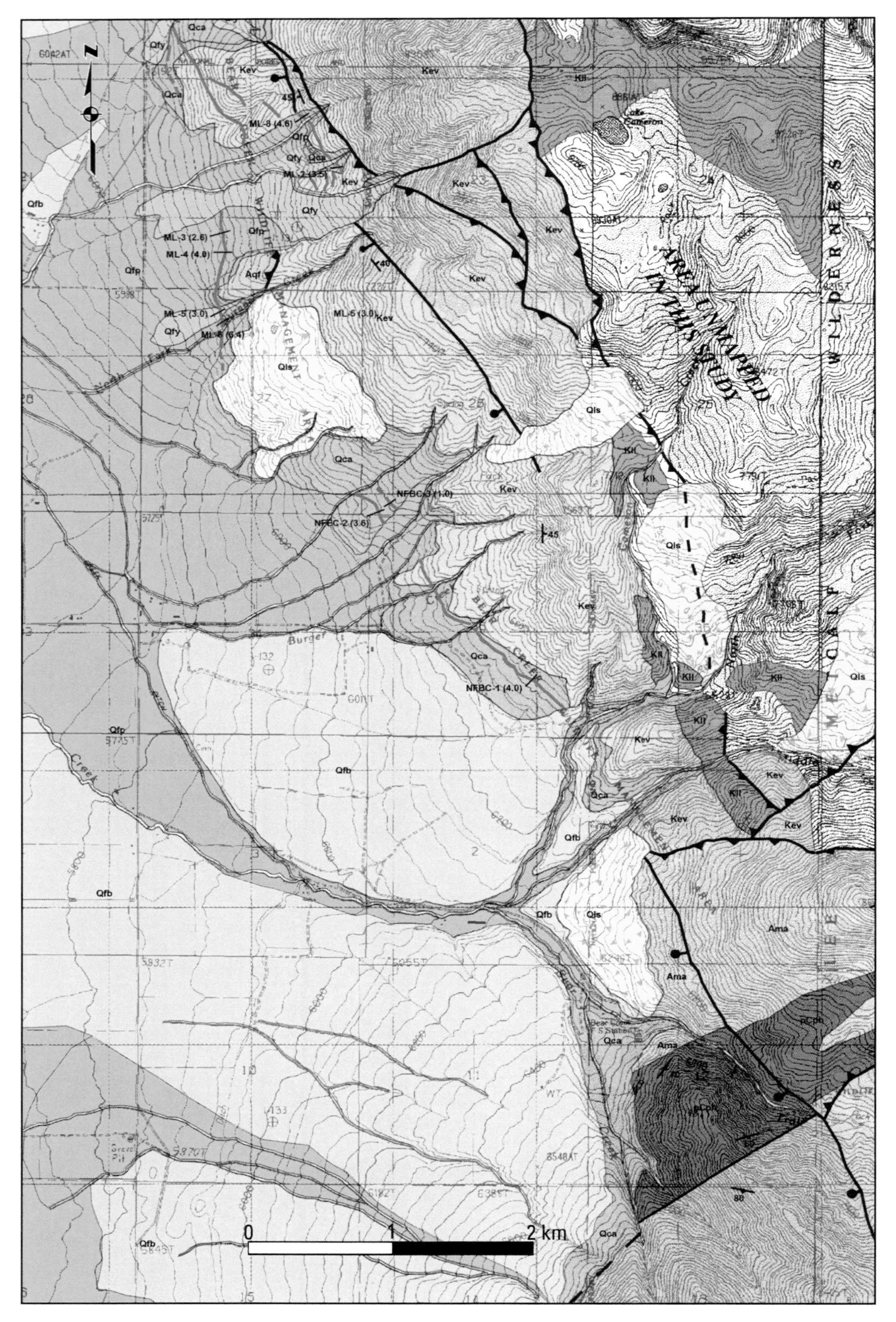

Figure 8. Geologic map of the Mill to Bear Creek range front modified from Kellogg and Williams (2000). Note the thrusting of Precambrian rocks over Cretaceous rocks (Kev–Everts Formation and Virgelle Sandstone, undivided) at the southeast corner of the map. Red lines show measurable late Pleistocene offset on the Madison Range fault system. Units are as follows: Aqf—Precambrian quartzofeldspathic gneiss; Ama—Precambrian marble; Kll—Livingston Formation, Lower Member; pCph—Precambrian phyllite; Qfb—Bull Lake outwash deposits; Qfp—Pinedale outwash deposits; Qca—late Pleistocene and Holocene colluvium alluvium undivided; Qls—Pleistocene landslide deposits; and Qfy—latest Pleistocene and Holocene debris-flow fan deposits.

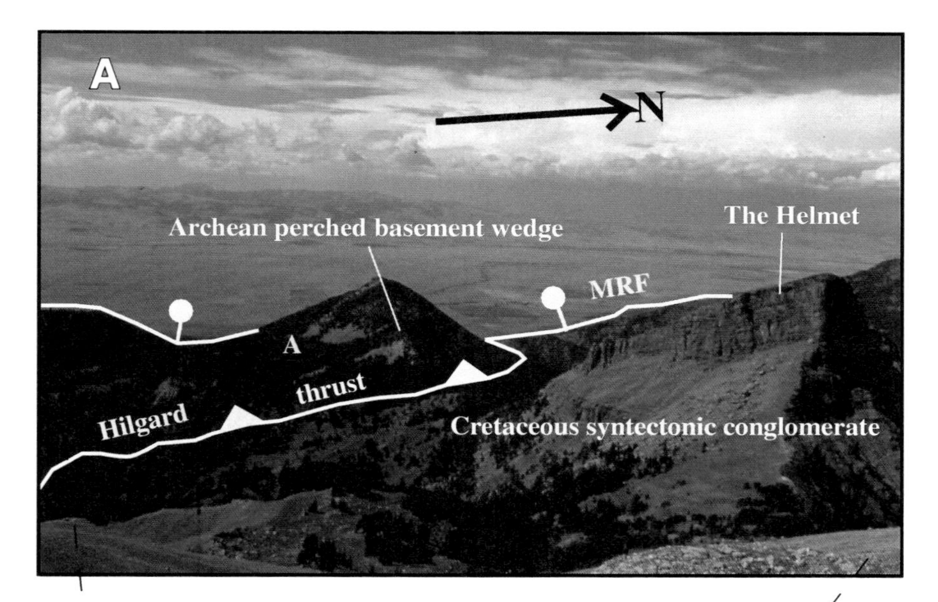

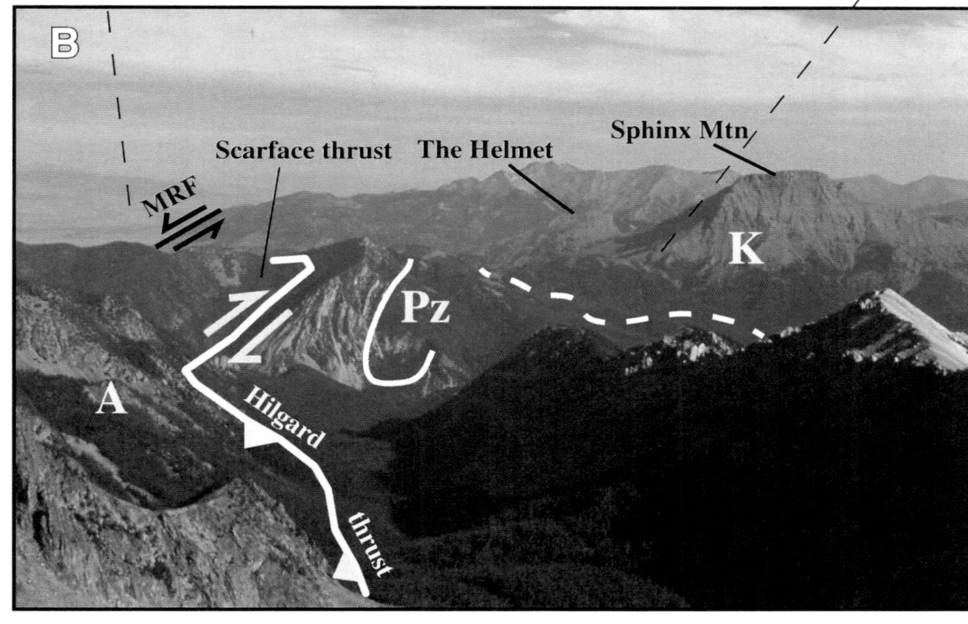

Figure 9. (A) View looking west from the top of Sphinx Mountain showing Precambrian rocks thrusted over an upper Cretaceous syntectonic conglomerate. This structure has been beheaded by Neogene extension resulting in a "perched basement" in the footwall of the normal Madison Range fault system. A—Archean metamorphic rocks and MRF—Madison Range fault. (B) View looking north along the crest of the Madison Range showing the overturned footwall syncline and the "perched basement wedge." A—Archean metamorphic rocks; Pz—Paleozoic sedimentary rocks; K—Cretaceous syntectonic conglomerate; and MRF—Madison Range fault.

potentially five segments (Fig. 15). Figure 15A shows the heights of basal faceted spurs measured along the fault system. Spur height is interpreted to be a product of fault separation and spur heights taper in height at section boundaries. Figure 15B is a plot of the valley-floor width to valley height ratio measurements (Vf) along the fault system. The higher ratios equate to more lateral erosion within a given drainage and less tectonically induced vertical incision (Keller, 1986). Vf ratios increase at proposed section boundaries. Finally, Figure 15C shows amounts of late Pleistocene displacement from fault scarp profiles measured along the Madison Range fault system and the tapering of late Pleistocene displacement at segment boundaries.

STOP 6—Madison Canyon Landslide Overview, Historic and Recent Seismicity
(UTM 465909E; 4963977N; elev. 6568 ft)

From Stop 5, return to U.S. Hwy 287 and drive north to junction with State Hwy 87. Continue southeast on U.S. Hwy 287 for 3.2 mi. Turn left toward Earthquake Lake Visitor Center and keeping to the left, proceed 0.4 mi to the upper parking lot near the western crest of the Madison Canyon slide. This is a fee area and you may need to stop at the Visitor Center (0.1 mi past highway turnoff) to pay a modest entrance fee.

This is the top of the Madison Canyon landslide, the largest seismically triggered landslide in North America to occur during

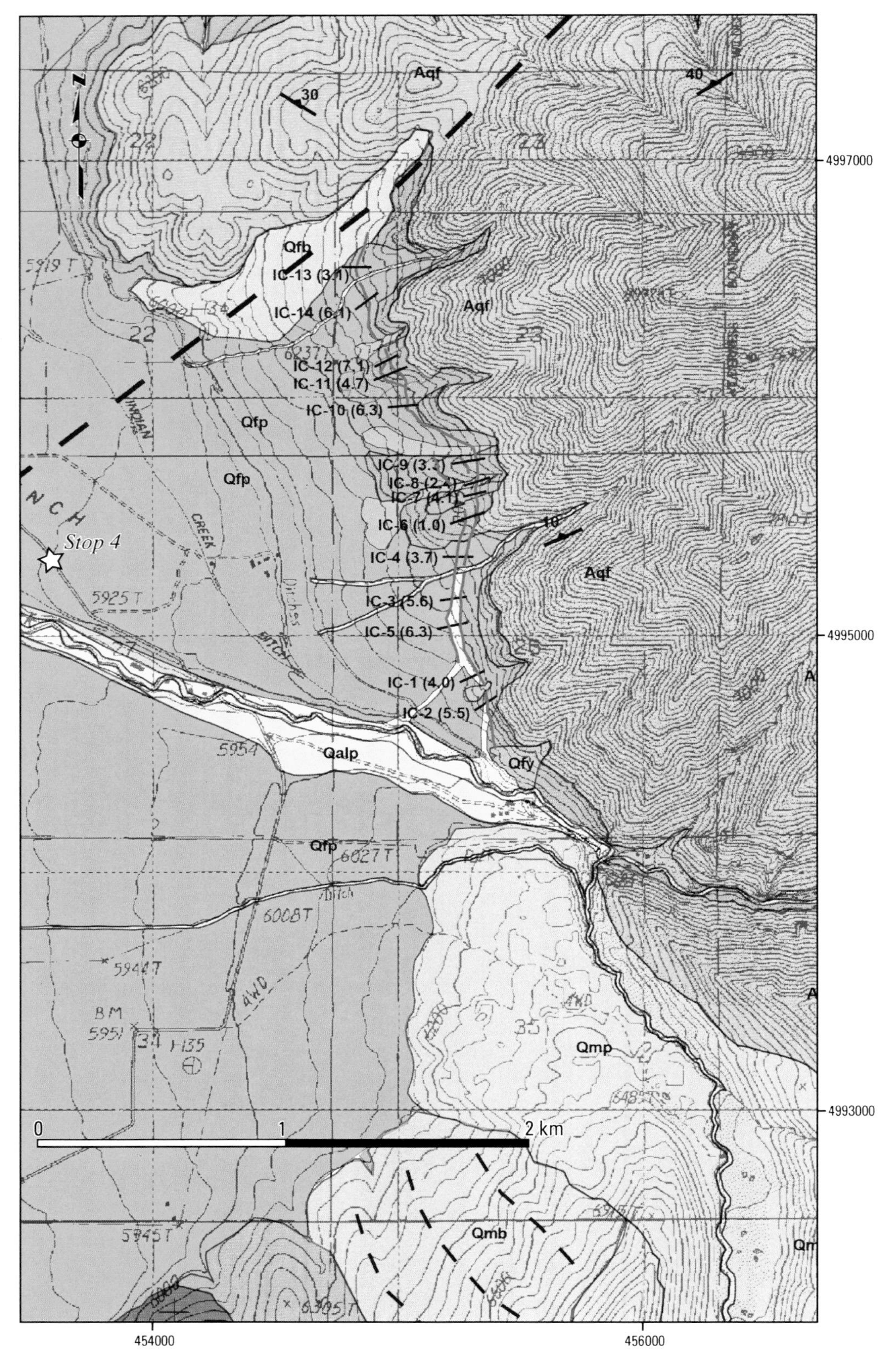

Figure 10. Geologic map of the mouth of Indian Creek. Late Pleistocene surface ruptures shown in red. Dashed line in northwest corner of map is an inferred fault based on sheared bedrock preserved in the footwall. Units are as follows: Aqf—Archean quartzofeldspathic gneiss; Qfb—Bull Lake outwash; Qfp—Pinedale outwash; Qmb—Bull Lake till deposits; Qmp—Pinedale till deposits; Qalp—latest Pleistocene and Holocene alluvium; and Qfy—Holocene fan deposits. Numbers in parentheses are amounts of surface offset in meters. Note the varying amounts of offset measured along this short stretch of the Madison Range fault system.

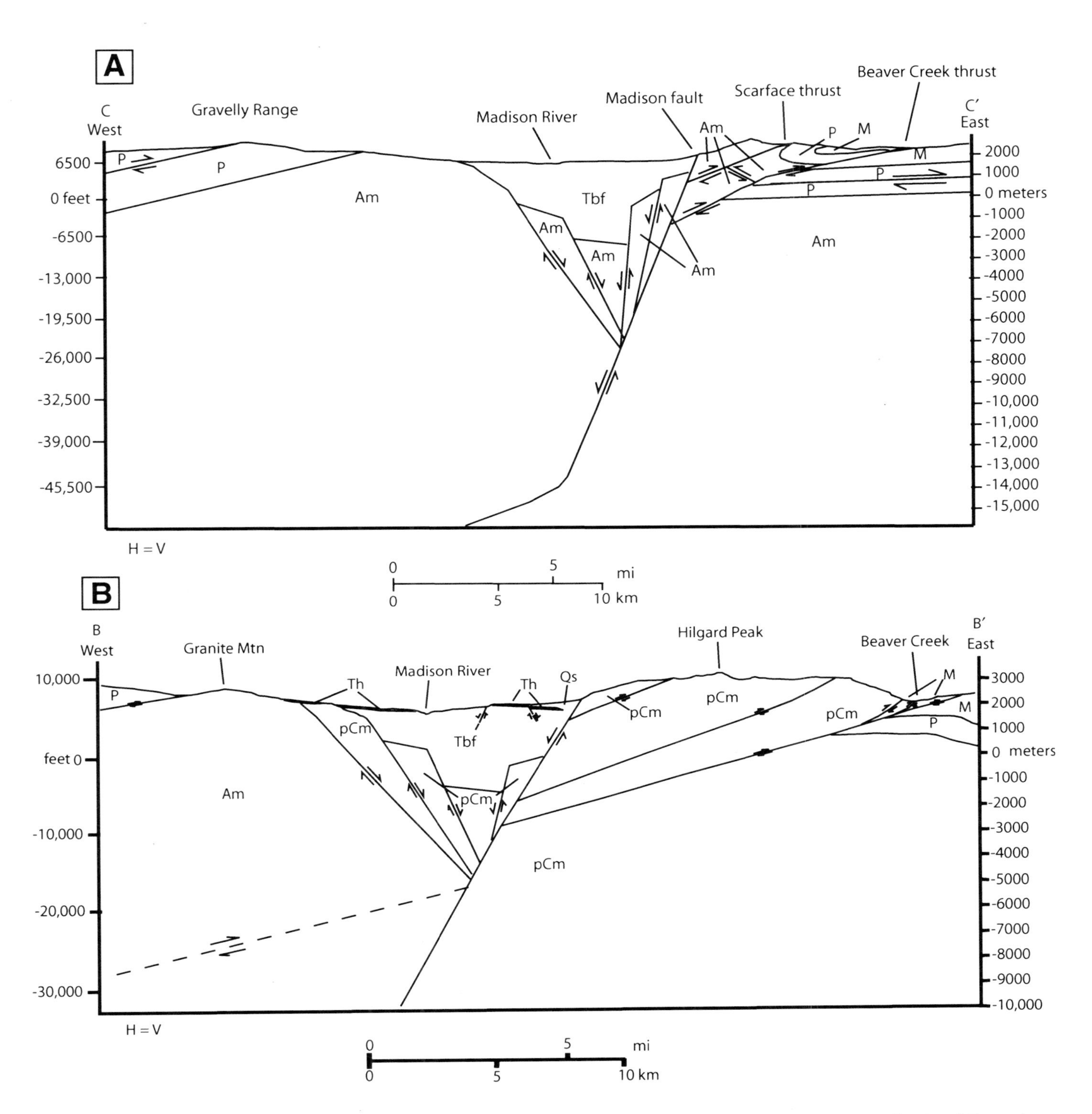

Figure 11. (A) Simplified cross section C–C′ (Fig. 2), Indian Creek section of the Madison Range fault. Rock units are listed as follows: Am—Precambrian metamorphic rocks; P—Paleozoic sedimentary rocks; M—Triassic to Cretaceous sedimentary rocks; and Tbf—Tertiary basin fill. Modified from Schmidt et al. (1993). (B) Cross section B–B′, Taylor-Hilgard segment of the Madison fault zone. Maximum basin fill estimated as 3300 m (Gary, 1980) and 4500 m (Rasmussen and Fields, 1983). Units listed as follows: pCm—Precambrian metamorphic rocks; P—Paleozoic rocks; M—Mesozoic rocks; Tbf—Tertiary basin fill; Th—Huckleberry Ridge Tuff (2.0 Ma); and Qs—Quaternary sediments.

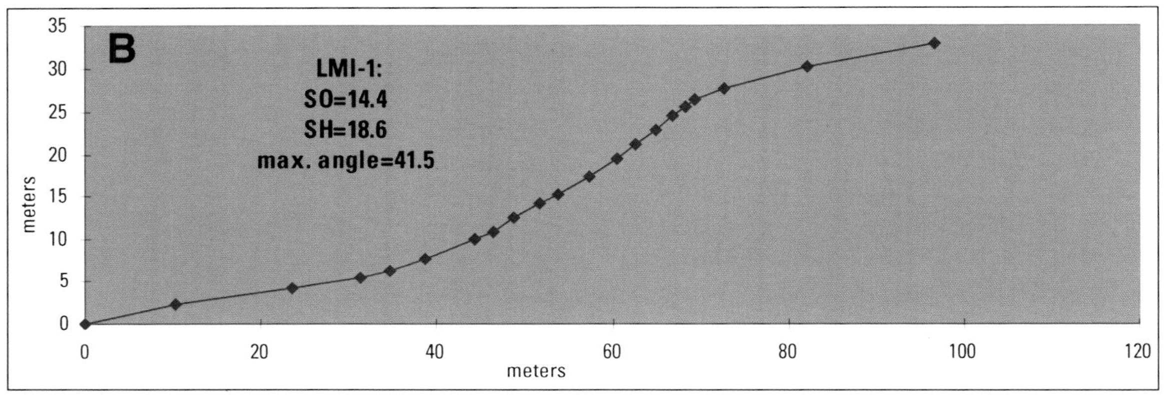

Figure 12. (A) View looking east at the mouth of Little Mile Creek. Late Pleistocene outwash deposits have a surface offset of 14.4 m. (B) Profile of the maximum surface offset at Little Mile Creek. SH is scarp height and SO is surface offset.

historic times. This 28 million cubic yard slide mass, with an estimated weight of 60 million tons, cataclysmically slid from the south wall into Madison Canyon only a few tens of seconds after the strongest seismic shaking generated by the M_W 7.3 Hebgen Lake earthquake, which occurred at 11:37 p.m. on 17 August 1959. The leading edge of the slide mass crossed the Madison River and came to rest ~120 m up the north canyon wall 150 m north of this spot. Hadley (1964) estimated that the emplaced slide mass had a total volume of 37 million cubic yards. While in motion, the slide mass reached an estimated velocity of 100 mph, which generated hurricane force winds around the perimeter of the slide. With a depth of more than 60 m, the aptly named Earthquake Lake formed behind the landslide. Immediately following the earthquake, Montana's governor requested the U.S. Army Corps of Engineers to excavate an outlet channel in the slide mass to reduce the chances of a catastrophic flood if the lake were to overtop and breach the landslide dam.

The eastern edge of the landslide partly buried the overflow section of the Rock Creek Campground, killing 26 campers. Nineteen of the victims were never found, and their remains are entombed in this landslide. A bronze plaque on the enormous boulder 550 ft northeast of this location memorializes these landslide casualties. Two additional victims died from rock fall at a small campground near Wade Lake, 6.5 mi southwest of here. A lone mountaineer who had apparently bivouacked near the summit of Granite Peak, 82 mi northeast of here, the night of the earthquake was killed by seismic-shaking induced rock fall. His fate remained a mystery until his remains were discovered after emerging from melting glacial ice at the base of Granite Peak 40 years after the Hebgen Lake earthquake.

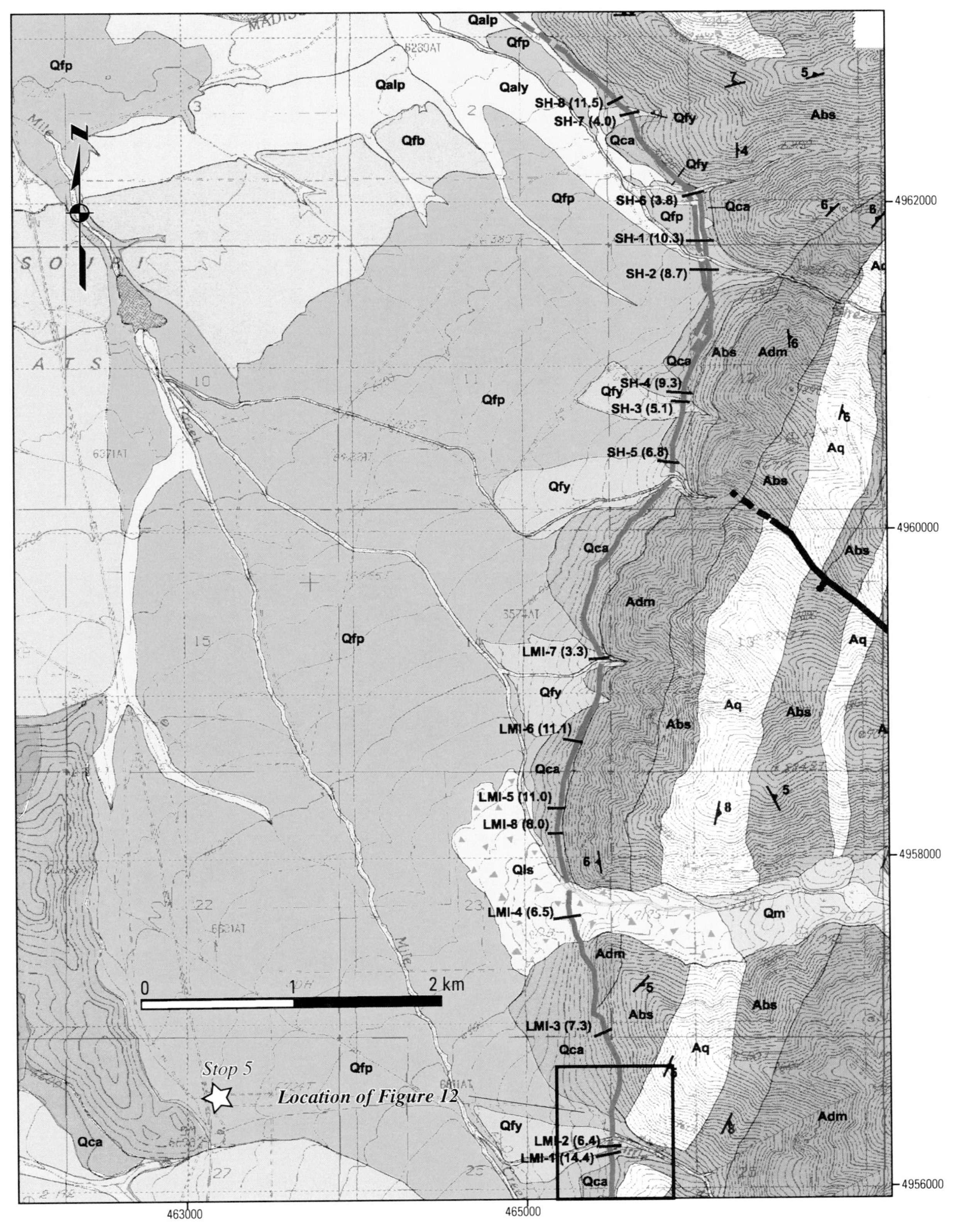

Figure 13. Geologic map of the Missouri Flats region along the southern Madison Range. Bedrock geology modified from Erslev (1982). Rock units listed as follows: Abs—Precambrian biotite schist; Adm—Precambrian dolomite; Aq—Precambrian quartzite; Qfb—late middle Pleistocene fan deposits; Qfp—late Pleistocene fan deposits; Qm—middle to late Pleistocene till; Qls—late Pleistocene landslide deposits; Qalp—latest Pleistocene and Holocene alluvium; Qaly—Holocene stream alluvium; Qfy—Holocene fan deposits; and Qca—late Pleistocene and Holocene colluvium alluvium undivided. LMI-# and SH-# are fault scarp profile numbers from Ruleman (2002b).

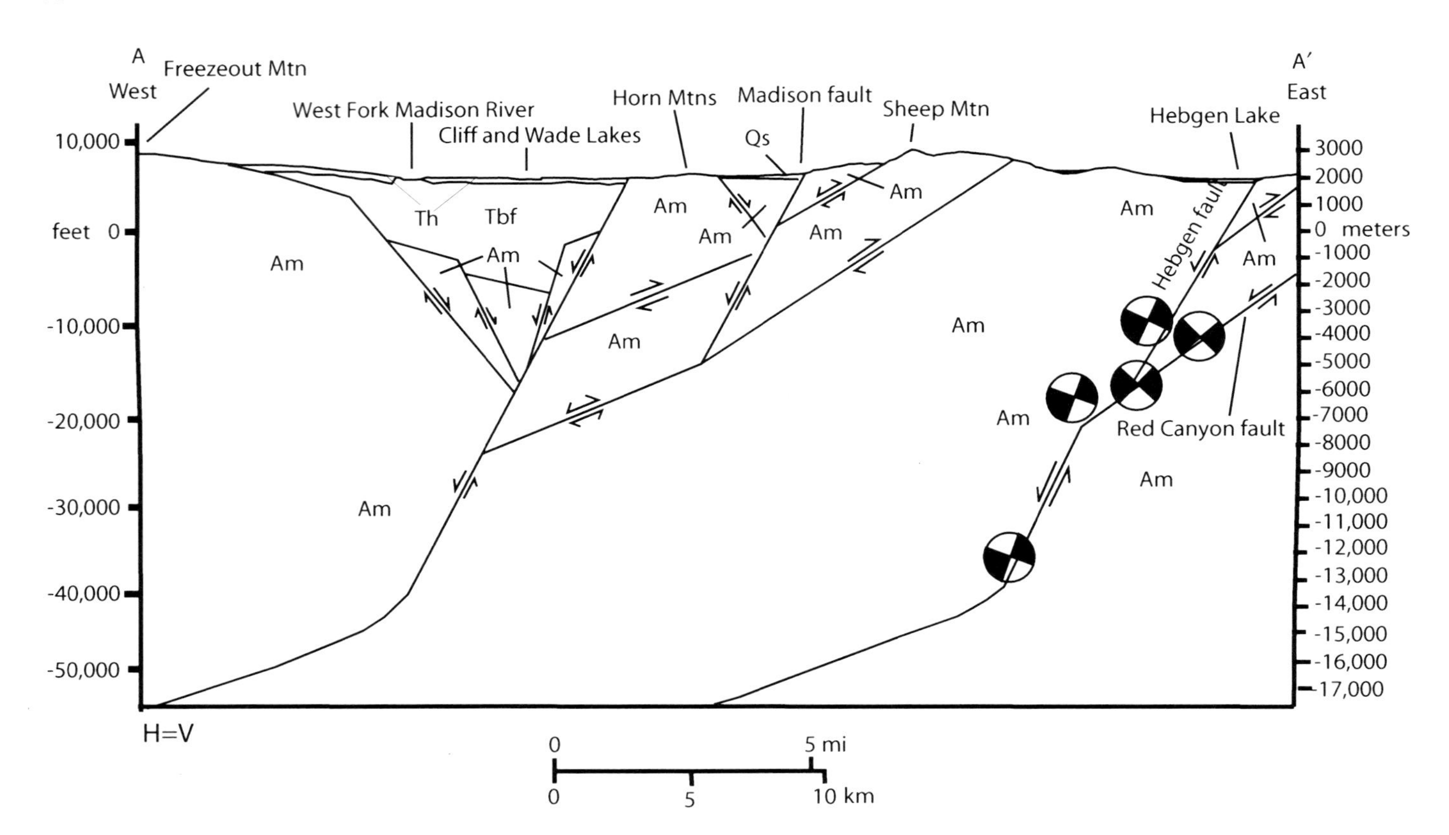

Figure 14. Simplified cross section line A–A′ (Fig. 2) of the southern Madison Range fault. Focal mechanisms and Hebgen Lake–Red Canyon fault system from Doser (1985). Map units listed as follows: Am—Precambrian metamorphic rocks; Tbf—Tertiary basin fill; Th—Huckleberry Ridge Tuff (2.0 Ma); and Qs—Quaternary sediments. Cliff and Wade Lakes are the location of a paleodrainage flowing north from the Centennial Valley to the Madison Valley.

To the southwest, across the Madison River is a layer of light yellow-ish brown Precambrian dolomite that dips steeply to the north. Prior to the landslide, this dolomite layer extended eastward along the base of the impending slide and apparently served as a buttress, retaining the stratigraphically lower but topographically higher schist, which extends up to the ridge crest. The schist foliation dips 60° north, into Madison Canyon and was predisposed for failure once the strong seismic shaking had compromised the integrity of the dolomite buttress below. The slide mass was apparently emplaced more or less as a coherent unit because the remnants of the dolomite buttress were deposited at the leading edge of the slide deposit and trees that were growing on the surface of the landslide remained on the surface.

The Madison Range and Hebgen Lake region lie within the northern part of the Intermountain Seismic Belt (Smith and Sbar, 1974; Smith and Arabasz, 1991), which is a zone of shallow seismicity and late Quaternary faulting that extends from northwestern Montana to southern Utah. Similar to many late Quaternary range-bounding normal faults in the belt, the Madison Range fault system has been aseismic during historic times—except for the 15-km-long section near the southern end that ruptured during the 1959 Hebgen Lake earthquake (discussed below and at Stop 5). Over the past three decades, regional seismograph networks within the belt have published catalogs of well-located hypocenters for small- and moderate-magnitude earthquakes. These catalogs document patterns of seismicity that commonly show little to no seismicity occurring along the down-dip extent of large, late Quaternary, range-bounding normal faults such as the Madison Range fault system (Fig. 16). Instead of occurring along range-bounding faults at depth, routine background seismicity occurs in the footwall block beneath the range, or in the hanging wall block, typically more than 10 km beyond the surface trace of the fault, or off either end of the fault. Examples of this conundrum of routine seismicity occurring near—but not on—range-bounding Intermountain Seismic Belt normal faults include the northern 65 km of the Madison Range fault system; the Mission, Canyon Ferry, and Red Rock faults in Montana; the Teton fault in Wyoming; and the Wasatch fault in Utah.

The 18 August 1959 Hebgen Lake earthquake had a moment magnitude of 7.3 and is the largest historic earthquake in the northern Rocky Mountains. The Hebgen Lake quake generated surface rupture along 26 km of the Hebgen Lake–Red Canyon fault system (Stops 7 and 8), which has an overall northwesterly strike. Using a global set of seismograms, Doser (1985) determined that the Hebgen Lake earthquake was a complex event that consisted of two sub-events separated by 5 seconds. The

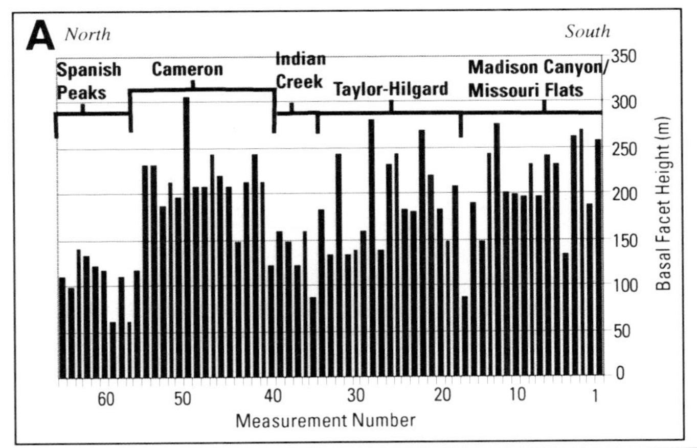

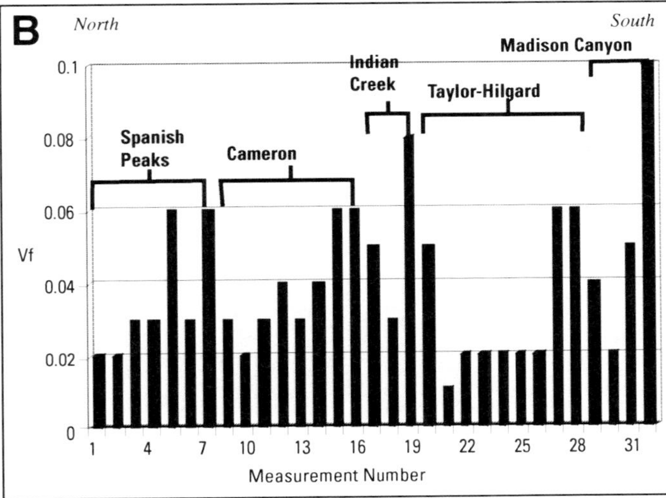

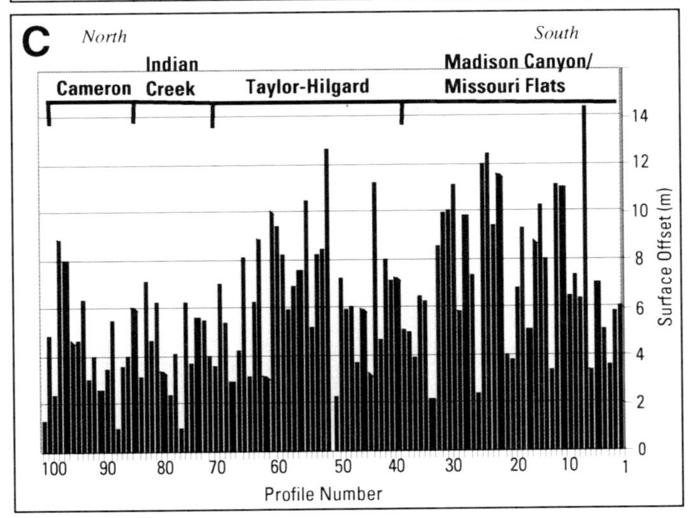

Figure 15. Comparative morphometric analyses measured along the Madison Range fault system. (A) Heights of basal faceted spurs measured at the mountain-piedmont junction. Note lower height magnitudes at section boundaries. (B) Valley-floor width to valley height ratios (Vf) measured at major drainages along the fault system. Note the increase in the ratio at section boundaries. (C) Measured surface offset along the fault system. Note a tapering of displacement at section boundaries and a general northward decrease in seismogenic offset.

primary seismic energy release occurred along a south-dipping normal fault striking N80°W and extending to a depth of 15 ± 3 km below the surface. Large early aftershocks extended from the Cliff Lake area west of the Madison Valley, 80 km eastward, to the Norris Geyser Basin in north-central Yellowstone National Park. Thus, the spectacular surface faulting occurred along relatively shallow, preexisting structures that were reactivated by slip at depth along an east-west-trending, south-dipping normal fault that extends well to the east and west of the surface ruptures.

A remarkable aspect of the 1959 earthquake was the region of coseismic subsidence (Fig. 17), which Myers and Hamilton (1964) documented by measuring flooding and emergence of shorelines on Hebgen and Cliff lakes and from highway releveling surveys. A maximum of 6.5 m of subsidence occurred along the northeast shore of Hebgen Lake and significant subsidence extends across the southern Madison Valley, as far as 20 km west of the westernmost surface rupture on the Hebgen Lake–Red Canyon fault system. The long axis of subsidence trends WNW, in general agreement with the trend and western extent of the seismogenic fault inferred from seismological data. Three short sections of the north-trending Madison Range fault system within the inferred subsidence area sustained up to ~1 m of surface rupture during the 1959 earthquake (Stop 5). The relatively minor surface ruptures along the Madison Range fault system and subsidence in the Missouri Flats area of the southern Madison Valley led Fraser et al. (1964) to propose the dual basin concept. They proposed that subsidence in the Madison Valley and in the Hebgen Lake basin were independent of each other and resulted solely from rupture of the adjacent faults. In contrast, Myers and Hamilton (1964) proposed the single basin concept, whereby subsidence in both valleys and across the intervening portion of the Madison Range resulted from movement along a deep-seated, east-west-trending normal fault. This inferred that a deep-seated fault controlled the overall pattern of subsidence, while optimally oriented bedding planes and faults accommodated surface ruptures. Areas lacking these bedding planes and faults experienced broad warping of the land surface. The seismologically determined orientation of the 1959 fault plane and pattern of subsidence seem to favor the single basin concept.

As noted above, the Missouri Flats area of the Madison Valley is very seismically active in marked contrast to the much larger portion of the valley to the north and a smaller portion to the south (Fig. 16). An interesting aspect of this ongoing seismicity is its spatial relationship with the 1959 coseismic subsidence. The majority of well-determined hypocenter locations lie within the area of the 1-ft subsidence contour determined by Myers and Hamilton (1964). Both to the north and south of this contour line, seismicity drops off to very low levels typical of other Intermountain Seismic Belt normal fault-bound valleys. This close association between 1959 subsidence and modern seismicity in an otherwise virtually aseismic valley suggests the possibility that the ongoing seismicity may somehow be related to the 1959 earthquake. Over the three decades for which detailed seismicity data are available, there is no decreasing trend that would be expected

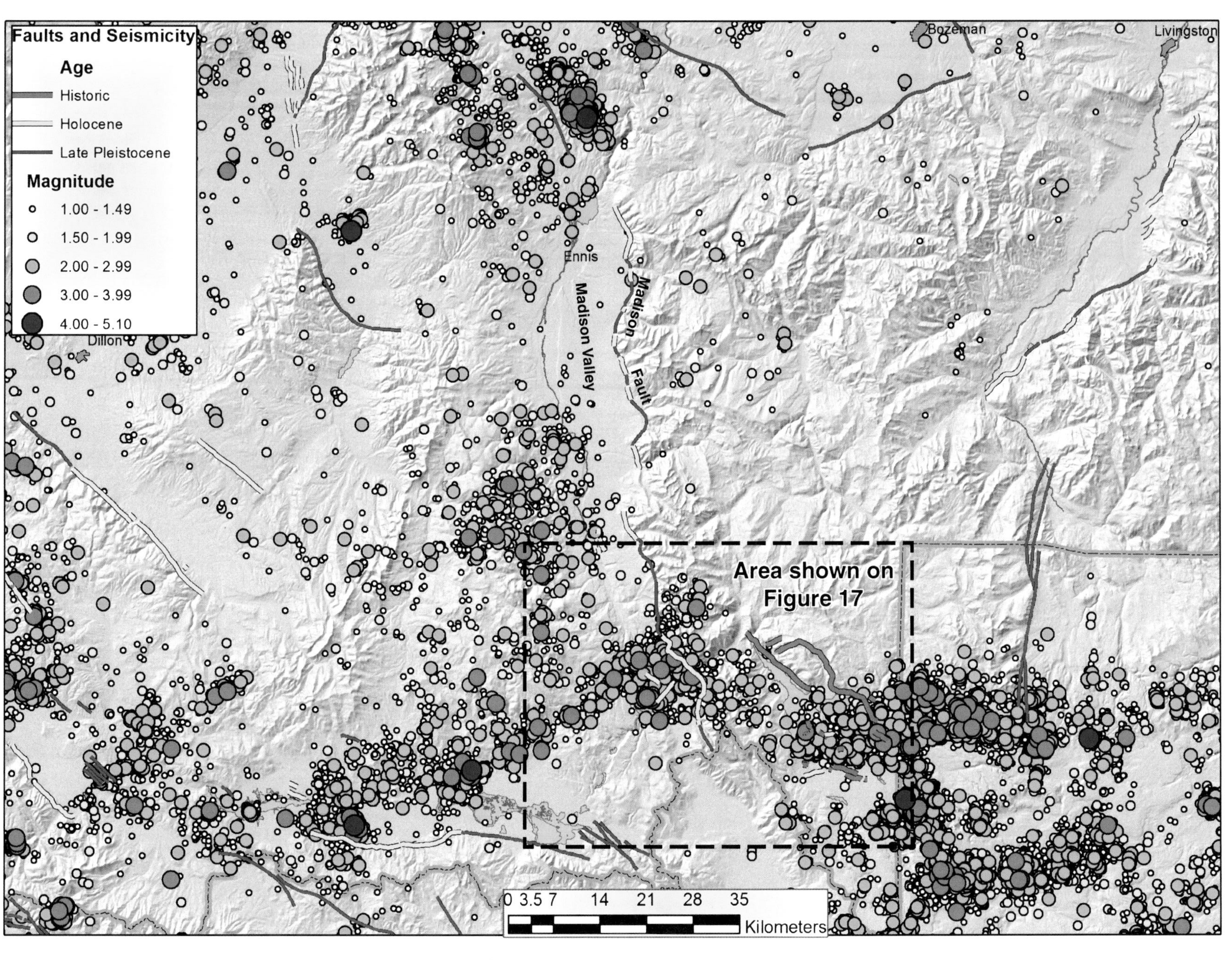

Figure 16. Map showing earthquake epicenters recorded by the Montana Bureau of Mines and Geology and the University of Utah Seismograph Stations since 1985.

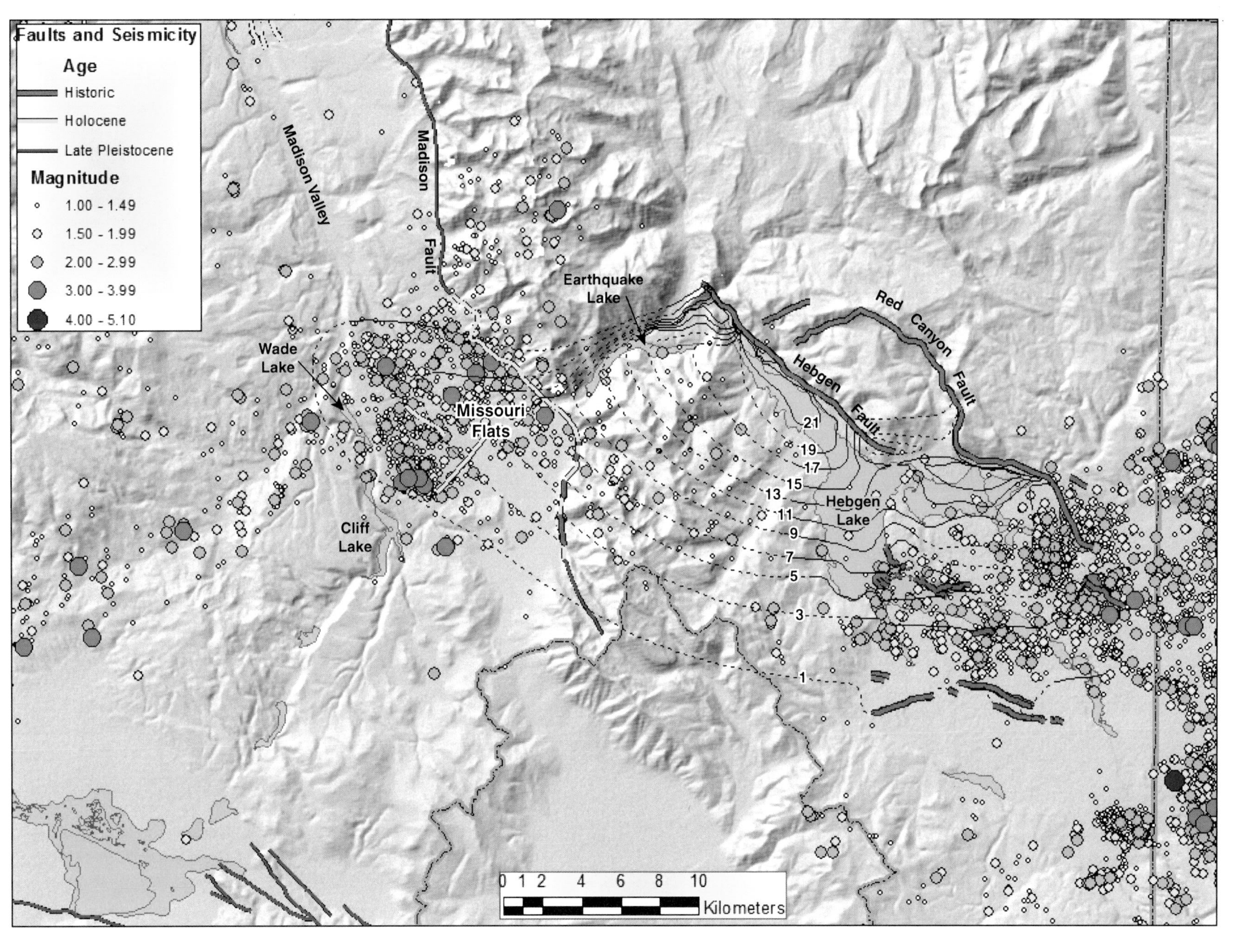

Figure 17. Seismicity in the Missouri Flats–Hebgen Lake region since 1985. Contours (dashed where inferred) show the amount of subsidence in feet that resulted from the 1959 Hebgen Lake earthquake. Note that the majority of seismicity in the southern Madison Valley and Hebgen Lake basin occurs within the area of subsidence.

of typical aftershock activity. Also, the 55 years that have elapsed since the 1959 earthquake extend well beyond the time of a classical aftershock sequence. It is noteworthy that there is no evidence to suggest that Missouri Flats seismicity is associated with the Madison Range fault system at depth. There is no tendency for the hypocenters to cluster along a west-dipping plane that may plausibly represent the down-dip projection of the fault system. Also, none of the 45 fault plane solutions determined for Missouri Flats earthquakes (Stickney and Smith, 2009) are consistent with slip at depth on a west-dipping normal fault. These fault plane solutions do exhibit a consistent N15°E, nearly horizontal T-axis orientation similar to the Hebgen Lake area deformation documented by Savage et al. (1993). The Missouri Flats seismicity is best explained as some sort of decades-long response of the hanging wall block above the 1959 slip plane rather than slip at depth along the Madison Range fault system.

STOP 7—Cabin Creek Fault Scarp
(UTM 472961E; 4968721N; elev. 6500 ft)

Proceed east on U.S. Hwy 287 for ~5.4 mi. Cross Cabin Creek and turn left into Cabin Creek fault scarp exhibit.

At this stop, we will see evidence of the 1959 event and one pre-1959 event on the Hebgen normal fault. For the two terraces at Cabin Creek, the lower terrace, dated at 800–1400 years (Van de Woerd et al., 2000), has 3.1 m of surface offset from the 1959

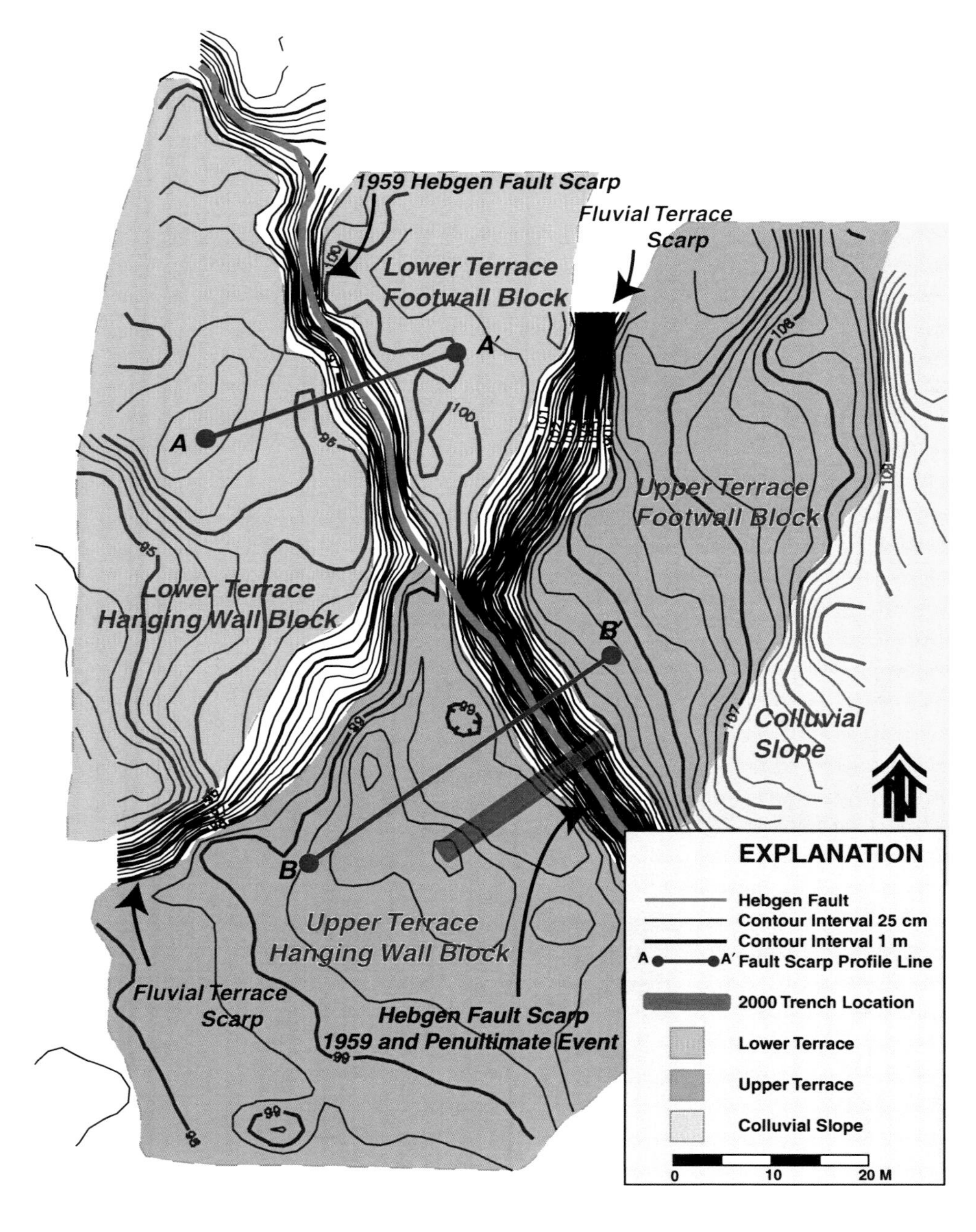

Figure 18. Total-station survey map of the Cabin Creek trench site, showing the trace of the Hebgen fault, location of the trench, the upper and lower terrace, and fault scarp profile locations. The site is near the northwest end of the 1959 surface rupture zone of the Hebgen normal fault.

event, whereas the upper terrace, dated at 6000–11,000 years from preliminary cosmogenic data, has 5.3 m of surface offset, indicating the higher terrace experienced an additional 2.2 m of offset since deposition (Figs. 18 and 19). In the summer of 2000, a trench was excavated across the upper terrace. The trench revealed one pre-1959 event (Fig. 20), indicating 2.2 m of displacement post-dating terrace deposition and predating the 1959 event.

STOP 8—Red Canyon and Red Canyon Fault Scarp *(UTM 483849E; 4968994N; elev. 7083 ft)*

From Stop 7, continue east on U.S. Hwy 287 for ~9.3 mi. Drive north for ~2.7 mi to the Red Canyon trailhead.

At this location we see the Red Canyon fault scarp 0.5 mi up Red Canyon trail. The Red Canyon fault closely follows the overturned limb of the Laramide fold and Divide Thrust fault to the east (Fig. 21). Where the Red Canyon fault crosses Red Canyon Creek, the fault scarp has 5.8 m of displacement (Fig. 22).

ACKNOWLEDGMENTS

We would like to thank Margaret Berry (USGS) and Jim Schmitt (Montana State University) for their thoughtful reviews. We would also like to thank the Cedar Creek Ranch for permitting access for the field trip.

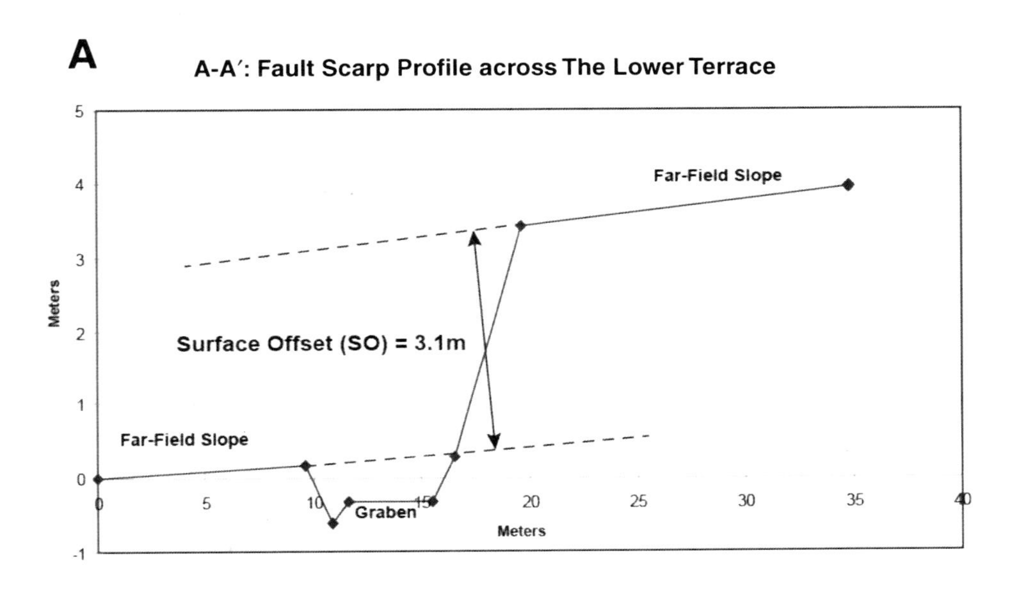

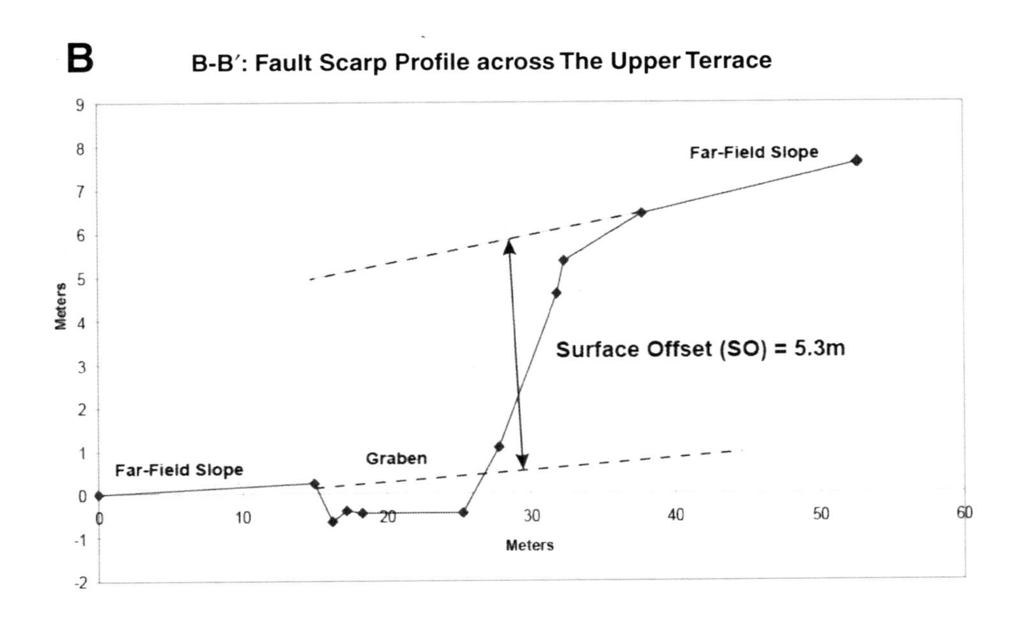

Figure 19. Fault scarp profiles of the lower and upper terraces at Cabin Creek. (A) Scarp profile (A–A′) across the lower terrace indicates only 3.1 m of surface offset. The surface offset was accomplished solely by the 1959 Hebgen Lake earthquake. The lower terrace predates the 1959 event, but post-dates the penultimate event. Cosmogenic dating from Van de Woerd et al. (2000) places the lower terrace in the 800–1400 year range. (B) Fault scarp profile (B–B′) across the upper terrace revealed 5.3 m of surface offset, which is the result of two earthquake events: the 1959 temblor (3.1 m measured from the lower terrace) and the penultimate event (2.2 m). The total vertical throw on the main fault (Tm) at this locality is 7.7 m based on the original terrace top.

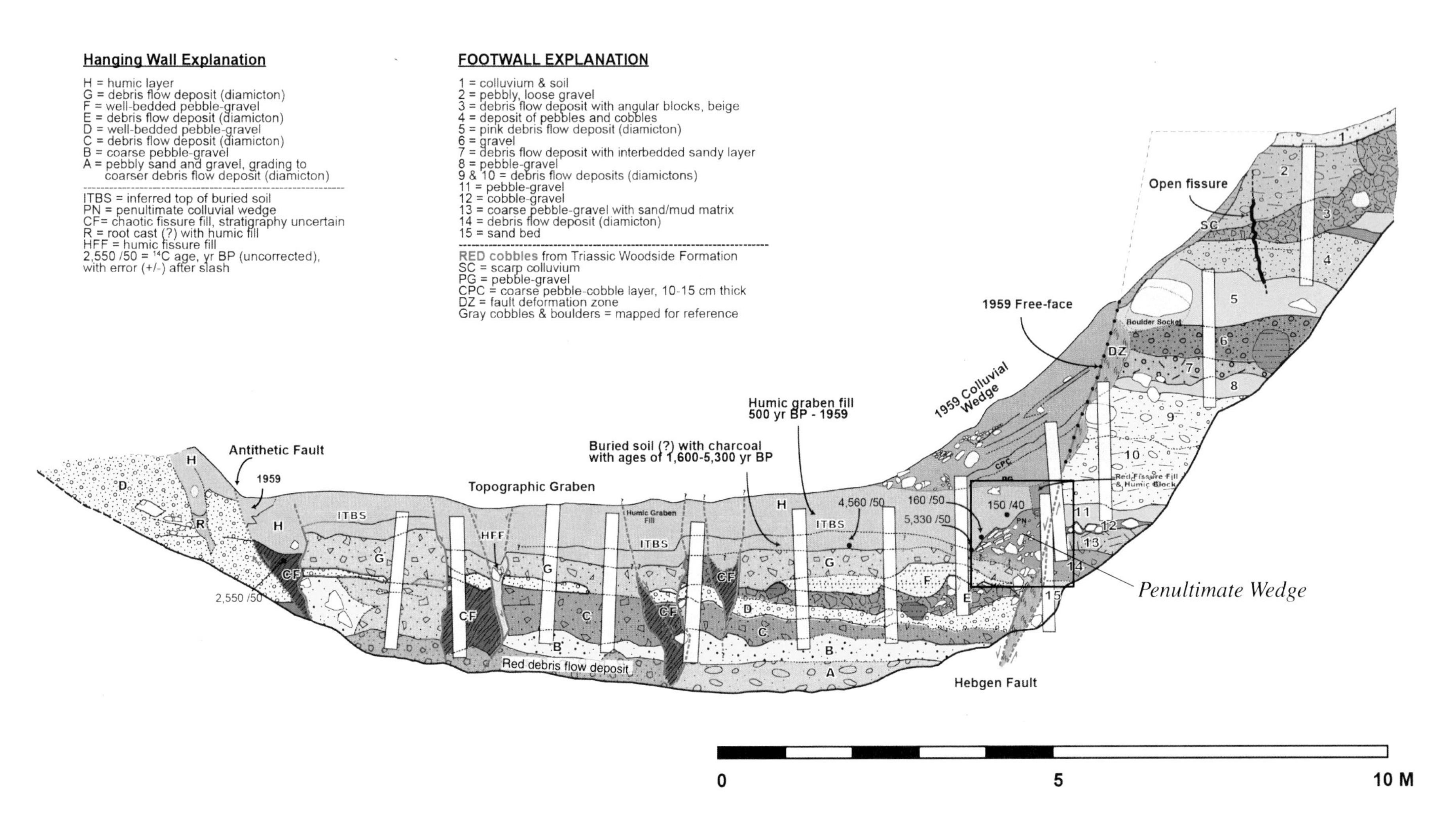

Figure 20. Log of the north wall of the Cabin Creek trench. The trench across the upper terrace revealed only the 1959 colluvial wedge and a penultimate wedge, totaling 7.4 m (Tm), consistent with 7.7 m Tm of the original terrace top. Following the penultimate event, 1 m of fine-grained humic sediment containing charcoal accumulated and eventually buried the penultimate colluvial wedge. Radiocarbon ages and cosmogenic data indicate that the penultimate event occurred during the Holocene, but prior to 1000 yr B.P.

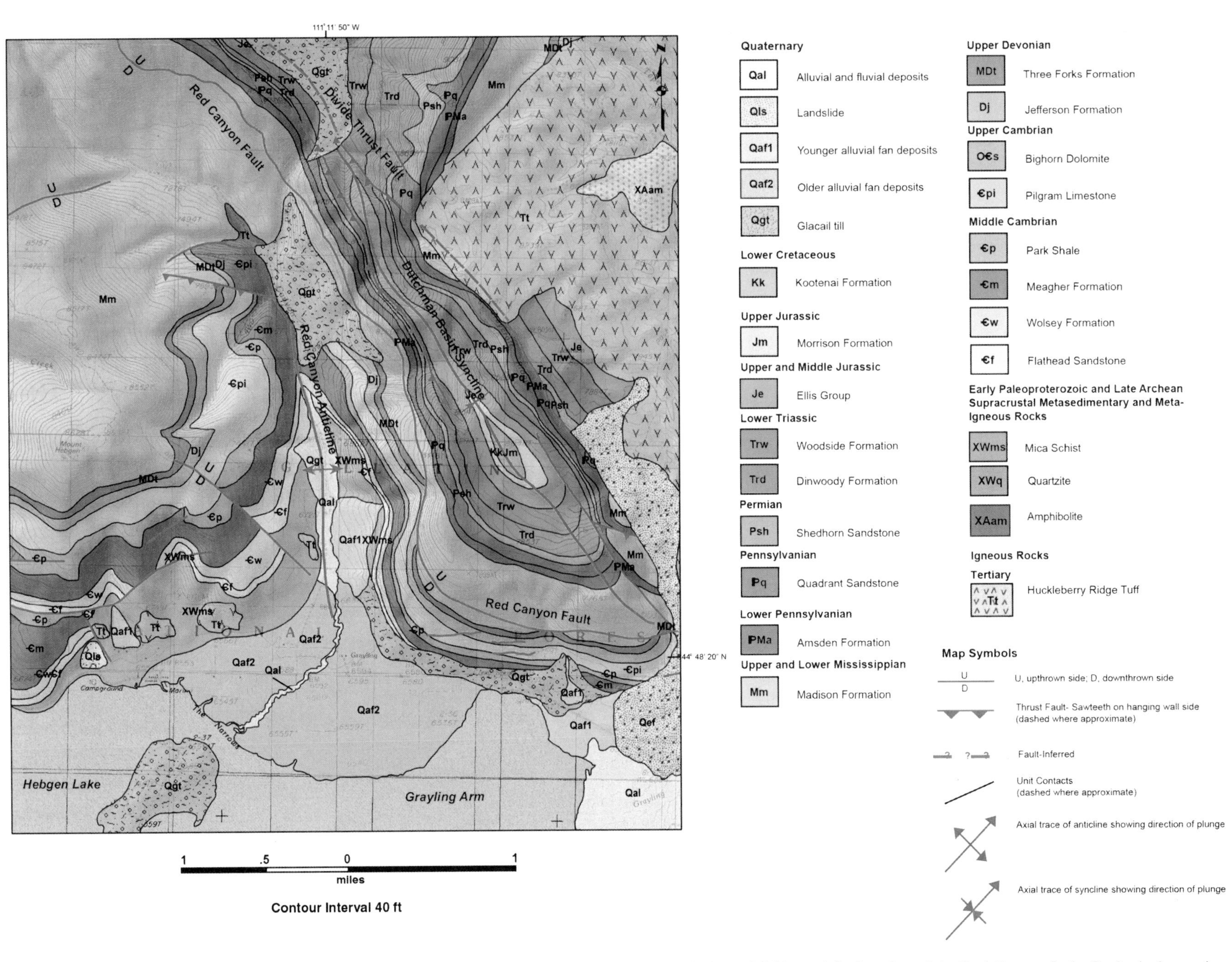

Figure 21. Geology of the Mount Hebgen 7.5′ Quadrangle showing bedrock units, surficial deposits, faults and folds, and the location of the Red Canyon fault. Geological mapping was done by Larsen and Lageson (2011).

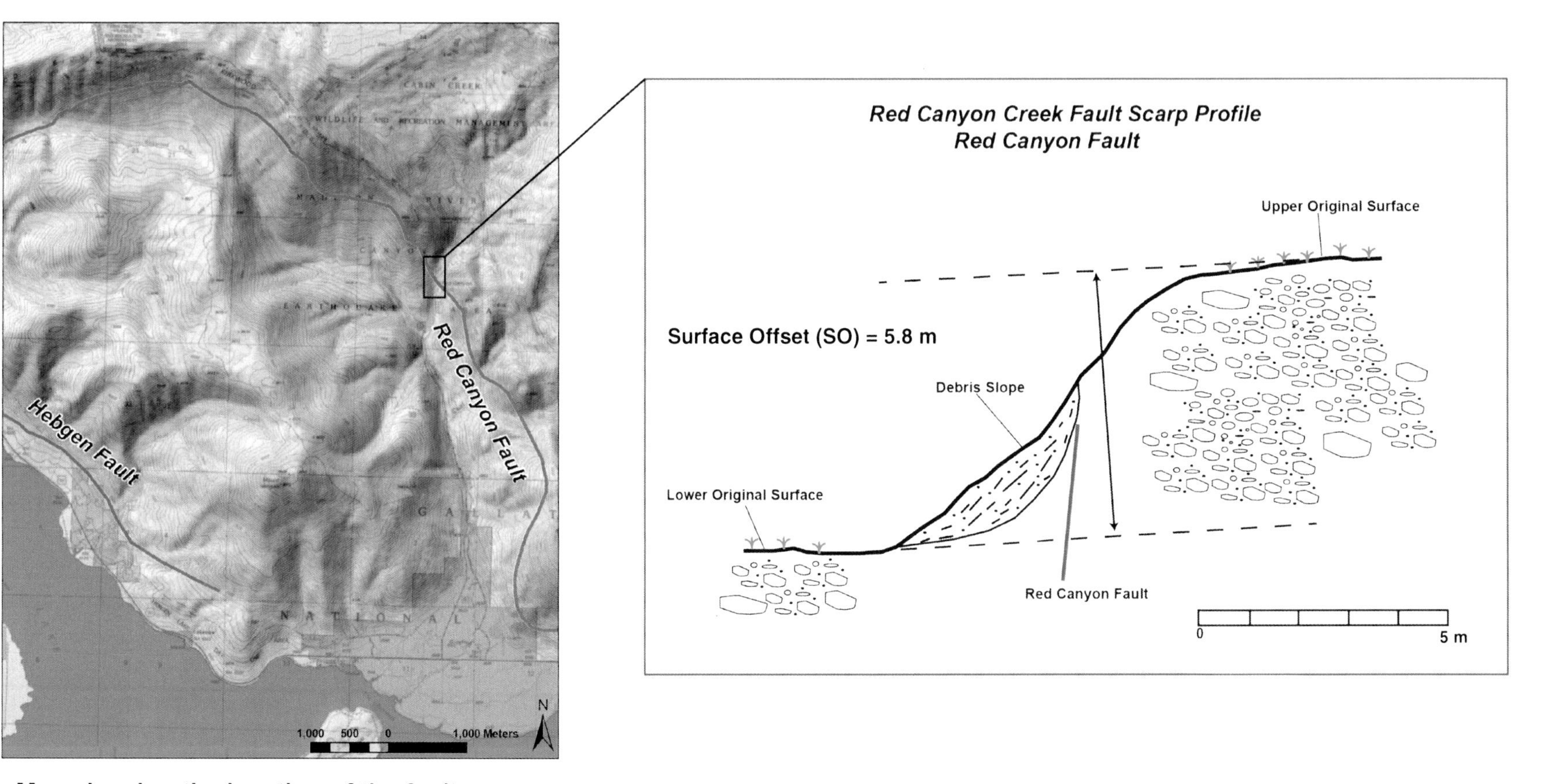

Figure 22. Fault scarp profile of the Red Canyon fault and location map. Profile of the Red Canyon fault was measured ~60 m northwest of Red Canyon Creek. The scarp is predominately coarse to fine-grained alluvium and colluvium. The fault scarp profile indicates a 5.8-m surface offset.

REFERENCES CITED

Anders, M.H., Geissman, J.W., Piety, L.A., and Sullivan, J.T., 1989, Parabolic distribution of circumeastern Snake River Plain seismicity and latest Quaternary faulting: Migratory pattern and association with the Yellowstone hotspot: Journal of Geophysical Research, v. 94, no. B2, p. 1589–1621, doi:10.1029/JB094iB02p01589.

Doser, D.I., 1985, Source parameters and faulting processes of the 1959 Hebgen Lake, Montana, earthquake sequence: Journal of Geophysical Research, v. 90, p. 4537–4555, doi:10.1029/JB090iB06p04537.

Erslev, E.A., 1981, Petrology and Structure of the Precambrian Metamorphic Rocks of the Southern Madison Range, Southwestern Montana [Ph.D. dissertation]: Cambridge, Massachusetts, Harvard University, 133 p.

Erslev, E.A., 1982, The Madison Mylonite Zone: A major shear zone in the Archean basement of southwestern Montana: Wyoming Geological Association 33rd Annual Field Conference Guidebook, Geology of Yellowstone Park Area, p. 213–221.

Erslev, E.A., 1993, Archean rocks: Beartooth Mountains and southwest Montana, *in* Reed, J.C., Jr., ed., Precambrian Conterminous U.S.: Boulder, Colorado, Geological Society of America, Geology of North America, v. C-2, p. 123–132.

Fraser, G.D., Witkind, I.J., and Nelson, W.H., 1964, A geological interpretation of the epicentral area—The dual-basin concept, *in* The Hebgen Lake, Montana, Earthquake of August 17, 1959: U.S. Geological Survey Professional Paper 435, p. 99–106.

Gary, S.D., 1980, Quaternary Geology and Geophysics of the Upper Madison Valley, Madison County, Montana [Master's thesis]: Missoula, Montana, University of Montana, 76 p.

Hadley, J.B., 1964, Landslides and related phenomena accompanying the Hebgen Lake earthquake of August 17, 1959, *in* The Hebgen Lake, Montana, Earthquake of August 17, 1959: U.S. Geological Survey Professional Paper 435-K, p. 107–138.

Keller, E.A., 1986, Investigations of active tectonics: Use of surficial earth processes, *in* Wallace, R.E., ed., Studies in Geophysics—Active Tectonics: Washington, D.C., National Academy Press, p. 136–147.

Kellogg, K.S., and Williams, V.S., 2000, Geologic Map of the Ennis 30′ × 60′ Quadrangle, Madison and Gallatin Counties, Montana, and Park County, Wyoming: USGS Geologic Investigations Series I-2690, scale 1:100,000.

Kellogg, K.S., Schmidt, C.J., and Young, S.W., 1995, Basement and cover-rock deformation during Laramide contraction in the northern Madison Range Montana and its influence on Cenozoic basin formation: American Association of Petroleum Geologists Bulletin, v. 79, no. 8, p. 1117–1137.

Lageson, D.R., 2009, Hebgen Lake-Red canyon normal fault system, MT: An example of early-stage structural inversion: Geological Society of America Abstracts with Programs, v. 41, no. 7, p. 54.

Lageson, D.R., and Zim, J.C., 1985, Uplifted basement wedges in the northern Rocky Mountain foreland: Geological Society of America Abstracts with Programs, v. 17, no. 4, p. 389.

Larsen, M.C., and Lageson, D.R., 2011, Geologic Map of the Mount Hebgen Quadrangle, Gallatin County, Montana: U.S. Geological Survey EDMAP grant, scale 1:24,000.

Lundstrom, S.C., 1986, Soil Stratigraphy and Scarp Morphology Studies Applied to the Quaternary Geology of the Southern Madison Valley, Southwestern Montana [M.S. thesis]: Humboldt, California, Humboldt State University, 53 p.

Mathieson, E.L., 1983, Late Quaternary Activity of the Madison Range Fault along Its 1959 Rupture Trace, Madison County, Montana [M.S. thesis]: Palo Alto, California, Stanford University, 169 p.

McCalpin, J.P., and Warren, G.A., 1993, Quaternary faulting on the Rock Creek fault, Overthrust Belt, western Wyoming: Geological Society of America Abstracts with Programs, v. 24, no. 6, p. 51.

Montana Bureau of Mines and Geology Geographic Information Center, 1:500,000 Faults of Montana: Montana.Gov Geographic Information, 1:500,000 Geology of Montana and 1:250,000 Digital Elevation Model, http://apps.msl.mt.gov/Geographic_Information/Data/DataList/datalist_ByCategories.aspx (accessed 2 December 2013).

Myers, W.B., and Hamilton, W., 1961, Deformation accompanying the Hebgen Lake, Montana, earthquake of August 17, 1959—Single-basin concept, *in* Geological Survey Research 1961: U.S. Geological Survey Professional Paper 424, p. D-168–D-170.

Myers, W.B., and Hamilton, W., 1964, Deformation accompanying the Hebgen Lake Earthquake of August, 1959, *in* The Hebgen Lake, Montana, Earthquake of August 17, 1959: U.S. Geological Survey Professional Paper 435, p. 37–98.

Pardee, J.T., 1950, Late Cenozoic block faulting in western Montana: Geological Society of America Bulletin, v. 61, p. 359–406, doi:10.1130/0016-7606(1950)61[359:LCBFIW]2.0.CO;2.

Peterson, D.L., and Witkind, I.J., 1975, Preliminary results of a gravity survey of the Henrys Lake quadrangle: Idaho and Montana: U.S. Geological Survey Journal of Research, v. 3, no. 2, p. 223–228.

Pierce, K.L., and Morgan, L.A., 1992, The track of the Yellowstone hotspot: Volcanism, faulting, and uplift, *in* Link, P.K., Kuntz, M.A., and Piatt, L.B., eds., Regional Geology of Eastern Idaho and Western Wyoming: Regional Geology of Eastern Idaho and Western Wyoming: Geological Society of America Memoir 179, p. 1–53.

Rasmussen, D.L., and Fields, R.W., 1983, Structural and depositional history, Jefferson and Madison basins, southwest Montana: American Association of Petroleum Geologists Bulletin, v. 67, p. 1352.

Ruleman, C.A., 2001, Late Quaternary tectonic activity along the Madison Range fault zone, northern arm of the Yellowstone tectonic parabola, southwest Montana: Association of Engineering Geologists Program with Abstracts, v. 44, no. 4, p. 74.

Ruleman, C.A., 2002a, Late Quaternary tectonic activity along the Madison fault zone, southwest Montana: Geological Society of America Abstracts with Programs, v. 34, no. 4, p. 37.

Ruleman, C.A., 2002b, Quaternary Tectonic Activity within the Northern Arm of the Yellowstone Tectonic Parabola and Associated Seismic Hazards, Southwest Montana [M.S. thesis]: Bozeman, Montana, Montana State University, 157 p. and 3 plates, scale 1:24,000.

Ruleman, C.A., 2009, Late Quaternary tectonic activity along the Madison Range fault zone, northern arm of the Yellowstone tectonic parabola: Geological Society of America Abstracts with Programs, v. 44, no. 7, p. 54.

Savage, J.C., Lisowski, M., Prescott, W.H., and Pitt, A.W., 1993, Deformation from 1973 to 1987 in the epicentral area of the 1959 Hebgen Lake, Montana, earthquake (Ms=7.5): Journal of Geophysical Research, v. 98, p. 2145–2153, doi:10.1029/92JB02410.

Schmidt, C.J., Evans, J.P., Harlan, S.S., Weberg, E.D., Brown, J.S., Batatian, D., Derr, D.N., Malizzi, L., McDowell, R.J., Nelson, G.C., Parke, M., and Genovese, P.W., 1993, Mechanical behavior of basement rocks during movement of the Scarface thrust, central Madison Range, Montana, *in* Schmidt, C.J., Chase, R.B., and Erslev, E.A., eds., Laramide Basement Deformation in the Rocky Mountain Foreland of the Western United States: Geological Society of America Special Paper 280, p. 89–105.

Schneider, N.P., 1985, Morphology of the Madison Range Fault Scarp, Southwest Montana: Implications for Fault History and Segmentation [M.S. thesis]: Miami, Florida, University of Miami, 131 p.

Schofield, J.D., 1981, Structure of the Centennial and Madison Valleys based on gravitational interpretation, *in* Tucker, T.E., ed., Montana Geological Society Field Conference and Symposium Guidebook, v. 5, p. 275–283.

Shelden, A.W., 1960, Cenozoic faults and related geomorphic features in the Madison Valley, Montana: Billings Geological Society 11th Annual Field Conference West Yellowstone-Earthquake Area, p. 178–184.

Smith, R.B., and Arabasz, W.J., 1991, Seismicity of the Intermountain Seismic Belt, *in* Slemmons, D.B., Engdahl, E.R., Zoback, M.D., and Blackwall, D.D., eds., Neotectonics of North America: Boulder, Colorado, Geology of North America, Decade Map, v. 1, p. 185–228.

Smith, R.B., and Sbar, M.L., 1974, Contemporary tectonics and seismicity of the western United States with emphasis on the Intermountain seismic belt: Geological Society of America Bulletin, v. 85, p. 1205–1218, doi:10.1130/0016-7606(1974)85<1205:CTASOT>2.0.CO;2.

Stickney, M.C., and Smith, D., 2009, Recent seismicity in the western part of the 1959 earthquake aftershock zone: Geological Society of America Abstracts with Programs, v. 41, no. 7, p. 54.

Tysdal, R.G., 1986, Thrust faults and back thrusts in Madison Range of southwest Montana foreland: American Association of Petroleum Geologists Bulletin, v. 70, p. 859–868.

Van de Woerd, J., Benedetti, L., Caffee, M.W., Finkel, R., and Hebgen Lake Paleoseismology Working Group, 2000, Slip-rate and earthquake recurrence time on the Hebgen Lake fault (Montana): Constraints from surface exposure dating of alluvial terraces and bedrock fault scarp: Eos (Transactions, American Geophysical Union), v. 81, no. 48, p. F1160.

Witkind, I.J., 1975, Preliminary Map Showing Known and Suspected Active Faults in Western Montana: U.S. Geological Survey Open-File Report 75-285, scale 1:500,000.

Manuscript Accepted by the Society 25 February 2014

Printed in the USA

The Geological Society of America
Field Guide 37
2014

Tracking a big Miocene river across the Continental Divide at Monida Pass, Montana/Idaho

James W. Sears
Department of Geosciences, 32 Campus Drive #1296, University of Montana, Missoula, Montana 59812, USA

ABSTRACT

Exotic Miocene and Pliocene river gravel lies on top of the Continental Divide along the Idaho-Montana border near Monida Pass. The gravel is interlayered with tuffs and basalt flows of the Heise volcanic field, which erupted from the site of the Yellowstone hotspot between 6.62 and 4.45 Ma. The gravel includes pebbles that may have been derived from bedrock outcrops in Nevada and Utah, implying a paleo-river with headwaters to the south of the modern Continental Divide and Snake River Plain. The river may have been a tributary of the pre–ice age Bell River of Canada.

The field trip examines evidence for the tectonic evolution of the Monida Pass area. The course of the Miocene river appears to have been diverted around growing mountain ranges, and then pinched off at Monida Pass on the northern shoulder of the Yellowstone hotspot track.

INTRODUCTION

This trip investigates field evidence for a recent proposal that a large Miocene river with headwaters in the southern Great Basin and Colorado Plateau may have flowed near the present location of Monida Pass on the Montana/Idaho segment of the Continental Divide, before being cross-cut and disrupted by the Yellowstone hotspot (Sears, 2013a). Before the ice age, the river may have joined the Bell River basin of Canada, which emptied into the Labrador Sea (Fig. 1, cf. Balkwill et al., 1990). The Miocene river valley may have followed the path of a Paleogene rift system along the east flank of the Nevadaplano plateau of DeCelles (2004).

The field trip begins near Spencer, Idaho, on the north edge of the Snake River Plain and proceeds north (Fig. 2). Stops 1–5 investigate evidence for the deflection and destruction of the Miocene paleovalley by tectonics and volcanism of the Yellowstone hotspot track along the present Continental Divide near Monida. Stop 6 examines a faulted remnant of the Miocene paleovalley in Sage Creek, near Dell, Montana, where Miocene basalt, tephra, and river gravel unconformably overlie the Paleogene Renova Formation. The final sites (Stops 7–12), along Red Rock graben and near Clark Canyon Reservoir, study effects of active faulting on the paleovalley in the Intermountain Seismic Belt (Stickney, 2007).

BELL RIVER BASIN

The Miocene paleovalley of SW Montana was likely a southern tributary of the Cenozoic Bell River basin (Fig. 1), which is thought to have drained most of Montana and Canada into the Labrador Sea, before being destroyed by Pleistocene continental glaciation (Balkwill et al., 1990). According to Howard (1958) during the ice age, Montana's tributaries to the Bell

Sears, J.W., 2014, Tracking a big Miocene river across the Continental Divide at Monida Pass, Montana/Idaho, *in* Shaw, C.A., and Tikoff, B., eds., Exploring the Northern Rocky Mountains: Geological Society of America Field Guide 37, p. 89–99, doi:10.1130/2014.0037(04). For permission to copy, contact editing@geosociety.org.

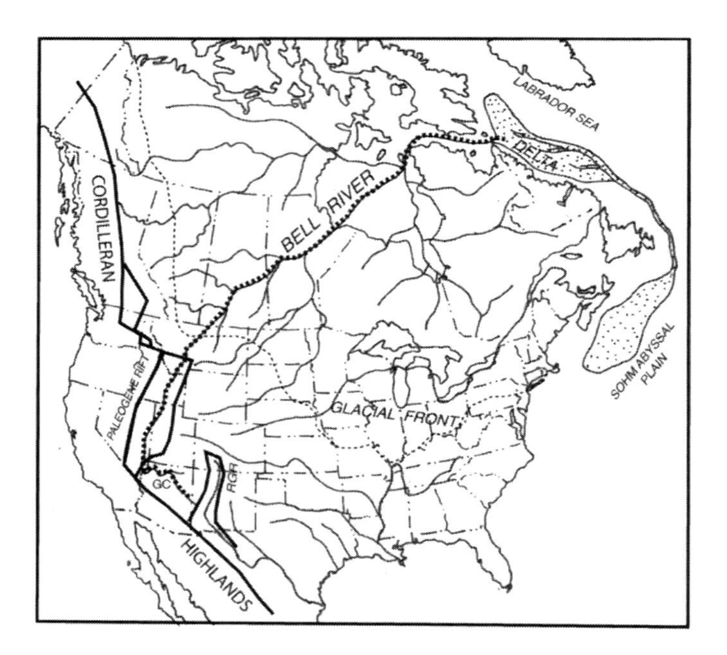

Figure 1. Bell River basin. GC—Grand Canyon; RGR—Rio Grande Rift. After Sears (2013a).

River basin—the upper Missouri, Musselshell, and Yellowstone Rivers—were deflected into the modern Missouri River along the southern front of the continental ice sheet.

Cenozoic deposits of the Bell River paleovalley that are preserved in southern Saskatchewan have provenances in the Montana Rockies, and record NE-directed paleoflow (Leckie et al., 2004). In Montana, the upper Missouri River valley follows segments of a Miocene paleovalley from the Continental Divide to Fort Benton. Fluvial gravel of the Miocene Sixmile Creek Formation in remnants of the paleovalley consistently exhibit southerly provenances, and paleocurrents indicate NE-directed paleoflow (Landon and Thomas, 1999).

The Sixmile Creek fluvial gravel includes distinctive pebbles that are exotic to Montana, but that lithologically match possible bedrock sources in the Great Basin of Nevada and Utah (Sears, 2013a). These include meta-chert and meta-quartzite of the Roberts Mountain allochthon and cherty litharenite of the Antler foredeep of Nevada. Pebbles of pink, white, and purple quartz-arenite may have been derived from the Cordilleran miogeocline of Utah.

The Sixmile Creek Formation contains detrital zircons that are diagnostic of the Antler belt (Beranek et al., 2006; Stroup,

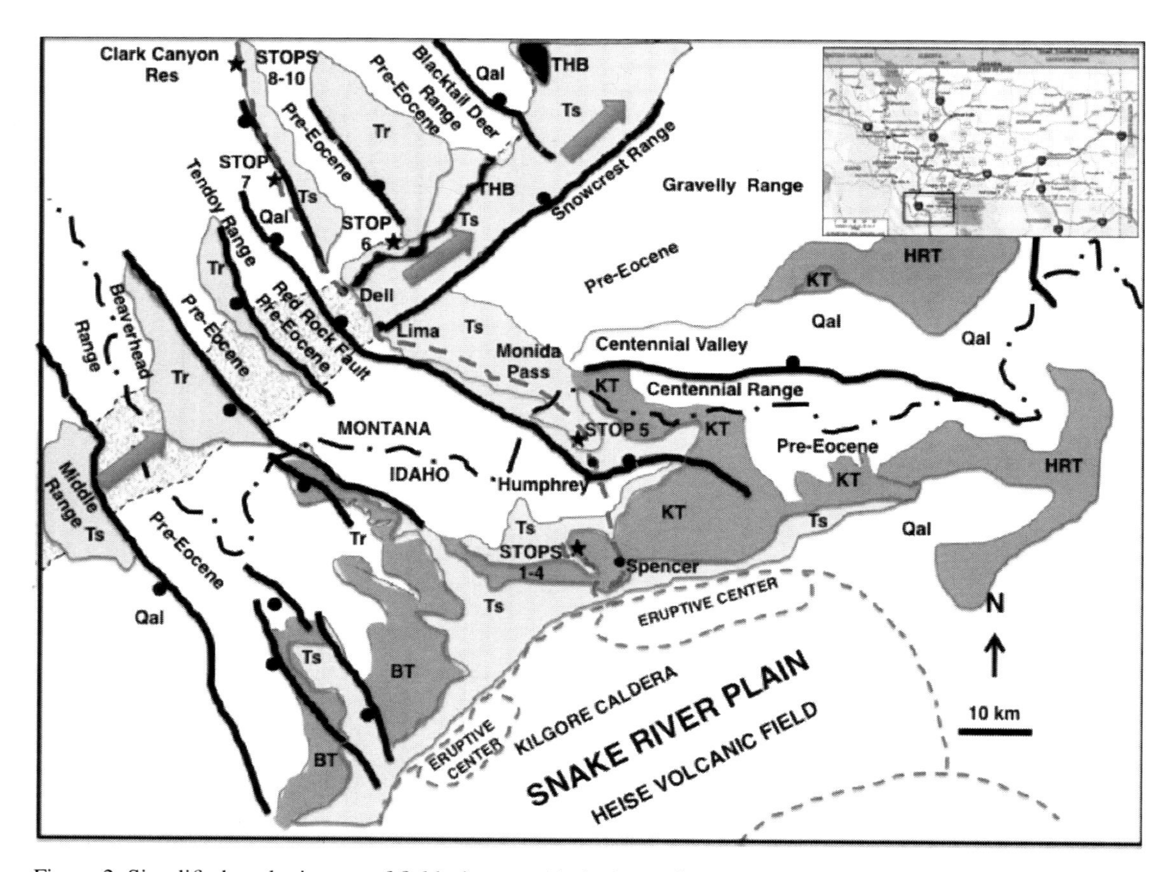

Figure 2. Simplified geologic map of field trip route (dashed gray line). Inset shows location. Arrows—interpreted original path of Miocene River. Stars—field trip stops. BT—Blacktail Tuff; HRT—Huckleberry Ridge Tuff; KT—Kilgore Tuff; THB—Timber Hill basalt; Qal—alluvium; Tr—Renova Formation; Ts—Sixmile Creek Formation. Compiled from Skipp et al. (1979), Lonn et al. (2000), and Lewis et al. (2012).

2008). Although Antler belt lithologies also occur in Idaho and may have provided some sources (Sears and Ryan, 2003), in Miocene time those localities were largely covered by Eocene Challis volcanics. While some Antler belt cobbles also occur in the Cretaceous Beaverhead conglomerate in SW Montana (Skipp, 1979), those localities are generally restricted to the footwalls of fault zones that post-date deposition of the Miocene gravel.

The Montana tributary of the Bell River may have had headwaters in Grand Canyon country (Sears, 2013a). The Miocene Colorado River may have flowed north from the southern Colorado Plateau into a rift system along the eastern margin of the Nevadaplano plateau of DeCelles (2004), and thence north through Nevada and Utah to Idaho and SW Montana.

TECTONIC SETTING

Paleogene Rift System

Paleogene rifting began at ca. 53 Ma along the Lewis and Clark Line in Montana, with extensional exhumation of metamorphic core complexes (Foster et al., 2007). The rifting is thought to have propagated to southern Nevada and Utah by 28 Ma, accompanied by volcanism (Mix et al., 2011), as the tectonically thickened axis of the Cordillera collapsed to the west (Coney and Harms, 1984; Hildebrand, 2009). The Paleogene rift system crosscut the Late Cretaceous–late Paleocene drainage of the Cordilleran foreland, when paleo-rivers flowed from Montana to NW Wyoming (Chetel et al., 2011).

Volcanics erupted across southwestern Montana between 50 and 46 Ma, concurrent with early phases of extensional faulting (Bausch, 2013). As the volcanic centers migrated south, the fine-grained, ash-rich Renova Formation was deposited in the northern part of the rift system in Montana. Its deposition (42–18 Ma) correlated with "super eruptions" from mega-caldera complexes in Nevada and Utah (Henry et al., 2012). The Renova accumulated in deep rift basins (Janecke, 1994; M'Gonigle and Dalrymple, 1996; Constenius, 1996).

The Renova may represent distal deposits of volcanic sediment derived from the calderas.

Fluvial sands have yielded syn-depositional detrital zircons that correlate with eruptions of the megacalderas of the Great Basin. The Renova has also yielded detrital zircon assemblages that closely match those from Paleozoic and Mesozoic strata of the Great Basin and southern Colorado Plateau (compare Stroup, 2008; Link et al., 2008; Barber et al., 2013; Rothfuss et al., 2012; Gehrels et al., 2011).

Basin and Range

In SW Montana, Basin-Range faulting appears to have diverted drainage from the north-trending Paleogene rift system into NE-trending grabens. The Miocene grabens radiated from the 17 Ma outbreak point of the Yellowstone hotspot in SE Oregon and some cut across the trends of the older rifts (Sears and Thomas, 2007; Sears et al., 2009).

In the field trip area, the Miocene disturbance faulted and tilted the Eocene Challis volcanics, Eocene-Oligocene Medicine Lodge lake beds, and Eocene–lower Miocene Renova Formation (Fields et al., 1985). The Eocene–lower Miocene units were mostly eroded off rising horsts, but were preserved in subsiding grabens (Fig. 3). The erosional period coincided with the Miocene climatic optimum, a time of unusually warm and wet global climate (Thompson et al., 1982; Zachos et al., 2001).

The Miocene grabens appear to have shunted heavy runoff into new rivers and lakes. Mega-debris flows cascaded down the old valleys on the flanks of the horsts and dumped bouldery diamictite into the Miocene grabens (Sears, 2007). These deposits typically comprise the basal part of the Sixmile Creek Formation, which was deposited from middle Miocene through late Pliocene time (Fields et al., 1985; Barnosky et al., 2007).

Gravel beds with distinctive pebbles suggest that the Miocene river threaded its way across SW Montana for 350 km, across the present Continental Divide to the Great Plains. The drainage flowed through wide valleys and lakes in Miocene grabens, and through deep canyons across the intervening horsts (Sears et al., 2009).

Dozens of active normal faults cross-cut the Miocene valleys in the Intermountain Seismic Belt of SW Montana, and have diverted drainage into new rivers and creeks. The younger faults

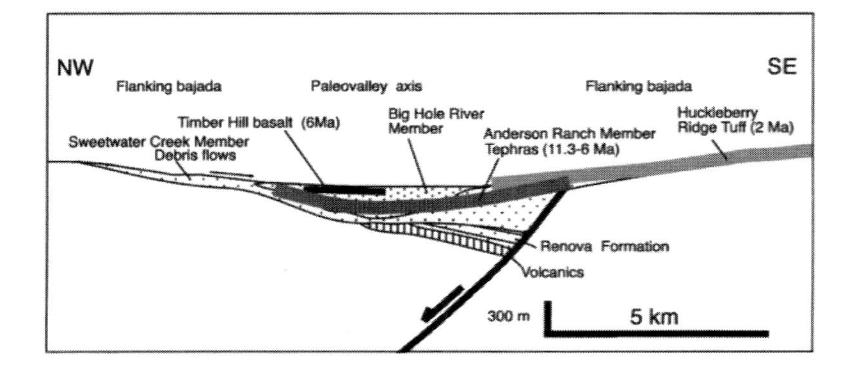

Figure 3. Schematic cross section of Ruby Graben, SW Montana. After Sears et al. (2009).

trend NW across the north flank of the Yellowstone hotspot track (Stickney, 2007).

Yellowstone Hotspot

In the field trip area, the paleovalley contains fluvially reworked tephra beds, ash-flow tuffs, and basalt that erupted from calderas along the Yellowstone hotspot track. These volcanics range in age from 16 to 2 Ma, and record the approach of the hotspot calderas toward the paleovalley. The hotspot calderas appear to have crossed the trace of the paleovalley between 10 and 6 Ma (cf. Pierce and Morgan, 1992).

The oldest tephra beds are thin and fine-grained, but by 10 Ma, as the eruptive sites reached the paleovalley, some tephra beds reached a thickness of 30 m and included pumice pebbles. The first welded tuff to appear in the paleovalley was the 6.62 Ma Blacktail Creek Tuff (Morgan and McIntosh, 2005). It was followed by the ca. 6 Ma Timber Hill basalt flow (Fig. 4), which is preserved along a 50 km reach of the paleovalley (Fritz et al., 2007). The paleovalley was overwhelmed by the thick 4.45 Ma Kilgore tuff in the area of the Continental Divide (Morgan and McIntosh, 2005).

Faulting in the shoulder zone of the hotspot began to displace segments of the paleovalley before 6 Ma, and by 4 Ma, the paleovalley appears to have been beheaded at the Continental Divide. Fluvial deposits of the paleovalley have been reworked and buried by alluvial fans derived from the horsts. The fans were capped by the 2 Ma Huckleberry Ridge Tuff, which was later offset by large-scale faults along the Centennial and Madison rangefronts (Pierce and Morgan, 1992).

NW-trending faults of the Intermountain Seismic Belt crosscut the NE-trending Miocene and N-trending Paleogene faults, creating complex sets of rectangular fault-block basins and ranges (Janecke, 1994; Sears and Fritz, 1998). The NW-trending faults diverted drainage into new grabens, leaving uplifted and tilted stream deposits in abandoned channels. These remnants of the older river deposits document the evolution of the paleovalley.

Detrital zircon data indicates that, between 7 Ma and 4 Ma, the Continental Divide migrated from the Idaho batholith near Boise, Idaho, to its present location along the Montana-Idaho border, and drainage was diverted west along the Snake River Plain (Beranek et al., 2006).

MONIDA PASS

Monida Pass occupies a normal fault transfer zone between the Tendoy and Centennial Ranges along the Continental Divide. The Centennial Range trends E-W, with a range-bounding fault on its north side. The fault dies out westward into a broad anticline that plunges west near Monida Pass. The fault and anticline deform the 4.45 Ma Kilgore tuff. Folds and faults at the east end of the Centennial Range involve the 2-Ma Huckleberry Ridge Tuff.

The Tendoy Range trends NW in near Dell, but bends to the east at Monida Pass. The Red Rock Fault bounds the Tendoy Range on the NE, and dies out into a SE-plunging anticline south of the Centennial Range. The fault and anticline both involve the Kilgore Tuff.

Monida Pass threads its way through the gap between the two ranges in non-resistant Miocene/Pliocene river gravel and weak, underlying Aspen Shale of Cretaceous age. Basalt from the Heise volcanic field flowed down the gap between the two ranges at ca. 4 Ma (Fritz et al., 2007).

The Kilgore Tuff and the Huckleberry Ridge Tuff dip southward into the Snake River Plain on the south flank of the Centennial Range. A canyon cuts through the tilted Kilgore Tuff at Spencer and is floored by a flat-lying Pleistocene basalt flow, part of the volcanic cover of the Snake River Plain. These relationships bracket the tilting of the Centennial Range to early Pleistocene.

The distribution of river gravel in the Monida Pass area traces the eastward migration of the paleovalley in concert with the eastward migration of the hotspot calderas. The upper Snake River may have flowed NW into the Miocene paleovalley along Medicine Lodge graben at ca. 6 Ma (Fig. 5). It was then progressively deflected around the SE-plunging end of the growing Tendoy Range (Fig. 6), until pinched off between the Tendoy and Centennial ranges, before eruption of the Kilgore Tuff at 4.45 Ma (Fig. 7). The Snake River then flowed down the eastern Snake River Plain to Lake Idaho in the Boise area, and the beheaded paleovalley became the headwaters of the Missouri River (Beranek et al., 2006).

ROAD LOG

Spencer, Idaho, to Clark Canyon Reservoir, Montana

Mi	*Description*
0.0	**Spencer, Idaho.** Begin at T-intersection of Opal Way and Old Highway 91 in Spencer. Proceed south on Old Highway 91.
0.2	Flat-lying Pleistocene basalt occupies valley beside Old Highway 91. It probably flowed into the canyon from shield volcano on Snake River Plain ahead to south. Angular unconformity exists between basalt and gently south-tilted Pliocene Kilgore Tuff (4.45 Ma), which forms dip-slope seen to east.
1.7	Sharp curve to right, proceed under I-15.
2.0	Curve left. South-tilted dip-slope of Kilgore Tuff seen to west. Tuff overlies river gravel layer on horizon.
3.7	Cross shallow canyon cut in Pleistocene basalt.
3.8	**Turn right onto Dry Creek Road** (called Skyline Road on some maps) **and proceed north.**
4.4	Pleistocene basalt on right.
5.5	Road proceeds up dipslope of Kilgore Tuff.
5.6	**Stop 1. Bouldery outcrop of Kilgore Tuff on right.**

View of tilted Kilgore Tuff to east. Morgan and McIntosh (2005) proposed that the Kilgore Tuff, a densely welded, crystal-poor, rhyolitic ignimbrite, erupted over a subdued landscape

Figure 4. Ten-meter-thick Timber Hill basalt flow (6 Ma) armors topographically inverted valley, offset by 250 m by Sweetwater normal fault, Ruby Range, SW Montana. View north.

from multiple source vents along the caldera margin. An elongate, 30-km-long vent zone on the north edge of the Kilgore caldera was adjacent to Spencer, where the tuff is thicker than 120 m (Morgan and McIntosh, 2005). The Kilgore Tuff has been mapped to Jackson Hole, Wyoming, and its original volume is estimated at 1800 km^3 (Morgan and McIntosh, 2005). The south dip of the tuff may be due to thermal subsidence of the hotspot track, or to evacuation of magma.

Continue north along Dry Creek Road.

6.3 **Stop 2. Park near crossing of small drainage.**

Kilgore Tuff overlies roundstone cobble gravel. The Miocene/Pliocene gravel is younger than the 6.62 Ma Blacktail Creek Tuff, but older than 4.45 Ma Kilgore Tuff. It contains exotic rock types that may have originated in SE Idaho, Utah, and Nevada.

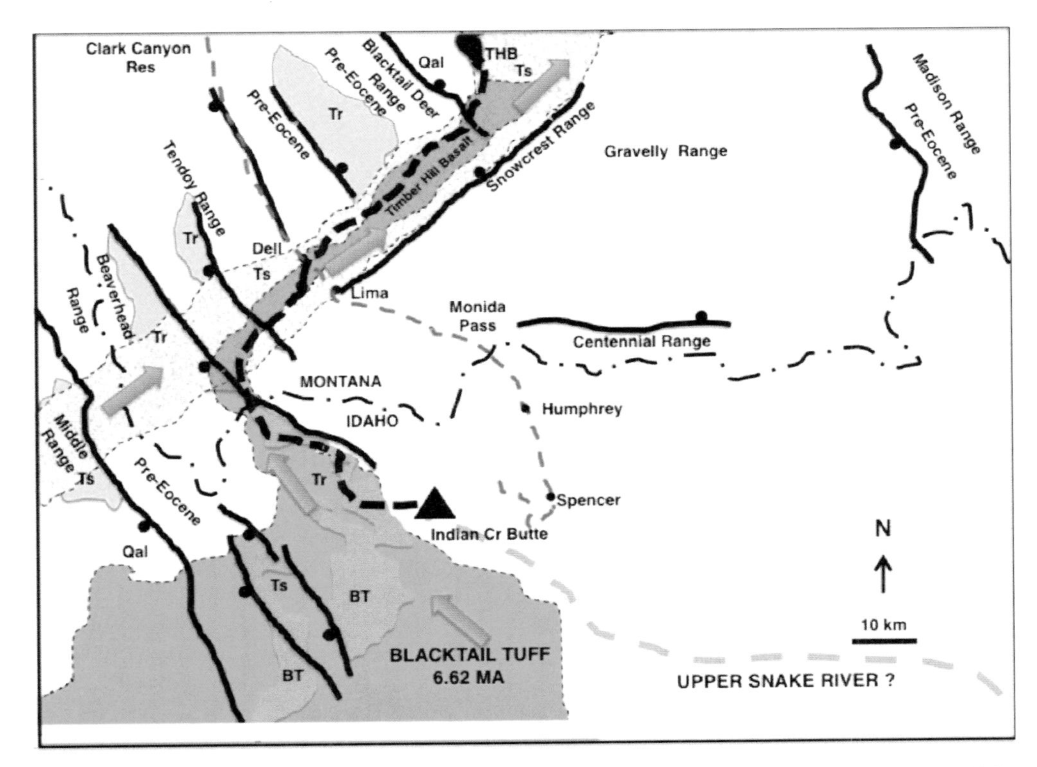

Figure 5. Volcanics of Heise field flowed down graben and into Miocene paleovalley at 6.62 and 6 Ma. Upper Snake River may have flowed into paleovalley on east flank of volcanic field. BT—Blacktail Tuff; Qal—alluvium; THB—Timber Hill basalt; Tr—Renova Formation; Ts—Sixmile Creek Formation.

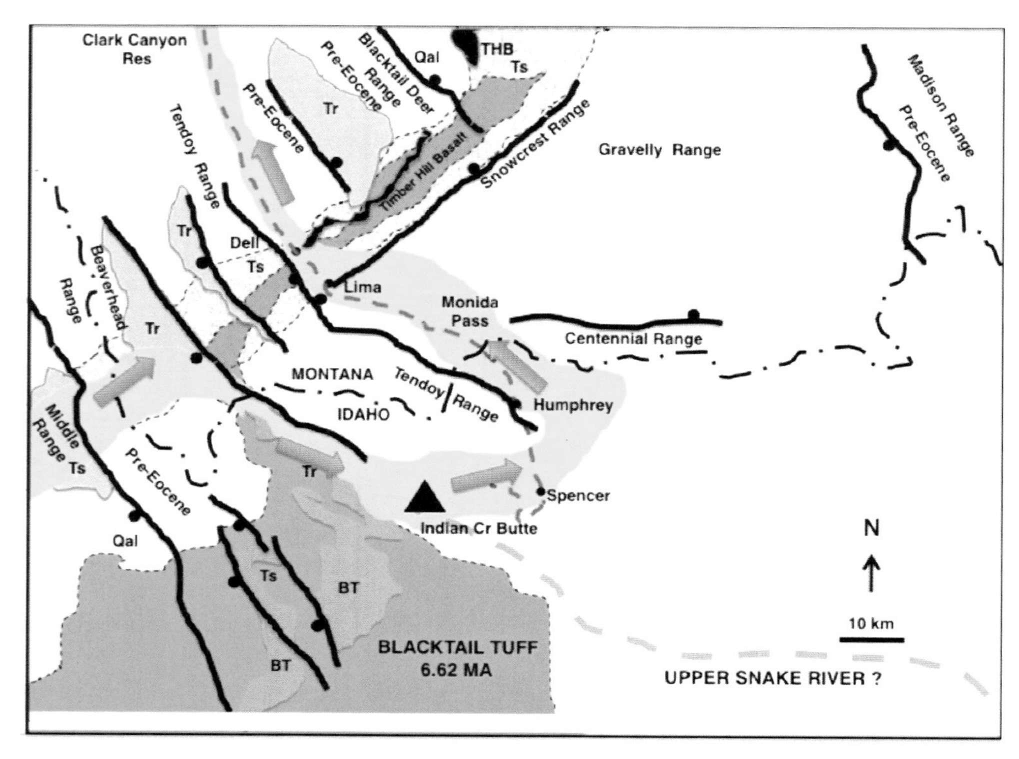

Figure 6. Faulting raised Tendoy Range horst, diverting paleovalley to Monida Pass and NW to Clark Canyon Reservoir. Paleovalley crossed its own path near Dell. BT—Blacktail Tuff; Qal—alluvium; THB—Timber Hill basalt; Tr—Renova Formation; Ts—Sixmile Creek Formation.

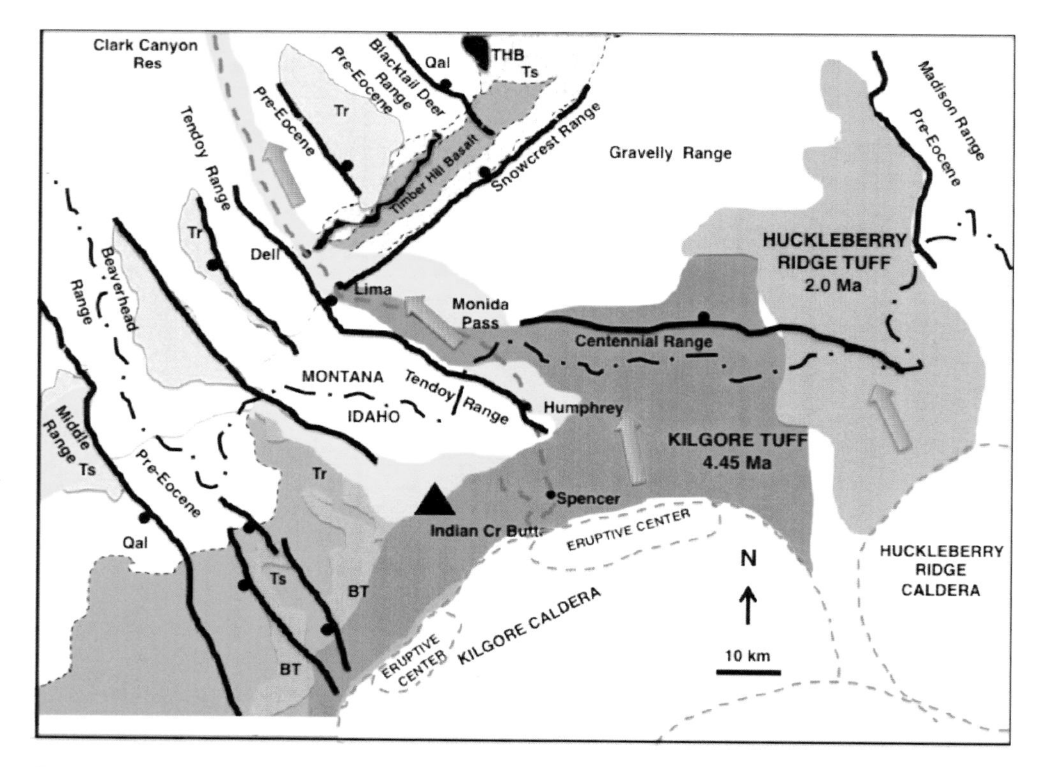

Figure 7. Kilgore Tuff overwhelmed paleovalley. Main vents were south of Monida Pass. Miocene/Pliocene river was pinched off, with Continental Divide forming headwaters of Missouri. BT—Blacktail Tuff; Qal—alluvium; THB—Timber Hill basalt; Tr—Renova Formation; Ts—Sixmile Creek Formation.

These include pink, white, and purple quartz-arenite, that may have come from Precambrian, Cambrian, and Ordovician quartzite beds in SE Idaho and Utah, and meta-chert, meta-quartzite, and cherty litharenite that may have come from the Roberts Mountain allochthon and the Antler foredeep in central Nevada. Cobbles have pressure-solution pits (Fig. 8), perhaps due to folding of gravel layer over anticline at plunging SE end of Tendoy Range.

Continue NW along Dry Creek Road. Note rounded river cobbles along road side. Road follows old river bed.

7.3 Cattle guard.
7.8 Vesicular lava.
8.9 **Stop 3. Intersection of Dry Creek and Cottonwood Creek Roads.**

Miocene/Pliocene river gravel of Sixmile Creek Formation. It appears to overlie Blacktail Creek Tuff, and underlie Kilgore Tuff. Walk SW along road beside fence line for ~1000 feet, to top of hill, for view of Snake River Plain and south-dipping Kilgore Tuff. Hill is underlain by river cobble gravel, as seen at Stop 2. This gravel may represent the course of the Miocene/Pliocene paleoriver around the SE end of the Tendoy Range, between 6 and 4.45 Ma.

Take Forest Road 200 to right. Cobble gravel overlies 6.62 Ma Blacktail Creek Tuff.

9.2 Cattle guard. Boundary of Targhee National Forest. Blacktail Creek Tuff (?) forms gentle south-sloping dipslope.
10.1 **Stop 4. Top of rise.**

Figure 8. Quartzite cobble with pressure-solution pits from Sixmile Creek gravel near Stop 2. Approximately 15 cm wide.

Although mapping is not yet complete in this area, the boulders seen here may be float of the 6.62 Ma Blacktail Creek Tuff, which was the first large-volume ignimbrite of the Heise field, at ~1200 km^3 (Morgan and McIntosh, 2005). It forms the long, gentle dipslope that the road follows here. It is a single outflow outflow facies less than 13 m thick along the Continental Divide and contains abundant lithophysal cavities (Morgan and McIntosh, 2005).

Return to Spencer.

Restart odometer at T-intersection in Spencer.
0.0 T-intersection of Opal Way and Old Highway 91 in Spencer.
Proceed south on Old Highway 91.
0.2 **Proceed west under I-15, then north on Old Highway 91.**
1.2 Cattle guard. Enter Targhee National Forest. Outcrops and rubble of Kilgore Tuff to left.
1.8 Approximate contact of Kilgore Tuff and underlying cobble gravel, covered by talus from Kilgore Tuff.
2.7 Pleistocene terraces cut on Miocene/Pliocene cobble gravel on left.
4.2 Stop sign at Stoddard Creek Road. Reclaimed gravel pit in Miocene/Pliocene cobble gravel. View east of Kilgore Tuff on skyline ridge, gently dipping to south. Continue north on Old U.S. Highway 91. Road approximately follows top of Humphrey Basalt, which underlies Miocene/Pliocene gravel.
7.0 Go under I-15.
7.6 **Stop 5. High bridge over canyon.** Stop at north end of bridge.

Humphrey Basalt overlain by Miocene/Pliocene gravel. Basalt may have erupted from Heise field. The basalt dips beneath the stream bed south of here.

Table Mountain, a basalt flow seen on skyline on Continental Divide to the NE, is folded over the broad anticline at the west end of the Centennial Range. **Continue north on Old U.S. Highway 91.**

9.8 **Humphrey. Take I-15 N.**

Road log now refers to mileposts (MP) along I-15 N.

MP	*Description*
186	Anticline on skyline to east where Centennial fault dies out westward.
190	Cretaceous Aspen Formation in road cuts.
195	Cretaceous Aspen Formation in road cuts.
0	Enter Montana at Monida Pass, Continental Divide. Cretaceous Aspen Formation in road cuts.
2	Cretaceous Aspen Formation in road cuts.
3	Pliocene volcanics K-Ar dated at ca. 4 Ma (Fritz et al., 2007). These probably flowed down the graben valley from the Kilgore caldera.
7	Tilted Pliocene volcanics on ridge.

8 Cretaceous Beaverhead Conglomerate in road cuts.
12 Cretaceous Beaverhead Conglomerate in road cuts.
14 Pliocene basalt caps buttes.
15 Lima, Montana.
18 Broad valley to east is course of Miocene paleoriver. Gravel is capped by 6 Ma Timber Hill basalt flow. Gravel and basalt are tilted into Red Rock Valley due to faulting along Red Rock Fault on SW side of valley. Faceted spurs can be seen along the front of the Tendoy Range to the west (Stickney and Bartholomew, 1987).
22 Cretaceous Beaverhead Conglomerate forms Red Butte on east.

Take Exit 23 to Dell. Reset odometer at stop sign.
0.0 Go straight east on Main Street.
0.1 Cross railroad. Turn left on Oregon Shortline Road. Pass "Calf-A" restaurant.
0.5 Turn right on Sage Creek Road.
6.6 Cattle guard. Timber Hill basalt on river gravel to right.
6.8 "Federal and State Lands" sign.
8.4 **Stop 6. Entrance to Matador Cattle Company.**

Renova Formation and Sixmile Creek Formation. The Renova Formation forms the gullied gray mudstone/siltstone at the base of the butte to the west of the road. The Sixmile Creek Formation caps the butte. This location was studied extensively by Fields et al. (1985). To the SE, the Timber Hill basalt caps a series of buttes and mesas that are gently tilted and cut by numerous NW-striking, NE-side down, normal faults (Fig. 9).

Return to Dell.

Proceed North on I-15.

Road log now refers to mileposts (MP) along I-15 N.

MP	*Description*
26	Pinedale glacial outwash and alluvial fans fill valley.
29	Kidd.
34	**Stop 7. Parking area on right.**

View of Red Rock Hills fault block. The fault block is tilted to the SW, toward the Red Rock Fault. The crest of the hill is Cretaceous Beaverhead conglomerate. The flank has two main alluvial fan complexes that have recycled fluvial gravel of the Sixmile Creek Formation as the range has been uplifted and tilted. It was probably a Miocene/Pliocene river valley diverted into Red Rock graben from the Miocene path of the paleovalley. Kidd is where fault shifts from SW to NE side of valley. A magnitude 5.3 earthquake, with a focal depth of 12.4 km occurred at the transfer point in 1999 (Stickney and Lageson, 2002).

38 Devonian Jefferson Formation forms foothills on west.
42 Sixmile Creek gravel in this area has been tectonically crushed.

Take Exit 44 to Clark Canyon.
0.0 **Restart odometer at stop sign. Cross interstate to west. Proceed west on MT 324.**
3.6 **Stop 8. Entrance to Cameahwait West Campground.** Stops 8–12 are adapted from Sears (2013b).

A low-angle normal fault places Lombard Limestone of the Mississippian/Pennsylvanian Snowcrest Group on the west against Cambrian Hasmark Dolomite, Wolsey Shale, Flathead Sandstone, and Archean basement on the east. These rocks are on the west limb of the Archean-cored Armstead anticline, a late Cretaceous Laramide structure. In this region, the basement-cored anticlines of the foreland (thick-skinned structures) were

Figure 9. Faulted and tilted 6 Ma Timber Hill basalt flow.

overprinted by low-angle thrusts of the Cordilleran fold-thrust belt (thin-skinned structures).

The normal fault dips gently west, synthetic with the main fault on the east side of the Medicine Lodge graben. As the fault traces north, displacement increases so that the Snowcrest and Quadrant are juxtaposed against the Archean. North of here, the fault is crossed by a paleovalley that is not faulted, and that contains Miocene Sixmile Creek debris flows derived from west of the graben. Those debris flows crossed the Armstead anticline and continued to the Beaverhead River at Stop 11.

Turn around, proceed east on MT 324.

4.9 Archean basement in core of Armstead anticline passes into a vertical Paleozoic section on the east limb of the anticline. The Armstead anticline was deeply eroded and was overlain with angular unconformity by conglomerate of the Campanian Beaverhead Group. Low-angle faulting then thrust the anticline eastward over the Beaverhead Group along the Armstead thrust.

Proceed up-section through Paleozoic formations from Flathead Sandstone through Wolsey Shale, Hasmark Dolomite, Jefferson Dolomite, Three Forks Shale, and Madison Group (Lodgepole Limestone, Mission Canyon Limestone, and Mackenzie Canyon Limestone).

6.0 Turn right on Shoshoni Road into marina area, and drive to Camp Fortunate overlook.

6.6 **Stop 9. Camp Fortunate Overlook.**

The Red Rock Hills lie to the SE, on the far side of the Interstate Highway. The hills comprise a fault block that is tilted to the SW against the Red Rock normal fault zone. The Red Rock fault zone trends NW and may be related to crustal extension on the shoulder of the Yellowstone hotspot track. A second normal fault zone trends across the lower slope of the Red Rock Hills, parallel to the Red Rock fault and 10 km to its NE. The second fault zone has the SW side downthrown, making part of the Red Rock valley a full graben. The Red Rock fault appears to die out to the NW as the second fault gains displacement. The second fault zone trends directly under the Clark Canyon dam. It may be exposed in a cliff face on the east side of the reservoir, where Miocene gravel beds have been crushed and faulted (Stop 10).

Alluvial fans cover most of the SW slope of the Red Rock Hills fault block, and these were reworked from Miocene/Pliocene Sixmile Creek gravel and Beaverhead Group conglomerate. Directly across the reservoir to the east, the rounded brown hills are remnants of thick Sixmile Creek river gravel (Fig. 10). The gravel composition indicates that the fluvial system had headwaters in the Lemhi Group of Idaho. The faulting that created the Red Rock valley graben diverted the drainage from its Miocene configuration, probably between 7 and 4 Ma.

Return to Montana 324.

7.0 Turn right on Montana 324.

7.9 Cross Clark Canyon Dam and 45th Parallel.

8.1 Turn right into picnic area and proceed to south parking lot.

8.5 **Park for Stop 10. Fractured Pliocene cobbles.** Walk along jeep trail next to the shore to a long cliff-cut of gravel beside the reservoir.

This conglomerate is part of the Miocene/Pliocene Sixmile Creek Formation. The bedding, shown by a thin sand bed, dips ~80 degrees to the east in a fault zone that trends NW, parallel with the Red Rock graben. The equivalent Miocene/Pliocene beds to the NE across I-15 are gently dipping. The cobbles are severely fractured and sheared, mostly at cobble-cobble contacts (Fig. 11). Note pressure-solution pits with radiating fractures, small normal-faulted cobbles, micro-grabens, and slickensides. Similarly fractured cobbles occur in the big railroad cut to the east, across I-15 from this site. The cobbles were faulted while buried, and then were re-cemented, uplifted on the fault zone, and exposed by erosion. The deposit was overlapped by Pleistocene fans, across which some normal fault scarps have been mapped. The fault is part of a zone that drops the Red Rock graben, and is seismically active. Several small, steep, NW-trending faults are marked by thin breccia, gouge, and offset cobbles.

Return to Montana 324.

9.0 Cross I-15 and turn left on High Bridge Road.

9.2 Bear left and continue down hill to north on frontage road. Follow Beaverhead River.

Figure 10. Remnant of Pliocene Sixmile Creek river gravel near Clark Canyon Reservoir. Gravel is ~150 m thick. View east.

Figure 11. Faulted river cobble (~20 cm across) in Pliocene Sixmile Creek Formation at Clark Canyon Reservoir. Cobbles are in place.

10.2 **Stop 11. Pull out on left for boulder of unusual size (Fig. 12).**

This boulder of Bonner Quartzite was derived from at least 35 km to the west, in the Beaverhead Mountains on the far side of the Oligocene Horse Prairie graben—the nearest source for the Bonner Quartzite. It traveled in a debris flow down a paleovalley that crossed the low angle fault seen at Stop 8. A similar paleovalley contains a 27 Ma basalt flow north of Stop 8. The debris flow includes clasts of Eocene basalt and comprises basal Sixmile Creek Formation. The valley incises 47 Ma Dillon volcanics.

Continue north on the frontage road.

10.7 **Stop 12. Railroad cut of Miocene unconformity.**

Stop in the parking area below I-15 overpass (High Bridge) and walk up the road to the east to near the railroad tracks. The road crosses an Eocene lava flow which exhibits crude columnar jointing and flow banding.

Figure 12. Large boulder of Precambrian Bonner Quartzite, Belt Supergroup, from debris flow in basal Sixmile Creek Formation, Stop 11.

Figure 13. Sixmile Creek Formation bouldery debris flows in channel cut on Eocene tuff. Large railroad cut at Stop 12.

View unconformity from the SW side of the railroad.

The railroad cut exposes the angular unconformity between the Eocene Dillon volcanics and the Miocene/Pliocene Sixmile Creek Formation (Fig. 13). This is the NW edge of a channel cut through the volcanics and filled with numerous debris flows. This exposure is near where the side valley entered the main Beaverhead graben. Across the river to the west, the bouldery deposit rises up the side of the valley to the skyline. The paleovalley has been tilted to the east into the Beaverhead graben. The tilting may have helped to trigger the debris flows.

END OF ROAD LOG.

REFERENCES CITED

Balkwill, H.R., McMillan, N.J., MacLean, B., Williams, G.L., and Srivastava, S.P., 1990, Geology of the Labrador shelf, Baffin Bay, and Davis Strait, *in* Keen, M.J., and Williams, G.L., eds., Geology of the Continental Margin of Eastern Canada: Geological Society of America, Geology of North America, v. I-1, p. 293–348.

Barber, D.E., Schwartz, R.K., Weislogel, A.L., Schricker, L., and Thomas, R.C., 2013, Implications for tectonic control on paleogeography and sediment dispersal pathway: Integrated U-Pb detrital zircon age-analysis of the Paleogene Missouri River headwater system, SW Montana: American Association of Petroleum Geologists, Search and Discovery Article #50863.

Barnosky, A.D., Bibi, F., Hopkins, S.S.B., and Nichols, R., 2007, Biostratigraphy and magnetostratigraphy of the mid-Miocene Railroad Canyon Sequence, Montana and Idaho, and age of the mid-Tertiary unconformity west of the continental divide: Journal of Vertebrate Paleontology, v. 27, p. 204–224, doi:10.1671/0272-4634(2007)27[204:BAMOTM]2.0.CO;2.

Bausch, W.G., 2013, Petrology, geochemistry, and age of exrtension in the Lost Trail Pass dike swarm, southwest Montana [unpublished MS thesis]: University of Montana.

Beranek, L.P., Link, P.K., and Fanning, C.M., 2006, Miocene to Holocene landscape evolution of the western Snake River Plain region, Idaho: Using the SHRIMP detrital zircon record to track eastward migration of the Yellowstone hotspot: Geological Society of America Bulletin, v. 118, p. 1027–1050, doi:10.1130/B25896.1.

Chetel, L.M., Janecke, S.U., Carroll, A.R., Beard, B.L., Johnson, C.M., and Singer, B.S., 2011, Paleogeographic Reconstruction of the Eocene Idaho River, North American Cordillera: Geological Society of America Bulletin, v. 123, p. 71–88, doi:10.1130/B30213.1.

Coney, P.J., and Harms, T.A., 1984, Cordilleran metamorphic core complexes: Cenozoic extensional relics of Mesozoic compression: Geology, v. 12, p. 550–555, doi:10.1130/0091-7613(1984)12<550:CMCCCE>2.0.CO;2.

Constenius, K., 1996, Late Paleogene extensional collapse of the Cordilleran foreland fold and thrust belt: Geological Society of America Bulletin, v. 108, p. 20–39, doi:10.1130/0016-7606(1996)108<0020:LPECOT>2.3.CO;2.

DeCelles, P.G., 2004, Late Jurassic to Eocene evolution of the Cordilleran thrust belt and foreland basin system, western USA: American Journal of Science, v. 304, p. 105–168, doi:10.2475/ajs.304.2.105.

Fields, R.W., Rasmussen, D.L., Nichols, R., and Tabrum, A.R., 1985, Cenozoic rocks of the intermontane basins of western Montana and eastern Idaho, *in* Flores, R.M., and Kaplan, S.S., eds., Cenozoic Paleogeography of the West-Central United States: Denver, Colorado, Rocky Mountain Section of the Society of Economic Paleontologists and Mineralogists, p. 9–36.

Foster, D.A., Doughty, P.T., Kalakay, T.J., Fanning, C.M., Coyner, S., Grice, W.C., and Vogl, J.J., 2007, Kinematics and timing of exhumation of Eocene metamorphic core complexes along the Lewis and Clark fault zone, northern Rocky Mountains, USA, *in* Till, A., Roeske, S., Sample, J., and Foster, D.A., eds., Exhumation along Major Continental Strike-Slip Systems: Geological Society of America Special Paper 434, p. 205–229, doi:10.1130/2007.2343(10).

Fritz, W.J., Sears, J.W., McDowell, R.J., and Wampler, J.M., 2007, Cenozoic volcanic rocks of SW Montana, *in* Thomas, R.C., and Gibson R., eds., Geologic History of Southwest Montana: Northwest Geology, v. 36, p. 91–110.

Gehrels, G.E., Blakey, R., Karlstrom, K.E., Timmons, J.M., Dickinson, B., and Pecha, M., 2011, Detrital zircon U-Pb geochronology of Paleozoic strata in the Grand Canyon, Arizona: Lithosphere, v. 3, no. 3, p. 183–200, doi:10.1130/L121.1.

Henry, C.D., Hinz, N.H., Faulds, J.E., Colgan, J.P., John, D.A., Brooks, E.R., Cassel, E.J., Garside, L.J., Davis, D.A., and Castor, S.B., 2012, Eocene–Early Miocene paleotopography of the Sierra Nevada–Great Basin–Nevadaplano based on widespread ash-flow tuffs and paleovalleys: Geosphere, v. 8, p. 1–27, doi:10.1130/GES00727.1.

Hildebrand, R.S., 2009, Did Westward Subduction Cause Cretaceous-Tertiary Orogeny in the North America Cordillera?: Geological Society of America Special Paper 457, 71 p., doi:10.1130/2009.2457.

Howard, A.D., 1958, Drainage evolution in northeastern Montana and northwestern North Dakota: Geological Society of America Bulletin, v. 69, p. 575–588, doi:10.1130/0016-7606(1958)69[575:DEINMA]2.0.CO;2.

Janecke, S.U., 1994, Sedimentation and paleogeography of an Eocene to Oligocene rift zone, Idaho and Montana: Geological Society of America Bulletin, v. 106, p. 1083–1095, doi:10.1130/0016-7606(1994)106<1083:SAPOAE>2.3.CO;2.

Landon, S.C., and Thomas, R.C., 1999, Provenance of gravel deposits in the mid-Miocene Ruby Graben, southwest Montana: Geological Society of America Abstracts with Programs, v. 37, no. 7, p. 292.

Leckie, D.A., Bednarski, J.M., and Young, H.R., 2004, Depositional and tectonic setting of the Miocene Wood Mountain Formation, southern Saskatchewan: Canadian Journal of Earth Sciences, v. 41, p. 1319–1328, doi:10.1139/e04-076.

Lewis, R.S., Link, P.K., Stanford, L.R., and Long, S.P., 2012, Geologic Map of Idaho: Idaho Geological Survey, Map M-9, scale 1:750,000.

Link, P.K., Fanning, C.M., and Stroup, C.S., 2008, Detrital zircon U-Pb geochronologic data for selected Cretaceous, Paleogene, Neogene and Holocene sandstones and river sands in southwest Montana and east-central Idaho: Montana Bureau of Mines and Geology Open-File Report MBMG-569.

Lonn, J.D., Skipp, B., Ruppel, E.T., Janecke, S.U., Perry, Jr., W.J., Sears, J.W., Bartholomew, M.J., Stickney, M.C., Fritz, W.J., Hurlow, H.A., and Thomas, R.C., 2000, Geologic Map of the Lima 30′ by 60′ Quadrangle, Southwest Montana: Montana Bureau of Mines and Geology-Open File 408, scale 1:100,000.

M'Gonigle, J.W., and Dalrymple, G.B., 1996, ^{40}Ar/^{39}Ar ages of some Challis Volcanic Group rocks and the initiation of Tertiary sedimentary basins in southwestern Montana: U.S. Geological Survey Bulletin 2132, 17 p.

Mix, H.T., Mulch, A., Kent-Corson, M.L., and Chamberlain, C.P., 2011, Cenozoic migration of topography in the North American Cordillera: Geology, v. 39, p. 87–90, doi:10.1130/G31450.1.

Morgan, L.A., and McIntosh, W.C., 2005, Timing and development of the Heise volcanic field, Snake River Plain, Idaho, western USA: Geological Society of America Bulletin, v. 117, p. 288–306, doi:10.1130/B25519.1.

Pierce, K.L., and Morgan, L.A., 1992, The track of the Yellowstone Hotspot: Volcanism, faulting, and uplift, *in* Link, P.K., Kuntz, M.A., and Piatt, L.B., eds., Regional Geology of Eastern Idaho and Western Wyoming: Geological Society of America Memoir 179, p. 1–54, doi:10.1130/MEM179-p1.

Rothfuss, J.L., Lielke, K., and Weislogel, A.L., 2012, Application of detrital zircon provenance in paleogeographic reconstruction of an intermontane basin system, Paleogene Renova Formation, southwest Montana, *in* Rasbury, E.T., Heming, S.R., and Riggs, N.R., eds., Mineralogical and Geochemical Approaches to Provenance: Geological Society of America Special Paper 487, p. 63–95, doi:10.1130/2012.2487(04).

Sears, J.W., 2007, Middle Miocene karst and debris flows and neotectonic faults near Clark Canyon Reservoir, Southwest Montana, *in* Thomas, R.C., and Gibson R., eds., Geologic History of Southwest Montana: Northwest Geology, v. 36, p. 251–260.

Sears, J.W., 2013a, Late Oligocene-Early Miocene Grand Canyon: A Canadian Connection?: GSA Today, v. 23, no. 11, p. 4–10, doi:10.1130/GSATG178A.1.

Sears, J.W., 2013b, A big river ran through it: Reconstructing the Miocene drainage of SE Idaho and SW Montana: Northwest Geology, v. 42, p. 325–336.

Sears, J.W., and Fritz, W.J., 1998, Cenozoic tilt-domains in southwest Montana: Interference among three generations of extensional fault systems, *in* Faulds, J.E., and Stewart, J.H., eds., Accommodation Zones and Transfer Zones: The Regional Segmentation of the Basin and Range Province: Geological Society of America Special Paper 323, p. 241–248, doi:10.1130/0-8137-2323-X.241.

Sears, J.W., and Ryan, P., 2003, Cenozoic evolution of the Montana Cordillera: Evidence from paleovalleys, *in* Raynolds, R., and Flores, J., eds., Cenozoic Paleogeography of Western US: Denver, Colorado, Rocky Mountain Section, Society of Exploration Paleontologists and Mineralogists, p. 289–301.

Sears, J.W., and Thomas, R.C., 2007, Extraordinary Middle Miocene crustal disturbance in southwestern Montana: Birth of the Yellowstone hot spot? *in* Thomas, R.C., and Gibson, R., eds., Geologic History of Southwest Montana: Northwest Geology, v. 36, p. 133–142.

Sears, J.W., Hendrix, M.S., Thomas, R.C., and Fritz, W.J., 2009, Stratigraphic record of the Yellowstone hotspot track: Neogene Sixmile Creek Formation grabens, Southwest Montana: Journal of Volcanology and Geothermal Research, v. 188, p. 250–259, doi:10.1016/j.jvolgeores.2009.08.017.

Skipp, B., Prostka, H.J., and Schleicher, D.L., 1979, Preliminary Geologic Map of the Edie Ranch Quadrangle, Clark County, Idaho, and Beaverhead County, Montana: U.S. Geological Survey Open-File Report 79-845.

Stickney, M.C., 2007, Historic earthquakes and seismicity in southwestern Montana: Northwest Geology, v. 36, p. 167–186.

Stickney, M.C., and Bartholomew, M.J., 1987, Seismicity and Late Quaternary faulting of the northern Basin and Range Province, Montana and Idaho: Bulletin of the Seismological Society of America, v. 77, p. 1602–1625.

Stickney, M.C., and Lageson, D.R., 2002, Seismotectonics of the 20 August 1999 Red Rock Valley, Montana, earthquake: Bulletin of the Seismological Society of America, v. 92, p. 2449–2464, doi:10.1785/0120010297.

Stroup, C.N., 2008, Provenance of Cenozoic continental sandstones in southwest Montana: Evidence from detrital zircon [M.S. thesis]: Pocatello, Idaho, Idaho State University, 117 p.

Thompson, G.R., Fields, R.W., and Alt, D., 1982, Land-based evidence for Tertiary climatic variations: Northern Rockies: Geology, v. 10, p. 413–417, doi:10.1130/0091-7613(1982)10<413:LEFTCV>2.0.CO;2.

Zachos, J., Pagani, M., Sloan, L., Thomas, E., and Billups, K., 2001, Trends, rhythms, and aberrations in global climate 65 Ma to present: Science, v. 292, p. 686–693, doi:10.1126/science.1059412.

Manuscript Accepted by the Society 24 February 2014

Printed in the USA

The Geological Society of America
Field Guide 37
2014

A slice through time: A Hyalite Canyon soil lithosequence

J.C. Sugden*
A.S. Hartshorn*
Department of Land Resources & Environmental Sciences, 334 Leon Johnson Hall, Montana State University, Bozeman, Montana 59717-3120, USA

J.L. Dixon*
Department of Earth Sciences, 112 Traphagen Hall, Montana State University, Bozeman, Montana 59717-3480, USA

C. Montagne*
Department of Land Resources & Environmental Sciences, 334 Leon Johnson Hall, Montana State University, Bozeman, Montana 59717-3120, USA

ABSTRACT

On this field trip, participants will get their hands dirty while characterizing soils formed on five different rock types: Archean Gneiss, Flathead Sandstone, Wolsey Shale, Meagher Limestone, and Absaroka Volcanics (a basaltic andesite rock). We first recap prior soil survey efforts across the Gallatin National Forest in southwestern Montana and introduce a state factor approach to understanding soils. For over 50 years, Montana State University faculty have explored parts of this lithosequence, using it as a natural laboratory for thousands of students. We continue this tradition with this field guide, emphasizing how the combination of field and laboratory data can enrich our understanding of soil processes. We will observe and measure striking differences in soils; these differences in physical and chemical properties, from textures and colors to pH and elemental composition, are discussed in the context of quantifying the influence of the underlying rock on soil properties. We use these differences to ask whether heterogeneity in soil properties justifies the inference that soil properties are dominated by the underlying lithology. We conclude that the underlying rock strongly influences soil properties, but in variable ways across this lithosequence. This influence is both direct and indirect: chemical weathering of the rock leads to compositional changes in overlying soil, but rock weathering also leads to coarse fragments in the soil profile, which alters soil hydrology.

*E-mails: john.sugden@msu.montana.edu; anthony.hartshorn@montana.edu; jean.dixon@montana.edu; montagne@montana.edu.

Sugden, J.C., Hartshorn, A.S., Dixon, J.L., and Montagne, C., 2014, A slice through time: A Hyalite Canyon soil lithosequence, *in* Shaw, C.A., and Tikoff, B., eds., Exploring the Northern Rocky Mountains: Geological Society of America Field Guide 37, p. 101–114, doi:10.1130/2014.0037(05). For permission to copy, contact editing@geosociety.org.

INTRODUCTION

This field trip will examine the role of parent material in soil formation in Hyalite Canyon. Just ~12 miles south of the Montana State University (MSU) campus in Bozeman, a sequence of five stratigraphic units will enable participants to appreciate the multiple roles lithology can play in soil development. The field trip learning outcomes include:

- how to characterize a soil profile,
- an introduction to geochemical fingerprinting of soils, and
- an understanding of the state factor and lithosequence approaches.

First we review historical MSU educational and research activities in Hyalite Canyon. Then we introduce the state factor approach (Jenny, 1941), and apply it across a lithosequence we will recharacterize as part of this field trip. We conclude with a discussion of future opportunities.

Hyalite Canyon as an MSU Field Trip Destination

Our field trip destination has roots in early efforts at MSU to take advantage of the natural laboratory represented by the Gallatin National Forest in general, and sites near Langohr Campground in Hyalite Canyon in particular (Figs. 1 and 2).

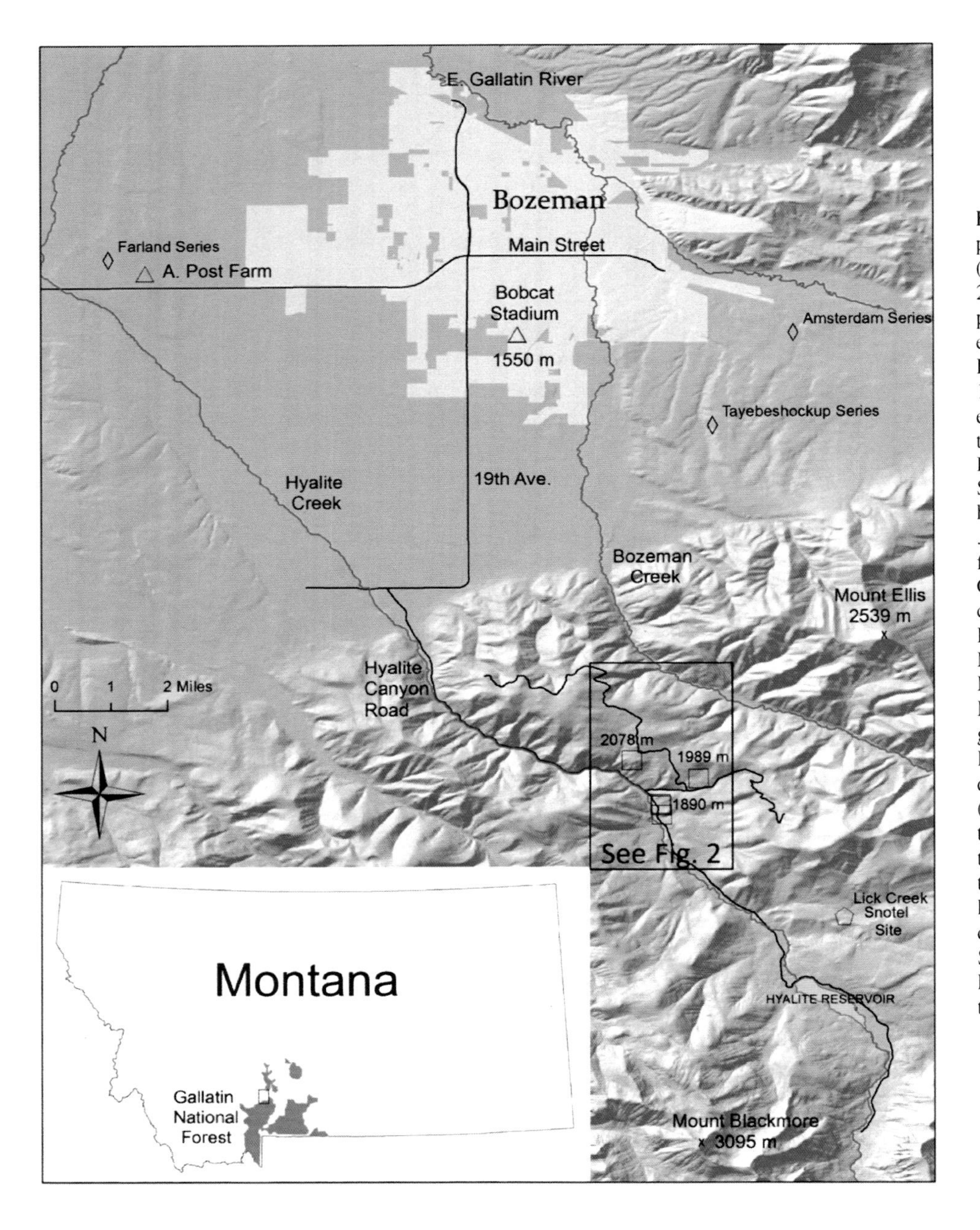

Figure 1. Hyalite Canyon lithosequence project area. Area of Figure 2 is inset. (Data from (1) U.S. Geological Survey, 2013, The national map viewer and data platform, http://nationalmap.gov/viewer.html [data used: National Elevation Datasets 46w111 and 46w112; accessed 15 November 2013]. (2) National Cooperative Soil Survey, National Cooperative Soil Survey Soil Characterization Data: Natural Resources Conservation Service, U.S. Department of Agriculture, http://ncsslabdatamart.sc.egov.usda.gov/ [data used: location information for Amsterdam and Farland Series, Lick Creek and Tayebeshockup soil pits; accessed December 2013]. (3) Montana Bureau of Mines and Geology, 2013, MBMG geologic research/mapping: Montana Tech of the University of Montana, www.mbmg.mtech.edu/gmr/gmr-statemap.asp#quadmaps [data used: Bozeman and Livingston 1:100,000 quads; accessed 2 December 2013]. (4) Montana Natural Resource Information System, Montana GIS portal: Montana's primary catalog of GIS data: Montana State Library, State of Montana, http://gisportal.msl.mt.gov/geoportal/catalog/main/home.page [data used: State of Montana Boundary, National Forests and Ranger Districts in Montana; accessed February 2014].

The Gallatin National Forest is a ~2 million-acre forest established in 1899 that stretches from just south of Bozeman to the northwestern corner of Yellowstone National Park and includes portions of seven mountain ranges. In 1998, a memorandum of understanding between MSU and the U.S. Forest Service (USFS, 1998; www2.montana.edu/policy/forest_service_use/mou.html) was developed to facilitate student and faculty engagement across the Gallatin and Custer National Forests. This area lies within the Montana metasedimentary subprovince of the Wyoming Province, and provides extensive exposures of Archean rocks uplifted during the Laramide Orogeny (Mueller and Frost, 2006).

More than 50 years ago, John Montagne, a faculty member at MSU hired by the Department of Earth Sciences in 1957, began exploring Hyalite Canyon as a field trip destination. John's teaching assignments included introductory geology, geomorphology, glacial geology, structural geology, and field geology, and he quickly recognized the value of Hyalite Canyon for pedagogic purposes. Together with his earth science colleagues, John developed teaching materials including an idealized stratigraphic column stretching from Archean gneiss to the Absaroka Volcanic caprock and younger strata (detail in Fig. 3). Although John passed away in 2008, he conveyed his passion for the site to his son Cliff at an early age. Cliff

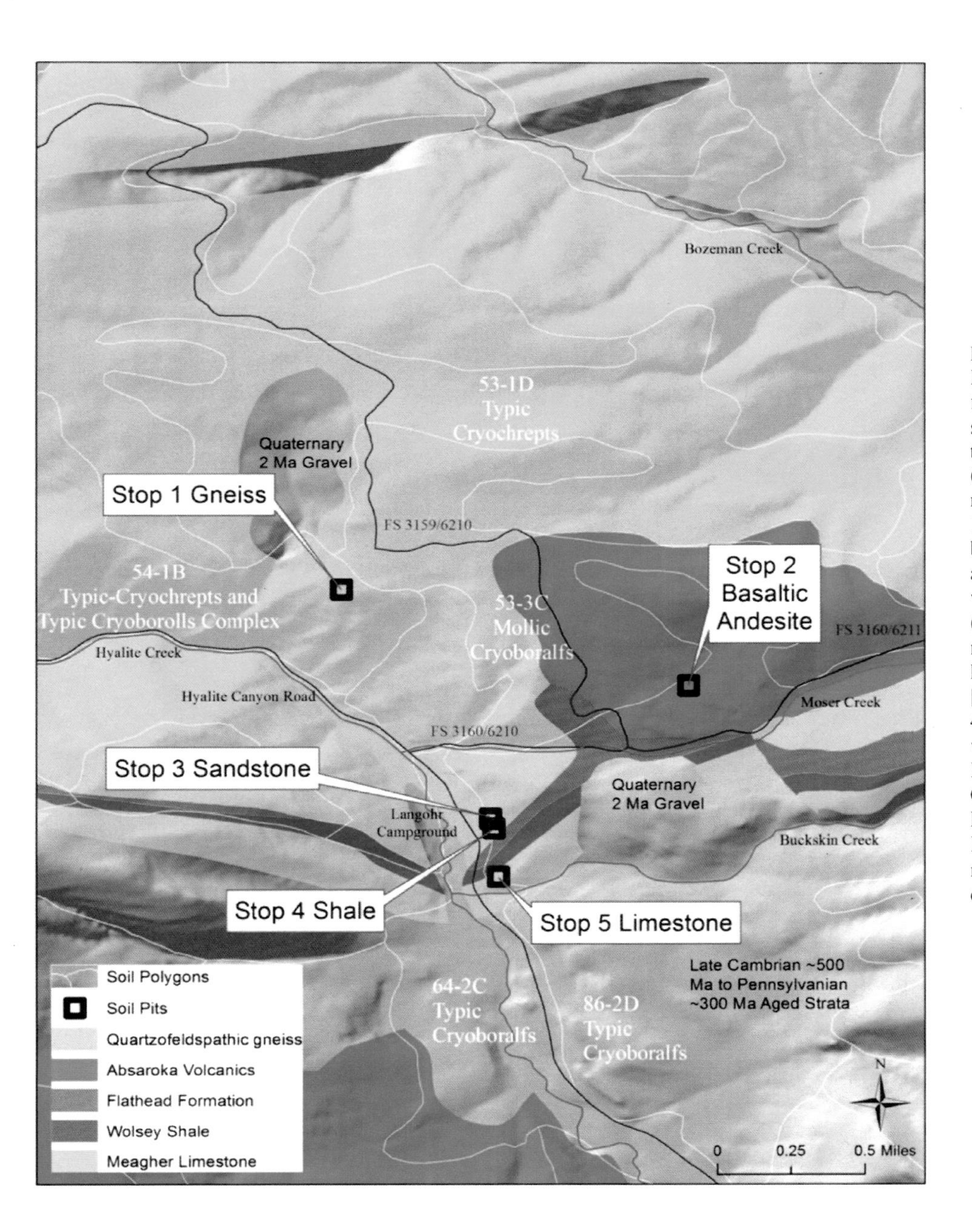

Figure 2. Soil and geologic map of the Hyalite Canyon lithosequence (soil profile locations are represented with bolded squares). Soil map units (polygons) from the 1996 survey are outlined in white (see text); geologic map units (colored regions; 1:100,000) are presented at 1:24,000 to match site details. As a result, boundaries between geologic map units appear to be shifted slightly north and west of their true locations. (Data from (1) U.S. Geological Survey, 2013, The national map viewer and data platform, http://nationalmap.gov/viewer.html [data used: National Elevation Datasets 46w111 and 46w112; accessed 15 November 2013]. (2) Gallatin County GIS Department, data available for download: Gallatin County, www.gallatin.mt.gov/Public_Documents/gallatincomt_gis/Data%20Download%20Page [data used: roads, soils, lakes and waterways; accessed 10 November 2013].)

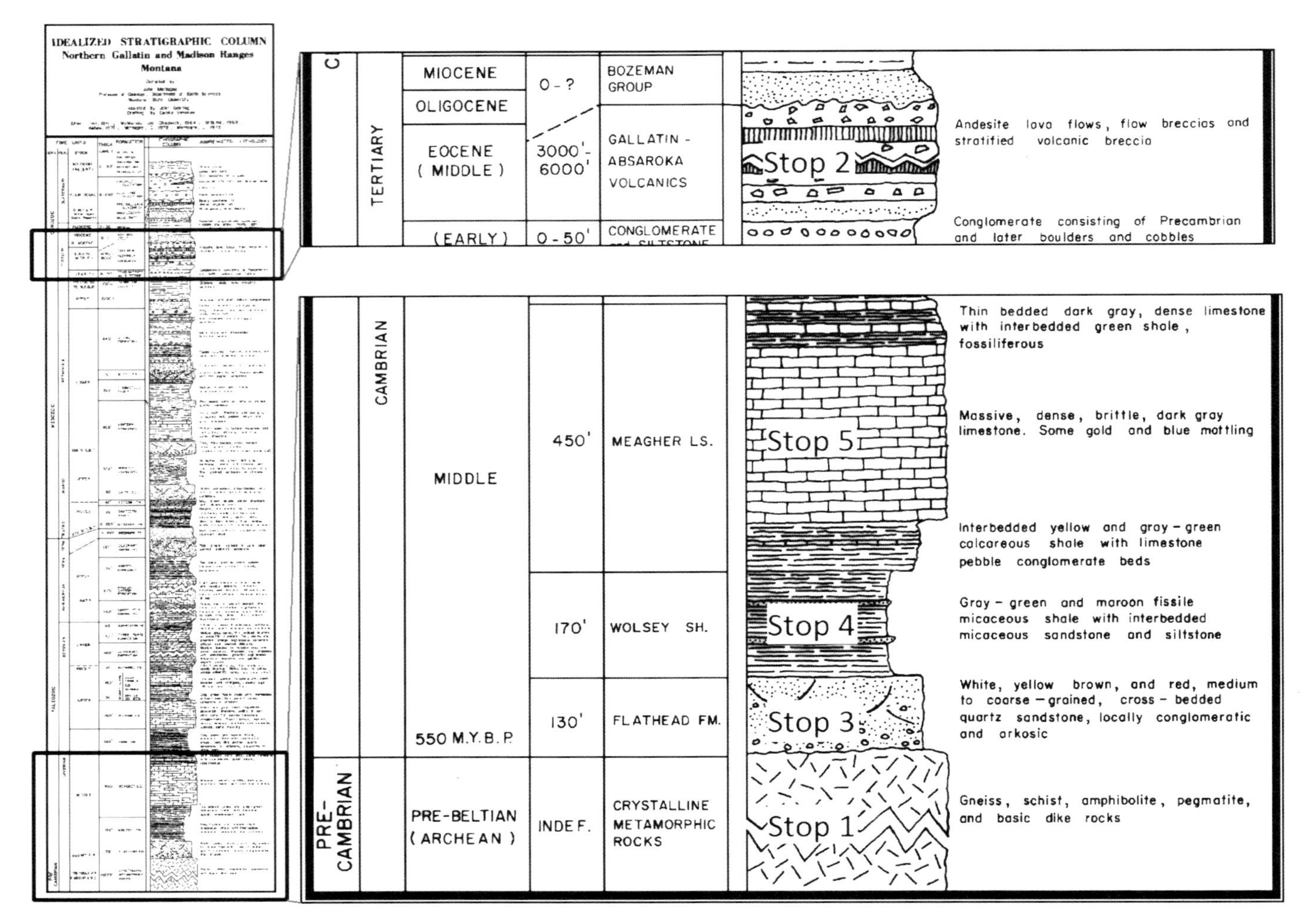

Figure 3. Detail from an idealized stratigraphic column for the northern Gallatin Range (Montagne, 1975; full version at www.montana.edu/wwwes/programs/Images/localstrat.pdf). The five stops on this field trip are indicated; for logistical reasons, we will visit the gneiss soil first, volcanic soil second, sandstone soil third, shale soil fourth, and limestone soil last, out of stratigraphic sequence.

accepted a teaching position at MSU in what was then called the Department of Plant and Soil Science in 1975; Cliff is currently an emeritus professor in what is now the Department of Land Resources and Environmental Sciences.

Cliff Montagne refined his father's field trip materials, and identified, characterized, and built the basis of this field trip for his introductory soils and pedology courses. An abridged version of this field trip still forms a core learning experience today, almost 40 years after it was first developed. (Decades of Langohr field trips have carved clearly visible trails in satellite imagery zoomed and centered on 45° 32′ [N], 111° 0.81′ [W].) Together with Jerry Nielsen and Larry Munn, faculty colleagues, Cliff has studied the influence of bedrock lithology, bedrock structure, and geomorphic process on landform, vegetation, and soil properties in this area (e.g., Montagne and Munn, 1980). How fire might interact with lithology to influence post-fire sediment yields remains unclear, but we will discuss recent Gallatin National Forest–specific findings quantifying forest vulnerability to climate change (Louie, 2013) as well as ongoing research into post-fire responses of volcanic- versus gneiss-dominated watersheds burned in August 2012 five miles west of this lithosequence. The City of Bozeman also has an interest in mitigating effects of future fires (and any subsequent sediment impairment of water quality) near the Hyalite Reservoir (optional roll-by Stop 6) because this reservoir supplies a large fraction of Bozeman's drinking water. Both the Earth Science and the Land Resources and Environmental Sciences Departments continue to bring students to Hyalite to bolster interest in—and expand our understanding of—soil-landscape relationships.

Prior Soil Survey Efforts across the Gallatin National Forest

Gallatin National Forest soils have been surveyed (Montagne and Munn, 1980; Davis and Shovic, 1996) and we summarize those findings here. In the mid-1970s, the USFS inventoried

TABLE 1. SELECTED PHYSICAL AND CHEMICAL PROPERTIES OF SELECTED SOIL HORIZONS IN THE GALLATIN NATIONAL FOREST (MONTAGNE AND MUNN, 1980: TABLE 49)

Geological group	Horizon	% clay	% plant-available water	pH	% base cation saturation	% organic matter
Granitic	A	15	11	5.9	47	2.6
	C	12	6	6.4	66	0.8
Sandstone	A	18	12	5.7	41	2.7
	C	–	–	–	–	–
Shale	A	30	11	6.1	58	4.7
	C	30	6	7.6	100*	1.3
Limestone	A	25	13	7.0	81	6.3
	C	18	11	8.1	100*	0.5
Volcanic	A	17	15	6.0	51	3.6
	C	39	14	6.6	12	0.5

Note: Samples represent <2 mm material. All values are averages of geological groups; groups are presented in inverted stratigraphic order.
*Reported as >100%.

soils across the Gallatin National Forest by collecting and characterizing samples from 98 soil profiles for a range of properties including coarse fragments (>2 mm), Munsell color (hue, value, chroma), texture (proportion sand, silt, clay [particles <2 µm]; by hydrometer), and structure (grade, size, type: "moderate medium subangular blocky"). Soil survey analyses also included additional physical (e.g., laboratory determination of plant-available water: -33 kPa > ψ > -1500 kPa) and chemical properties (e.g., pH, Bray phosphorus, base cation saturation, organic matter, and electrical conductivity [EC]).

These soil survey results were analyzed and relationships between soil properties and eight geological groups were summarized. Table 1 presents results for the five groups that most closely correspond to those we will visit on this field trip: A horizon clay content ranged from <15% in granitic soils to >30% in shale-derived soils; A horizon pH ranged from 5.7 (sandstone-derived soils) to 7.0 (limestone-derived soil); and A horizon organic matter ranged from 2.6% (granitic rock-derived soils) to 6.3% (limestone-derived soils; Table 1).

The five soils we will visit are mapped as components of three soil map units (Fig. 2; gneiss: 53-1D; sandstone/shale/limestone: 64-2C; and volcanic: 86-2D). Each map unit is composed of subgroup taxonomic classifications as well as unclassified areas mapped as rock. For example, the gneiss soil (Stop 1; Fig. 2) map unit is comprised of a complex of two subgroups, Typic Cryochrepts and Typic Cryoborolls; the sandstone/shale/limestone profiles (Stops 3–5) are mapped as two subgroups, Typic Cryoboralfs and Argic Cryoborolls, associated with floodplains and terraces; and the volcanic soil (Stop 2) map unit includes two subgroups, Typic Cryoboralfs and Mollic Cryoboralfs. While these classifications are no longer current, we have listed them here to match current online soil maps of this area (http://websoilsurvey.sc.egov.usda.gov/App/HomePage.htm or http://casoilresource.lawr.ucdavis.edu/gmap/). Our five soils (gneiss-sandstone-shale-limestone-volcanic) would currently be classified to the subgroup level as Typic Dystrocryepts, Typic Haplocryepts, Aquic Haplocryolls, Lithic Haplocryolls, and Typic Haplocryolls, respectively.

For those unfamiliar with hierarchical U.S. soil taxonomic nomenclature, we use two of these current soil classifications to provide examples of how soil properties are used to classify soils (Table 2). We echo recent calls to include soil taxonomic information across a broad spectrum of studies (e.g., Schimel and Chadwick, 2013). Subgroup classifications represent an intermediate level of detail in this taxonomic system, which includes orders, suborders, great groups, subgroups, families, and series. The series level, while the most restrictive taxonomic designation, is unlike the other classification levels because there is no informational value in a soil series name. One cannot tell if the soil mapped as a "Blackmore" under the MSU Strand Union

TABLE 2. CURRENT TAXONOMIC BREAKDOWN OF TWO LITHOSEQUENCE SOILS TO ILLUSTRATE HOW FORMATIVE ELEMENTS PROPAGATE INTO LOWER ORDER (e.g., SUBGROUP) CLASSIFICATIONS

Soil	Taxonomic classification	Taxonomic level	Formative element/explanation
Stop 3: Sandstone	Inc**ept**isol	Order	ept/Weakly developed soil
	Cry**ept**	Suborder	cry/Cold soil temperature regime
	*Haplo*cry**ept**	Great group	haplo/Relatively undistinguished; a *Dystro*cryept, by contrast, would refer to a soil with <50% base cation saturation
	Typic *Haplo*cry**ept**	Subgroup	typic/Relatively undistinguished at the subgroup level
Stop 4: Shale	M**oll**isol	Order	oll/Well-developed soil with a thick, dark surface horizon and generally base-cation rich and fertile
	Cry**oll**	Suborder	cry/Cold soil temperature regime
	*Haplo*cry**oll**	Great group	haplo/Relatively undistinguished; an *Argi*cryoll, by contrast, would imply the presence of translocated (illuvial) clay in a subsurface, argillic horizon
	Aquic *Haplo*cry**oll**	Subgroup	aquic/A wetter soil with reducing conditions

Note: Explanations are based on *Keys to Soil Taxonomy* (Soil Survey Staff, 2010).

Building, for example, is an Inceptisol or a Mollisol, although online tools (e.g., Soil Survey Division, Official Series Descriptions: https://soilseries.sc.egov.usda.gov/osdname.asp; accessed February 2014) enable the rapid decoding of any series name.

The Gallatin National Forest survey results (Table 1) were drawn from soils collected over millions of acres, multiple mountain ranges, and at least 10 landforms, whereas this field trip explores soils across a single ~2-mile (~3-km) transect (Fig. 2). Nevertheless, these results defined relationships between lithology and soil properties, in much the same way as has been done for the state of Montana (Veseth and Montagne, 1980). Soils over gneiss or sandstone are generally acidic and coarse-textured, whereas those formed over limestone, shale, or volcanics are generally less acidic and more fine-textured. Slope stability is influenced by these lithology-texture relationships (Montagne, 1976). For example, gneiss-derived soils weather to quartz-rich matrices with many coarse fragments, macropores, high infiltration, and limited plant-available water. Gneiss can also contribute high amounts of mica to the soil, leading to poor engineering properties and slope failure (Montagne and Munn, 1980). Shale soils are also often prone to slope failure as well as road or other construction problems, and most landslides in the Gallatin National Forest have been associated with shale or shale-derived, clay-rich soils. We may explore a slope failure (optional roll-by Stop 7) that employs hollow-bore soil nails for stabilization purposes (Lundgreen, 2013).

State Factor Approach

A state factor approach (Jenny, 1941) provides a strong foundation to explore soil processes. Five state factors clarify influences on soil development and soil processes: climate (**cl**), organisms (**o**), relief (hillslope or topographic position, **r**), parent material (**p**), and duration of soil-forming processes (commonly shortened to time [**t**]). "Clorpt" factors account for the diversity of soils on many scales. The soils beneath the MSU campus, for example, have a semiarid climate (cl), grassland vegetation (o), occupy a gently north-sloping interfluve (r), are built of alluvium and eolian inputs (p), and have been developing for thousands of years (t).

The field trip and this field guide consider how soil properties differ when only one factor is altered: parent material. Soils formed across parent material gradients are called lithosequences, across which soil properties can be measured and soil processes inferred. Although the importance of parent material to soils has been studied for more than a century (Dokuchaev, 1879 *in* Heckman and Rasmussen, 2011), there are relatively few "pure lithosequence studies in the literature" (Heckman and Rasmussen, 2011, p. 99). For example, an 8-fold difference in clay, 60-fold difference in acidity, and 6-fold increase in extractable aluminum were reported across a rhyolite-granite-basalt-dolostone lithosequence in Arizona (Heckman and Rasmussen, 2011). As another example, a 3-fold difference in clay, 100-fold difference in acidity, and 30-fold difference in extractable aluminum were reported for a granite-marble-dacite lithosequence in Greece (Yassoglou et al., 1969).

It is challenging, however, to identify locations where only one state factor is varied across space. As one example, climate (cl) is affected by hillslope position (r), which influences soil residence times (t). As a second example, plant communities (o) often reflect underlying parent material (p). Thus, while differences in lithosequence soils perhaps should not be unambiguously attributed to only varying parent materials when plant communities also differ (Jenny, 1941; Jenny, 1958; Druce, 1959), underlying lithologic differences influence plant communities indirectly through effects on soil texture and nutrient availability.

Langohr Campground Lithosequence: Climate, Relief, and Time Factors

Long-term climate data for our lithosequence can be inferred from the Lick Creek SNOTEL site (3 miles [5 km] southeast of Stop 3; 2090 m [6855 ft]; north northwest aspect; U.S. Department of Agriculture, Natural Resources Conservation Service, Lick Creek Snotel Site #578, no date; http://www.wcc.nrcs.usda.gov/nwcc/site?sitenum=578&state=mt [accessed February 2014]). Thirty-year (1983–2013) mean annual precipitation is 76 cm (30 in) and mean annual air temperature is –1.1 °C. From 2008 to 2013, volumetric soil moisture at 7.5 cm averaged 18%; at 100 cm, soil moisture averaged 46%. Over the same time period, soil temperatures at 7.5 cm depth averaged 3.9 °C and at 100 cm, 4.5 °C, suggesting a cryic soil temperature regime.

All five soils on today's field trip have similar aspects (west or northwest), elevations (1900–2100 m; 6200–6880 ft), and slopes (15–20%). These elements of the climate and relief factors are important, because in this area aspect and elevation interact to control plant community composition (Weaver and Perry, 1978). Although these topographic factors are relatively similar, upslope contributing areas could vary, influencing soil processes. While soil ages are not precisely known, advances by Wisconsinan Hyalite Canyon glaciers are unlikely to have reached these soils. The Bull Lake Glaciation may have extended to within 1 mile of the limestone soil (Stop 5; ~150–140 thousand years [ka] ago; Pierce, 2003) according to a map by Weber (1965). These five soils were not glaciated during the Pinedale Glaciation either (~21 ka; Pierce, 2003; Weber, 1965). Long-term denudation rates from Idaho (Kirchner et al., 2001) provide a means of estimating soil residence times; these catchment-averaged, ^{10}Be rates ranged from 55 to 327 Mg km^{-2} y^{-1}, with a median flux of 104 Mg km^{-2} y^{-1} (104 g m^{-2} y^{-1}). Assuming a rock density of 2.6 Mg m^{-3} enables the conversion of a denudation rate of 100 Mg km^{-2} y^{-1} to a volumetric flux of 3.8×10^{-8} km^3 km^{-2} y^{-1}, or a landscape lowering rate of 0.04 mm y^{-1}. The quotient of the average soil thickness (average depth to base of C horizons [Table 3]: 74 cm) and this estimated overall lowering rate (0.04 mm y^{-1}) yields an estimated soil residence time of ~19 ka.

Langohr Campground Lithosequence: Biota and Parent Material Factors

This compositionally broad set of rocks has been expressed as part of a stratigraphic sequence (Fig. 3). Like much of southwest Montana, this area is underlain by ~2750 million-year (Ma) Archean gneiss (Stop 1; James and Hedge, 1980). At the north end of Hyalite Canyon, this gneiss forms steep, sandy, scree-filled hillsides. Further south are three Cambrian (550–495 Ma) sedimentary rocks characteristic of shallow marine environments (Lebauer, 1964). The oldest is the Flathead sandstone (550 Ma), which uncomformably overlies Archean gneiss (and whose erosional remnants may have formed this sandstone [Middleton, 1980]). The Flathead sandstone (quartzite) is associated with sandy soils (Stop 3) that support forests dominated by lodgepole pine (*Pinus contorta*) and Douglas fir (*Pseudotsuga menziesii*).

Wolsey shale conformably lies on the Flathead sandstone and was formed as the Cambrian oceanic shoreline moved from west to east (Lebauer, 1964). The soil over Wolsey shale (Stop 4) is clay-rich with nonexpanding illite (Lebauer, 1964). This shale soil also contains sandstone colluvium from upslope and supports meadow-like grasslands containing timothy grass (*Phleum pratense*) and common snowberry (*Symphoricarpos albus*). Just a bit further south, Meagher limestone (Stop 5) lies conformably over Wolsey shale creating thin, fine-grained soils supporting mixed grasslands with islands of Douglas fir. Finally, Tertiary-aged Absaroka volcanics (basaltic andesite) lie unconformably on gneiss and sandstone ~1 mile (1.6 km) northeast of Langohr Campground (Stop 2). Absaroka volcanics in the northern Gallatin Range are generally potassium-poor (<~3% K_2O; Table 3) and composed of layered, andesitic flows interspersed with breccia (Chadwick, 1970). A pre-Wisconsin glacier may have moved these volcanic rocks underlying the Stop 2 soil profile to their current position (Weber, 1965). Their formation is attributed to lithospheric extension and resulting decompressional melting; rock exposures associated with the Tertiary are found throughout local mountain ranges (Feeley et al., 2002). Near Langohr Campground, fine-grained soils on this rock support lodgepole pine forests and aspen (*Populus tremuloides*) groves.

Parent materials can be either residual or transported (e.g., Schaetzl and Anderson, 2005). Residual parent materials underlie a soil and are best visualized as coherent bedrock. Transported parent materials, by contrast, include glacial moraines, alluvium, colluvium, and finer-grained eolian dust (e.g., loess), with the size of materials scaling with the viscosity of the transporting fluid. If that fluid is a glacier, the viscosity of ice enables the transport of large blocks of material. Lower viscosity fluids such as the water in Hyalite Creek can only transport smaller blocks like cobbles. Air is a relatively low viscosity fluid, and so can typically only transport silt-sized particles (2–50 mm). Wind-transported materials originate as either acute events (e.g., the eruptions of Glacier Peak in present-day Washington ~11,000 years ago [Harward and Borchardt, 1972] or Mazama in present-day Oregon ~7600 years ago [Zdanowicz et al., 1999]) or chronic events (continuous dustfall).

Here we consider the potential for eolian inputs to these lithosequence soils. Although there are no short- or long-term records of atmospheric inputs to the Langohr Campground area specifically, this site lies halfway between two National Atmospheric Deposition Program (NADP) sites: one in Yellowstone National Park ~60 miles (100 km) southeast (~1900 m elevation; http://nadp.sws.uiuc.edu/nadpdata/ads.asp?site=WY08) and another in Clancy, ~60 miles (100 km) northwest (~1400 m; http://nadp.sws.uiuc.edu/nadpdata/ads.asp?site=MT07). Wet deposition

TABLE 3. SELECTED FIELD-ESTIMATED PHYSICAL AND CHEMICAL PROPERTIES FOR THE FIVE SOIL PROFILES COMPRISING THE LITHOSEQUENCE, PRESENTED IN INVERTED STRATIGRAPHIC ORDER

Underlying Rock Lat./Long. *Tentative Tax. Subgroup*	Horizon	Lower depth (cm)	% coarse fragments	Moist Munsell color	Texture	% clay	Structure grade, size, type	pH
Gneiss	A/B	10	10	10 YR 2/2	sandy loam	10	2,m,gr	5.2
45.54535N /111.02085W	Bw	29	50	10 YR 3/4	sandy loam	10	1,f,sg	5.7
Typic Dystrocryept	C1	40	60	10 YR 4/4	loamy sand	5	sg	6.1
	C2	75	80	10 YR 6/4	sand	5	ma	5.9
	A	5	20	10 YR 4/4	sandy loam	10	1,m,gr	5.7
Sandstone	Bw1	30	35	10 YR 4/6	sandy loam	10	2,c,sbk	5.5
45.533843N /111.013541W	Bw2	50	35	7.5 YR 4/3	sandy loam	10	2,c,sbk	5.9
Typic Haplocryept	BC	70	40	5 YR 4/3	sand	5	sg	6.1
	C	90	50	5 YR 4/4	sand	5	sg	6.7
Shale	A	28	5	10 YR 3/2	silt loam	20	3,m,sbk	5.8
45.533363N /111.01337W	Bt	74	15	10 YR 5/4	silty clay loam	35	1,tn,pl	5.7
Aquic Haplocryoll	C	83	50	10 YR 5/3	silty clay	50	ma	5.8
Limestone	A1	25	80	10 YR 2/2	clay loam	35	2,f,gr	7.7
45.531488N /111.013577W	A2	35	20	7.5 YR 3/3	clay loam	35	3,m,gr	8.0
Lithic Haplocryoll	C	55	70	7.5 YR 4/4	clay loam	35	3,f,abk	8.2
	A	26	0	10 YR 3/1	clay	50	abk	6.2
Basaltic Andesite	Bt1	38	3	10 YR 4/2	clay loam	35	2,f,abk	6.2
45.530967N /111.01315W	Bt2	49	10	10 YR 4/4	clay loam	35	2,m,abk	6.4
Typic Haplocryoll	C	65	45	10 YR 5/4	clay loam	35	3,m,sbk	6.3

Note: Methods and nomenclature per Schoeneberger et al. (2012).

has been monitored at these sites since 1980 and 1984, respectively. From 2008 to 2012, combined inputs of calcium (Ca), magnesium (Mg), potassium (K), and sodium (Na) averaged 1.5 and 0.8 kg ha^{-1} y^{-1} for the two stations, respectively, with Ca constituting 68% of these inputs. If these average short-term inputs (~1 kg ha^{-1} y^{-1} or ~0.1 g m^{-2} y^{-1}) are representative of long-term inputs (~20 ka; cf. Neff et al., 2008), and we multiply this total by a factor of 6 to account for the fraction of dust comprised of these four elements (this study), this would translate to a time-integrated atmospheric mass flux to these soils of 120 Mg ha^{-1}, or 12 kg m^{-2}. Global annual suspended dust sediment estimates vary 50-fold (from 60 to 3000 Mg yr^{-1}; Tegen, 2003), with reported global dust fluxes (deposition to soils) varying >100-fold (from <1 to >100 g m^{-2} yr^{-1}; Lawrence et al., 2013). Estimated dust fluxes into the southern Rocky Mountains range from 10 to 22 g m^{-2} yr^{-1} (100–220 kg ha^{-1} yr^{-1}; Lawrence et al., 2013), about two orders of magnitude greater than our estimated NADP inputs. This discrepancy could reflect greater fluxes to the southern versus northern Rocky Mountains, uncertainties in the fraction of deposition in the form of other elements (e.g., quartz and feldspar components such as silica [Si], aluminum [Al], and oxygen [O]), or that wet deposition is only a fraction of total deposition. Long-term denudation rates from Idaho (~100 g m^{-2} y^{-1}; Kirchner et al., 2001) help contextualize these eolian inputs, since they are 5- to 10-fold greater than the southern Rocky Mountain eolian inputs.

We are not the first group to consider the potential impacts of eolian inputs on soil properties in the Rocky Mountains. John Retzer (1962, p. 32), for example, noticing silt deposited on snow in Colorado's Fraser Experimental Forest, commented, "It is not known to what degree this dust influences soil development, or to what extent it is responsible for the silt and clay measured in the soil profiles."

Field Trip Stop Descriptions

Mileage	*Directions (coordinates in Table 3)*
0.0	Depart from Strand Union Building (West Grant St. and South 8th Avenue) on the Montana State University campus, Bozeman. Proceed 5.7 miles south to Hyalite Canyon Road turnoff via 19th Avenue, which doglegs west ~4.5 miles south of Bozeman.
5.7	Turn left (south) on Hyalite Canyon Road. Proceed another 5.7 miles to Moser Creek Road (Forest Service [FS] road 6210/3160).
5.7	Turn left uphill (east) on Moser Creek Road.
0.5	Proceed uphill (east) on Moser Creek Road to the intersection with FS 6210/3159.
1.3	Turn left onto FS 6210/3159, proceed 1.3 miles, and park. Stop 1. Gneiss soil profile. *Location:* ~1000 ft (~300 m) west of parking area over a small hill.
1.3	Reverse direction and proceed back to intersection of FS 6210/3159 and FS 6210/3160.
0.2	Turn left uphill (east) onto FS 6211/3160, proceed 0.2 miles, and park. Stop 2. Basaltic andesite soil profile. *Location*: ~1000 ft (~300 m) north of parking area in a meadow.
0.7	Reverse direction to intersection of Moser Creek Road/ FS 6210/3160 and Hyalite Canyon Road.
0.3	Park near dumpsters. Stops 3–5. Sandstone, shale, and limestone soil profiles, respectively. *Locations*: We will first walk to Stop 3, just uphill (~1000 ft; ~300 m) and northeast of the parking area. Stop 4 is ~500 ft (~150 m) south of Stop 3; Stop 5 is ~1000 ft (~300 m) south of Stop 4.
3.9	Following examination of all five soil profiles, there may be an opportunity for two optional stops. The first (roll-by, Stop 6, Hyalite Reservoir) is accessed by continuing south of Hyalite Canyon Road from Langohr Campground.
8.6	Reversing direction from Hyalite Reservoir (north), a second (roll-by Stop 7) stop will be on our right (east): a slope failure.
1.3	Return to campus by proceeding north on Hyalite Canyon Road until its intersection with 19th Avenue.
5.7	Turn right and proceed east on 19th Avenue as it doglegs north, to intersection with Lincoln or Kagy to return to Strand Union Building.

Site 1 (the gneiss-derived soil) is the highest soil pit in the lithosequence; it is furthest from Hyalite Canyon Road on the edge of a plateau dividing Hyalite Creek from Bozeman Creek. On the drive or walk in, look south for views of Mount Blackmore and the Mummy. West of the parking area, ascend a short, steep hill and wander south to a treed summit. The soil pit we will characterize lies just west and a bit north of this summit on the edge of a treed meadow. This profile is poorly developed with very little clay formation and abundant coarse fragments (>50%) starting at just 10 cm; the relatively dark surface horizon is the most acid horizon we will see across the lithosequence.

Site 2 (the volcanic-derived soil) sits on a boundary between a narrow tight meadow and forest close to the intersection of FS 6210 and FS 6211. Soils here are finely textured, hold water, and boggy conditions may be found on the short walk to the soil pit. Relative to the gneiss-derived soil, this soil is much more finely textured and contains far fewer coarse fragments (<10% through 49 cm).

Site 3 (the sandstone-derived soil) is within a stone's throw of Langohr Campground. Cross the road and walk up hill through the meadow and past Stop 4 (shale), heading north just into the trees. Notice how the slopes steepen slightly after leaving the meadow. Relative to the volcanic soil, this sandstone-derived soil shows much coarser textures and strikingly red colors.

Site 4 (shale-derived soil) is just south of the sandstone pit, in the middle of a broad meadow. The slumping conditions, textural contrast, distinct horizonation, and coarse fragments are just some of the features distinguishing the shale pit from its sandstone neighbor to the north. We will discuss the pronounced vegetative and soil differences between the sandstone and shale soils.

To reach site 5 (limestone-derived soil) descend ~50 m to an obvious roadbed on the contour below the shale pit and walk south toward Buckskin Creek (Fig. 2). Observe the vegetative changes that shift from a distinct meadow to a forested meadow. This soil pit is ~30 m above the trails just before entering the Buckskin Creek drainage below a small rock outcrop. A limestone outcrop on the other side of Hyalite Creek is the same limestone that lies beneath this soil pit. This intermediately textured soil (clay loam) has an unusual configuration of coarse fragments with depth.

METHODS

Soils from the lithosequence were sampled and described following standard methods (Schoeneberger et al., 2012). We will have an opportunity to practice many of these characterization techniques on the field trip, including the definition of horizon boundaries, coarse fragments, soil colors, soil textures, soil structure, root densities, pore densities, pH and EC, and effervescence. For the analyses presented below, two liters (less in thin or very rocky horizons) of soil and rock from each horizon were collected. Soil pH and EC were determined using a portable combination pH/EC meter submerged in a 1:10 ratio of soil to deionized water. To support this study, we also characterized four samples to serve as proxies for dust inputs to the lithosequence: (*i*) a relatively unweathered, silt-rich, C-horizon sample of eolian loess from ~1 m depth at the Arthur M. Post Agricultural Experiment Station west of Bozeman ("PF4"; Fig. 1); (*ii*) a grab sample from a thick volcanic ash packet exposed in fall 2013 from Helena, ~100 miles north; (*iii*) a grab sample from the 5th floor roof of the MSU Bobcat Stadium ("BS5"; Fall 2013; Fig. 1); and (*iv*) a grab sample from the 6th floor roof of the same stadium ("BS6"; Fall 2013). Post Farm lies within a transect of five eolian soils characterized as a climosequence, with the most eastern and wettest soil characterized mineralogically (Bourne and Whiteside, 1962). Soil and dust samples were air-dried, and passed through a 2 mm sieve. Textures and clay estimates are from field-texturing.

Of the 38 samples collected to support this study, 28 were soil, six were rock and four were dust; ~50 g samples were sent to ALS Laboratories (Reno, Nevada) to determine the concentrations of eight major elements (Si, Al, iron [Fe], Ca, Mg, K, Na, and phosphorus [P], all expressed here on a percent oxide weight basis) by ICP-MS (inductively coupled plasma–mass spectrometry). Loss on ignition (LOI) was obtained after 20 min at 1000 °C. Because rare earth elements (REEs) can be sensitive indicators of material provenance, and so provide a means of chemically fingerprinting soils (Muhs et al., 2008; Nakase et al., 2014), all samples were also analyzed for 5 minor elements (all expressed on a mg kg^{-1} weight basis; europium [Eu], gadolinium [Gd], lanthanum [La], samarium [Sm], and ytterbium [Yb]), all by ICP-AES (inductively coupled plasma–atomic emission spectroscopy). Analytical uncertainties (as average coefficients of variation [CV] for five pairs of duplicate samples) were ± 7% for major elements except for Ca, Na, and P which were 20, 20, and 13% respectively, or ± 10% for the five REEs.

To chemically fingerprint the soils, we calculated two REE ratios. First, however, all soil, rock, and dust REE concentrations were normalized to REE concentrations in chondrite meteorites, which best represent unaltered igneous materials (Taylor and McLennan, 1995). We normalized REE using these corresponding chondrite concentrations: Eu 0.0563, Gd 0.199, La 0.237, Sm 0.148, and Yb 0.161 (McDonough and Sun, 1995). We then calculated the europium anomaly as $Eu^* = (Sm_N{*}Gd_N)^{0.5}$, where Sm_N and Gd_N represent chondrite-normalized Sm and Gd. Our first geochemical ratio is simply Eu_N/Eu^*. Our second geochemical ratio is chondrite-normalized lanthanum (La_N) to ytterbium (Yb_N), calculated as La_N/Yb_N. Positive Eu anomalies suggest oceanic crustal origins and mafic rock, whereas negative Eu anomalies are indicative of upper continental origins and felsic rock (Taylor and McLennan, 1995). La_N/Yb_N ratios determine the amount of fractionation between light REE and heavy REE. Higher La_N/Yb_N ratios (>20) can indicate an enrichment of light REE (Muhs et al., 2008).

RESULTS AND DISCUSSION

The five soil profiles we will characterize exhibit measurable differences in physical and chemical properties (Table 3). For example, coarse fragments, textures, and colors show large shifts both with depth and across rock types. On this field trip and in this field guide, we will discuss the processes most likely to have produced these patterns.

The soils on gneiss and sandstone have predominantly sandy or sandy loam textures. These coarse textures minimize soil surface area and reduce water holding capacity, which can ultimately slow soil development relative to the development of finer textured soils. In other settings, however, coarser textures could lead to greater translocation of materials, resulting in greater soil development (e.g., Schaetzl and Anderson, 2005). (Soil textural control of water holding capacity and plant-available water is well-illustrated through an online, interactive tool [Juma, 2012; www.pedosphere.ca/resources/texture/triangle_us.cfm].) An important difference between the gneiss and sandstone profiles is the distribution of coarse fragments with depth. Coarse fragments in the gneiss soil profile jump sharply from 10 to 80% by volume (A to C2 horizons), whereas in the sandstone soil profile, coarse fragments increase from 20 to 50% (A to C horizons). These differences may reflect greater bioturbation (vertical mixing by animals or plants) in the sandstone profile relative to the gneiss profile. In fact, we will observe differences in the forests between the gneiss and sandstone profiles: the forest on sandstone appears thicker and to have experienced more windthrow, where trees and their root plates have been blown over, bringing relatively unweathered bedrock and saprolite to the surface. Consistent with greater bioturbation, horizon boundaries are far less clear, and soil colors more uniform—and much redder in hue—in the sandstone soil than in the gneiss soil. Because greater windthrow delivers more weatherable primary minerals (e.g., feldspar) to the surface of the sandstone profile, where chemical weathering can be maximized, this could lead to more rapid and/or intense chemical weathering in this profile.

In the shale and limestone profiles, by contrast, the dominant soil textures are silty clay loams and clay loams, respectively. Generally, the clay-rich soils on shale are less rocky, although we will see evidence of colluvial processes in this profile—the Bt and C horizons contain coarse fragments consisting of unweathered sandstone. In fact, the sandstone rock sample characterized for this study was actually pulled from a depth of 80 cm in the shale profile (Fig. 4, Table 4). At Stop 4, we will see evidence of solifluction-type slumps within the shale meadow. As a result, roots of larger shrubs and trees may have greater difficulty establishing here. At Stop 5, the soil on limestone is situated below a small limestone outcrop that helps explain the anomalously high rock content (colluvium) throughout the profile, including the surface A horizon (80% coarse fragments). In contrast, the soil on shale has very little rock (5%) in the most mobile A horizon. An important take home message from this field trip will be the critical roles rock fragments play in altering hydrologic pathways and therefore extent and rates of chemical weathering. As Pavich (1986) has noted, "...how climate affects rock weathering is complexly dependent upon soil and rock structure."

Unlike the other two fine-textured lithosequence soils on shale and limestone, the soil on volcanic rock supports stands of Douglas fir, lodgepole pine, and aspen. This soil has relatively high clay content throughout the profile and an abrupt increase in coarse fragments at the transition between B and C horizons. Unlike the existing 1996 soil survey, our profile-specific classifications yielded only two taxonomic orders—Inceptisols (gneiss, sandstone) or Mollisols—and these generally reflected the sharp textural differences (Table 3).

These soils also differ in ways that we cannot observe directly; their elemental composition varies from the surface horizon to bedrock, and between the five profiles (Fig. 4, Table 4). Not surprisingly, Si represents the largest fraction of the rock samples except for the limestone, which was dominated by Ca (47%). Two soils show small, but clear, reductions in Si from rock (gneiss and sandstone); two soils show large gains in Si (shale and limestone); and the basaltic andesite soil shows a small gain in Si relative to the underlying rock. We interpret these patterns as suggesting differential chemical weathering of the rocks underlying each soil; for example, less weathering of the volcanic rock relative to the sandstone rock could have occurred because of the textural differences of the overlying soils. These patterns could also reflect the incorporation of chemically distinct material as dust or upslope colluvium.

The major elemental composition of the four dust samples also differed from one another (top row, Fig. 4; Table 4). These dust samples provide an additional end-member for geochemically fingerprinting A horizons, since these horizons must have formed from a combination of underlying rock and eolian inputs. We quantify the importance of dust to soils with three categories (Yaalon and Ganor, 1973): those composed entirely of eolian inputs (e.g., loess, as appears to be the case at the Post Farm [Fig. 1]); those reflecting comparatively less eolian inputs (e.g., no silt mantle is present); and, finally, those with "eolian contamination." Yaalon and Ganor define eolian-contaminated soils as those where eolian influence can be measured with mineralogical or geochemical fingerprinting techniques, but the eolian material is secondary in importance to the underlying rock material.

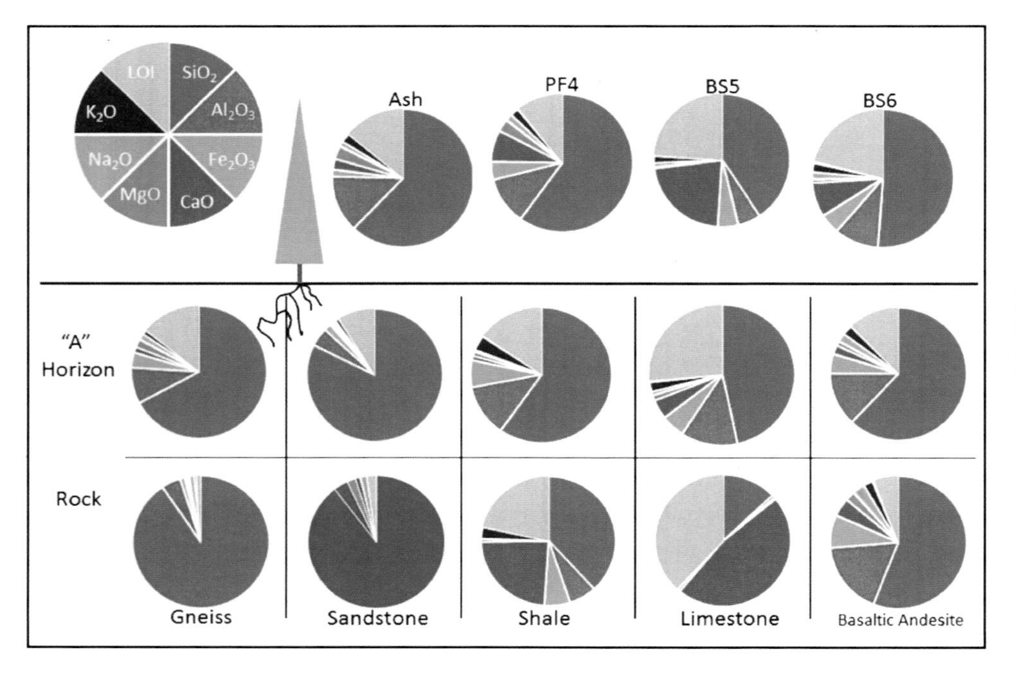

Figure 4. Major elemental composition for rock, soil A horizon, and dust across the lithosequence. Raw data are in Table 4. LOI—loss on ignition.

TABLE 4. MAJOR AND MINOR (NOT NORMALIZED) ELEMENTAL COMPOSITION OF LITHOSEQUENCE SOILS BY MORPHOLOGIC HORIZON (SEE TABLE 3), UNDERLYING ROCK, AND FOUR DUST SAMPLES

Site/sample	Horizon	SiO_2	Al_2O_3	Fe_2O_3	CaO	MgO	Na_2O	K_2O	P_2O_5	LOI	Total	Eu	Gd	La	Sm	Yb
Gneiss	A/B	68.1	8.8	3.77	1.44	2.05	1.51	1.18	0.07	14.4	102.0	0.41	1.48	10.9	1.68	0.96
	Bw	77.9	8.2	3.61	1.02	2.91	1.25	1.42	0.03	3.6	100.4	0.27	1.20	8.9	1.28	0.57
	C1	77.9	7.1	4.09	0.63	4.00	0.60	1.65	0.02	3.9	100.3	0.20	0.91	8.3	1.06	0.39
	C2	79.1	7.0	4.00	0.63	3.48	0.75	1.48	0.02	3.6	100.6	0.25	0.99	9.0	1.23	0.59
	Rock	90.3	4.6	1.15	0.54	0.50	1.28	0.54	0.01	1.0	100.0	0.30	0.96	11.6	1.70	0.29
	[C1 dup]	77.9	7.1	4.15	0.59	4.00	0.59	1.64	0.02	3.9	100.3	0.20	0.65	7.4	0.89	0.38
Sandstone	A	82.5	4.6	1.81	0.49	0.41	0.56	0.99	0.07	8.8	100.6	0.49	2.07	15.1	2.15	1.43
	Bw1	87.9	4.7	1.98	0.43	0.41	0.52	1.05	0.07	2.6	100.1	0.55	2.29	19.4	2.67	1.36
	Bw2	89.5	4.2	1.90	0.32	0.38	0.41	1.08	0.05	2.0	100.3	0.56	2.28	18.9	2.83	1.59
	BC	95.8	2.2	1.22	0.16	0.18	0.19	0.63	0.03	0.9	101.6	0.37	1.68	11.9	1.69	1.20
	C	97.6	0.9	1.04	0.03	0.05	0.02	0.27	0.01	0.6	100.7	0.23	1.23	7.2	1.06	0.62
	Rock	90.1	3.4	2.05	1.07	0.19	0.03	1.40	0.06	2.2	100.8	0.78	2.86	12.5	3.42	0.73
	[Bw1 dup]	87.6	5.3	2.28	0.38	0.46	0.59	1.16	0.06	2.5	100.9	0.52	2.59	20.5	3.16	1.68
	[C dup]	96.7	1.0	1.32	0.11	0.05	0.01	0.29	0.02	0.6	100.3	0.27	1.44	7.9	1.25	0.76
Shale	A	59.7	12.2	6.43	1.08	1.02	0.41	3.67	0.23	15.2	100.8	1.44	6.12	40.8	6.32	2.81
	Bt	68.3	12.0	7.16	0.64	0.97	0.33	3.82	0.10	6.7	100.8	1.41	6.00	41.5	6.44	2.88
	C	63.1	14.6	7.45	0.58	1.17	0.22	4.87	0.09	6.7	99.6	1.73	7.07	53.1	8.02	2.99
	Rock 1	37.7	6.5	6.12	23.40	0.62	0.06	2.90	0.08	21.2	99.7	2.10	7.81	26.8	8.58	2.51
	Rock 2	39.1	6.4	5.72	22.90	0.55	0.07	3.11	0.16	20.5	99.5	2.58	10.60	26.3	10.60	3.10
Limestone	A1	46.2	12.4	5.20	4.41	1.20	0.81	2.25	0.33	26.1	99.8	1.20	5.10	40.2	5.75	2.51
	A2	47.6	11.8	4.84	7.71	1.20	0.88	2.06	0.24	21.2	98.3	1.30	5.61	40.0	5.92	2.70
	C	43.4	11.2	4.65	11.15	1.18	0.80	1.94	0.26	23.0	98.3	1.20	5.47	36.6	6.12	2.76
	Rock	13.1	0.7	0.39	47.30	0.69	0.02	0.16	0.03	38.2	100.6	0.13	0.57	4.9	0.62	0.36
Volcanic	A	62.0	13.0	4.79	2.20	1.14	2.08	2.48	0.13	12.2	101.2	1.74	5.54	46.5	6.73	2.71
	Bt1	61.7	14.7	5.28	2.27	1.32	2.16	2.50	0.11	8.8	100.2	1.59	4.71	40.8	5.75	2.52
	Bt2	61.1	15.4	5.55	2.40	1.44	2.17	2.45	0.10	8.8	100.6	1.48	4.23	35.4	5.16	2.35
	C	58.6	15.7	5.77	2.39	1.56	2.11	2.38	0.11	9.2	99.0	1.53	4.14	35.7	5.09	2.31
	Rock	55.1	17.3	7.90	4.47	2.34	2.80	2.28	0.34	5.8	99.7	1.70	4.82	37.6	5.88	1.95
Dust	PF4	60.0	11.0	4.14	7.47	2.88	1.58	1.99	0.21	10.7	100.7	1.24	5.33	38.4	5.53	2.73
	BS6	50.2	9.9	4.62	7.78	0.83	1.57	2.16	0.42	20.6	98.8	0.72	2.90	25.6	3.96	1.39
	BS5	39.9	5.3	4.75	21.50	0.37	0.90	1.72	0.13	23.4	98.6	0.55	2.36	16.0	2.94	1.09
	Ash	62.2	13.2	1.59	2.56	3.13	1.15	2.26	0.03	13.9	100.1	0.07	16.50	18.4	14.30	8.53
	[PF4 dup]	59.4	10.5	3.87	7.26	2.71	1.54	1.90	0.19	10.7	98.8	1.20	5.25	38.0	5.51	2.65
	[BS6 dup]	51.7	10.0	4.90	7.62	0.85	1.59	2.18	0.42	19.8	99.7	0.77	3.18	27.0	4.40	1.56

Note: See text for element abbreviations. Major elements are presented as percent oxides; loss-on-ignition (LOI) is also in percent; minor elements (Eu, Gd, La, Sm, Yb) are in mg kg^{-1}. Total column includes eight oxide results and LOI, as well as an additional five elements (data not shown): chromium, titanium, manganese, strontium, and barium. Five duplicate samples are bracketed. Sites are listed in inverted stratigraphic order except for dust samples.

Our REE ratios show that these soils are comprised of varying mixtures of rock and dust (Fig. 5). For example, gneiss soil ratios plot far to the left of the La_N/Yb_N ratios for the underlying gneiss rock (plotted as off-scale in Fig. 5) but overlap with three dust ratios. Similarly, shale soil ratios overlap with dust ratios, but plot to the right of the two underlying shale rock La_N/Yb_N ratios. Limestone soils, by contrast, overlap with both the underlying limestone and dust, particularly the PF4 and BS5 dust samples (Fig. 5, Table 4). (Ratios for these same two dust samples plot most closely to soil ratios from all soil profiles except the volcanic soil profile.) The volcanic soil La_N/Yb_N ratios plot closer to the ratios for the underlying rock than to the dust, but also lie between them. This provides an opportunity to quantify—using Euclidean distances—the fraction rock, which we estimate at 0.6 for the volcanic A horizon soil. Finally, the sandstone soil La_N/Yb_N ratios plot to the left of *both* dust and rock La_N/Yb_N ratios, with minimal overlap, suggesting the potential for weathering-related shifts in these REE ratios.

Our combination of field observations; the elemental composition of the soil, rock, and dust; and our REE ratios together show pronounced differences in soils across this 2-mile lithosequence. These differences suggest the underlying rock drives soil differentiation, although further research is required to determine whether lithology-driven textural differences lead to selective retention of eolian materials. If so, this would imply that heterogeneity in both parent materials and their chemical weathering lead to complex soil development pathways over the very fine spatial scales of this lithosequence.

Future Directions

This field trip and field guide illustrate the influence of parent material on soil properties. It is, as with any field study, a Rumsfeldian effort: we go to the field with the tools (field and laboratory data, a state factor approach) and sites we have, not with the tools and sites we want or wish we had at a later time (e.g., Suarez, 2004). We look forward to building on this field trip in several ways and include these future directions to promote further discussion.

First, we are eager to explore ways of separating the biota and lithology factors. Despite the possibility of variable dust influence on soils, texture and vegetation community differences could affect whether and how soils incorporate long-term dust inputs. For example, a relatively high concentration of clays associated with weathering of shales could slow incorporation of dust inputs into the soils overlying shale; conversely, we could expect greater incorporation of dust into the more quartz-rich soils with high infiltration capacities associated with the gneiss and sandstone profiles. Textural control on the fate of eolian inputs—whether these are integrated into the soil or simply washed downslope—could derive from the underlying rock, weathered dust, or their combination. Differences between the forest-dominated sandstone and

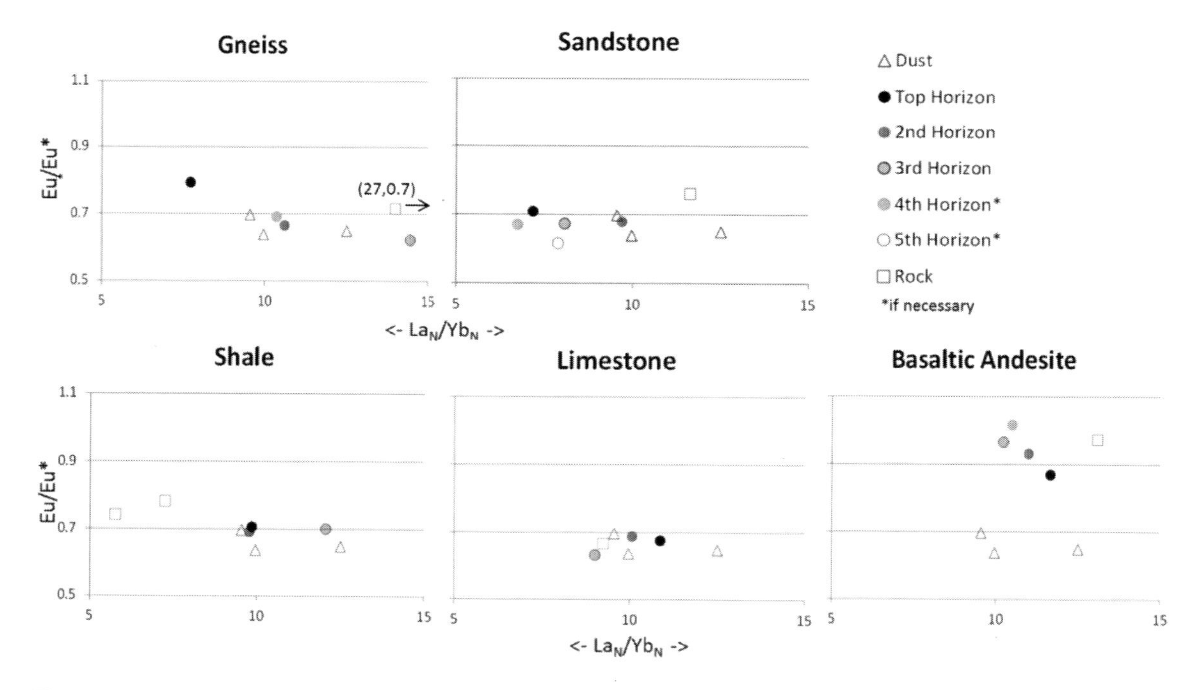

Figure 5. Chondrite-normalized europium to europium anomaly ratios (*y* axis; see text for details) versus chondrite-normalized lanthanum to ytterbium ratios (*x* axis) for the five soil profiles. Note two samples are omitted for clarity: the gneiss rock had corresponding values of 27 and 0.7 (indicated) and the Helena ash sample had corresponding x, y values of 1.5 and 0.01, respectively (not indicated).

grassland-dominated shale provide an additional complication to quantifying dust inputs, since the much greater leaf area associated with the forest could serve as a more effective dust trap, funneling eolian inputs into these soils.

Second, we look forward to determining particle size distributions and particle size-specific (e.g., silt and clay) mineralogy for these lithosequence soils as these physical and mineralogical fingerprinting approaches complement our geochemical fingerprinting method (e.g., Muhs et al., 2008). A useful target across this lithosequence could be Mazama glass. Third, once we have completed our measurements of soil bulk densities, we can calculate mass fluxes, both on an individual element basis and total mass basis. Finally, quantifying total denudation rates would enable the definition of both chemical weathering as well as physical erosion rates (Riebe et al., 2003). We recognize the lithologic limitations to this technique as only the gneiss and sandstone are likely to have the requisite quartz, but we could also use fallout nuclides to measure erosion via mixing rates (Dixon et al., 2009; Kaste et al., 2007). Either approach would build on prior regional efforts (^{10}Be: Licciardi and Pierce, 2008; ^{137}Cs: Arnalds et al., 1989).

CONCLUSIONS

This field trip to a lithosequence in the Gallatin National Forest continues a tradition begun decades ago with the efforts of John Montagne to use Hyalite Canyon for hands-on, in-field learning. For this field trip, we apply a state factor approach to soils formed from a spectacular stratigraphic sequence of five rock types: gneiss, sandstone, shale, limestone, and basaltic andesite. Field and laboratory data showed the influence of rock on soil properties, revealing clear differences. Although this heterogeneity could imply eolian materials are not an important influence on soil properties, lithology-driven vegetation and texture differences could lead to uneven retention (and consequent weathering) of any eolian materials. Where eolian inputs are sufficient in both duration and magnitude, a readily observable (and geochemically distinct) silt mantle can form, but we do not observe this across this lithosequence. The differences we observe in underlying parent materials appear to control the coarse fragment content of soil profiles, altering soil hydrology, and ultimately contributing to striking soil differences across this lithosequence.

ACKNOWLEDGMENTS

We thank the following colleagues for their constructive reviews of earlier versions of this manuscript: Cory Cleveland, Stephanie Ewing, Paul McDaniel, Larry Munn, Jerry Nielsen, Jay Norton, Todd Preston, and T. Weaver. We thank Tom Howes and Ron Wiens for their help collecting dust samples. These field trips would not have been possible without the cooperation and support of the Bozeman Ranger District, U.S. Forest Service. This work was funded in part by the Montana State University Vice President for Research, the MSU College of Agriculture, and a Faculty Excellence Grant (Hartshorn). We are also appreciative of over 40 years of students' and colleagues' questions, critiques, and discussions of portions of this field trip and the ideas they have sparked.

REFERENCES CITED

Arnalds, O., Cutshall, N.H., and Nielsen, G.A., 1989, Cesium-137 in Montana soils: Health Physics, v. 57, no. 6, p. 955–958, doi:10.1097/00004032-198912000-00010.

Bourne, W.C., and Whiteside, E.P., 1962, A study of the morphology and pedogenesis of a medial chernozem developed in loess: Soil Science Society of America Journal, v. 26, p. 484–490, doi:10.2136/sssaj1962.03615995002600050023x.

Chadwick, R.A., 1970, Belts of eruptive centers in the Absaroka-Gallatin volcanic province, Wyoming-Montana: Geological Society of America Bulletin, v. 81, p. 267–274, doi:10.1130/0016-7606(1970)81[267:BOECIT]2.0.CO;2.

Davis, C.E., and Shovic, H.F., 1996, Soil Survey of Gallatin Forest, Montana: Gallatin National Forest, Bozeman, Montana, USDA Forest Service, p. 9, 53–59, 78–79, 104–105.

Dixon, J.L., Heimsath, A.M., Kaste, J., and Amundson, R., 2009, Climate-driven processes of hillslope weathering: Geology, v. 37, p. 975–978, doi:10.1130/G30045A.1.

Dokuchaev, V.V., 1879, Short historical description and critical analysis of the more important soil classifications: Trav. Soc. Nat. St. Petersburg, v. 10, p. 64–67.

Druce, A.P., 1959, An empirical method of describing stands of vegetation: Tuatara, v. 8, p. 1–11, http://nzetc.victoria.ac.nz/tm/scholarly/tei-Bio08Tuat01-t1-body-d1.html.

Feeley, T.C., Cosca, M.A., and Lindsay, C.R., 2002, Petrogenesis and implications of calc-alkaline cryptic hybrid magmas from the Washburn volcano Absaroka volcanic province, USA: Journal of Petrology, v. 43, no. 4, p. 663–703, doi:10.1093/petrology/43.4.663.

Harward, M.E., and Borchardt, G.A., 1972, Mineralogy and trace element composition of ash and pumice soils in the Pacific Northwest of the United States, *in* Harward, M.E., ed., Mineralogy and Trace-Element Composition of Ash and Pumice Soils in the Pacific Northwest of the United States: Special Report No. 276: Corvallis, Oregon Agricultural Experiment Station, p. B.5.1–B.5.12.

Heckman, K., and Rasmussen, C., 2011, Lithologic controls on regolith weathering and mass flux in forested ecosystems of the southwestern USA: Geoderma, v. 164, p. 99–111, doi:10.1016/j.geoderma.2011.05.003.

James, H.L., and Hedge, C.E., 1980, Age of the basement rocks of southwest Montana: Geological Society of America Bulletin, v. 91, p. 11–15, doi:10.1130/0016-7606(1980)91<11:AOTBRO>2.0.CO;2.

Jenny, H., 1941, Factors of Soil Formation: A System of Quantitative Pedology: Mineola, New York, Dover Publications, Inc. (reprinted 1994).

Jenny, H., 1958, Role of the plant factor in the pedogenic functions: Ecology, v. 39, p. 5–16, doi:10.2307/1929960.

Juma, N., 2012, Pedosphere.ca: Empowering soil science students, educators & professionals, www.pedosphere.ca (accessed February 2014).

Kaste, J.M., Heimsath, A.M., and Bostick, B.C., 2007, Tracing short-term soil mixing with fallout radionuclides: Geology, v. 35, p. 243–246, doi:10.1130/G23355A.1.

Kirchner, J.W., Finkel, R.C., Riebe, C.S., Granger, D.E., Clayton, J.L., King, J.G., and Megahan, W.F., 2001, Mountain erosion over 10 yr, 10 k.y., and 10 m.y. time scales: Geology, v. 29, no. 7, p. 591–594, doi:10.1130/0091-7613(2001)029<0591:MEOYKY>2.0.CO;2.

Lawrence, C.R., Reynolds, R.L., Ketterer, M.E., and Neff, J.C., 2013, Aeolian controls of soil geochemistry and weathering fluxes in high-elevation ecosystems of the Rocky Mountains, Colorado: Geochimica et Cosmochimica Acta, v. 107, p. 27–46, doi:10.1016/j.gca.2012.12.023.

Lebauer, L.R., 1964, Petrology of the middle Cambrian Wolsey shale of southwest Montana: Journal of Sedimentary Petrology, v. 34, no. 3, p. 503–511.

Licciardi, J.M., and Pierce, K.L., 2008, Cosmogenic exposure-age chronologies of Pinedale and Bull Lake glaciations in greater Yellowstone and the Teton Range, USA: Quaternary Science Reviews, v. 27, p. 814–831, doi:10.1016/j.quascirev.2007.12.005.

Louie, J.Y., 2013, Pilot National Forest Reports: Gallatin National Forest, *in* Assessing the Vulnerability of Watersheds to Climate Change: Results of National Forest Watershed Vulnerability Pilot Assessments: USDA Forest Service General Technical Report PNW-GTR-884, p. 33–49, www.fs.fed.us/ccrc/wva/PilotNFWatershedVulnerabilityReport.pdf (accessed February 2014).

Lundgreen, C.A., 2013, Stabilization of an infinite slope failure utilizing hollow bar soil nails with long-term monitoring plan: Breckenridge, Colorado, Association of Conservation Engineers Conference, http://conservationengineers.org/conferences/2013.html (accessed February 2014).

McDonough, W.F., and Sun, S., 1995, The composition of the Earth: Chemical Geology, v. 120, p. 223–253, doi:10.1016/0009-2541(94)00140-4.

Middleton, L.T., 1980, Sedimentology of middle Cambrian Flathead sandstone, Wyoming [Ph.D. thesis]: Laramie, Wyoming, Department of Geology, University of Wyoming.

Montagne, C., 1976, Slope stability evaluation for land capability reconnaissance in the northern Rocky Mountains [Ph.D. thesis]: Bozeman, Montana, Montana State University.

Montagne, C., and Munn, L., 1980, Statistical summaries of soil characterization and site data: Summary report: Gallatin National Forest Contract No. R1-11-80-30.

Montagne, J., Goering, J., and Vaniman, C., 1975, compilers, Idealized stratigraphic column—northern Gallatin and Madison Ranges, Montana: Montana State University; www.montana.edu/wwwes/programs/Images/localstrat.pdf (accessed March 2014).

Mueller, P.A., and Frost, C.D., 2006, The Wyoming Province: A distinctive Archean craton in Laurentian North America: Canadian Journal of Earth Sciences, v. 43, p. 1391–1397, doi:10.1139/E06-075.

Muhs, D.R., Budahn, J.R., Johnson, J.L., Reheis, M., Beann, J., Skip, G., Fischer, E., and Joes, S.A., 2008, Geochemical evidence for airborne dust additions to soils in Channel Islands National Park, California: Geological Society of America Bulletin, v. 120, no. 1/2, p. 106–126, doi:10.1130/B26218.1.

Nakase, D.K., Hartshorn, A.S., Spielmann, K.A., and Hall, S.J., 2014, Eolian deposition and soil fertility in a prehistoric agricultural complex in central Arizona, USA: Geoarchaeology, v. 29, p. 79–97.

National Atmospheric Deposition Program, Annual data for site: MT07 (Clancy), http://nadp.sws.uiuc.edu/nadpdata/annualReq.asp?site=MT07 (accessed 20 November 2013).

National Atmospheric Deposition Program, Annual data summary for site: WY08 (Yellowstone National Park-Tower Falls), http://nadp.sws.uiuc.edu/nadpdata/ads.asp?site=WY08 (accessed 20 November 2013).

Neff, J.C., Ballantyne, A.P., Farmer, G.L., Mahowald, N.M., Conroy, J.L., Landry, C.C., Overpeck, J.T., Painter, T.H., Lawrence, C.R., and Reynolds, R.L., 2008, Increasing eolian dust deposition in the western United States linked to human activity: Nature Geoscience, v. 1, p. 189–195.

Pavich, M.J., 1986, Processes and rates of saprolite production and erosion on a foliated granitic rock of the Virginia Piedmont, *in* Colman, S.M., and Dethier, D.P., eds., Rates of Chemical Weathering of Rocks and Minerals: Orlando, Florida, Academic Press, p. 551–590.

Pierce, K.L., 2003, Pleistocene glaciations of the Rocky Mountains: Development in Quaternary Sciences, v. 1, p. 63–76, doi:10.1016/S1571-0866(03)01004-2.

Retzer, J.L., 1962, Soil Survey of Fraser Alpine Area, Colorado, Colorado Agricultural Experiment Station, U.S. Forest Service, U.S. Department of Agriculture, Series 1956, #20, p. 1–47, www.fs.fed.us/rm/pubs_other/rmrs_1962_retzer_j001.pdf (accessed March 2014).

Rhoades, C., Elder, K., and Greene, E., 2010, The influence of an extensive dust event on snow chemistry in the southern Rocky Mountains: Arctic, Antarctic, and Alpine Research, v. 42, no. 1, p. 98–105, www.fs.fed.us/rm/boise/AWAE/scientists/profiles/Rhoades/rmrs_2010_rhoades003.pdf, doi:10.1657/1938-4246-42.1.98.

Riebe, C.S., Kirchner, J.W., and Finkel, R.C., 2003, Long-term rates of chemical weathering and physical erosion from cosmogenic nuclides and geochemical mass balance: Geochimica et Cosmochimica Acta, v. 67, no. 22, p. 4411–4427, doi:10.1016/S0016-7037(03)00382-X.

Schaetzl, R.J., and Anderson, S., 2005, Soils: Genesis and Geomorphology: New York, Cambridge University Press, 827 p.

Schimel, J., and Chadwick, O.A., 2013, What's in a name?: Frontiers in Ecology and the Environment, v. 11, no. 8, p. 405–406, doi:10.1890/13.WB.016.

Schoeneberger, P.J., Wysocki, D.A., and Benham, E.C., and Soil Survey Staff, 2012, Field book for describing and sampling soils, Version 3.0: Lincoln, Nebraska, Natural Resources Conservation Service, National Soil Survey Center, ftp://ftp-fc.sc.egov.usda.gov/NSSC/Field_Book/FieldBookVer3.pdf (accessed February 2014).

Soil Survey Staff, 2010, Keys to Soil Taxonomy: 11th ed. U.S. Department of Agriculture. Natural Resources Conservation Service, ftp://ftp-fc.sc.egov.usda.gov/NSSC/Soil_Taxonomy/keys/2010_Keys_to_Soil_Taxonomy.pdf (accessed February 2014).

Suarez, R., 2004, Troops question secretary of defense Donald Rumsfeld about armor: PBS News Hour, www.pbs.org/newshour/bb/military/july-dec04/armor_12-9.html (accessed February 2014).

Taylor, S.R., and McLennan, S.M., 1995, Geochemical evolution of continental crust: Reviews of Geophysics, v. 33, no. 2, p. 241–265, doi:10.1029/95RG00262.

Tegen, I., 2003, Modeling the mineral dust aerosol cycle in the climate system: Quaternary Science Reviews, v. 22, p. 1821–1834, doi:10.1016/S0277-3791(03)00163-X.

U.S. Forest Service and Montana State University, 1998, Memorandum of Understanding between the Montana State University, Bozeman and Billings Campuses and the Gallatin and Custer National Forests, www2.montana.edu/policy/forest_service_use/mou.html (accessed February 2014).

Veseth, R., and Montagne, C., 1980, Geologic parent materials of Montana soils: Bulletin 721: Montana Agricultural Experiment Station and USDA-Soil Conservation Service, Bozeman.

Weaver, T., and Perry, D., 1978, Relationship of cover type to altitude, aspect, and substrate in the Bridger Range, Montana: Northwest Science, v. 52, no. 3, p. 212–219.

Weber, W.M., 1965, General geology and geomorphology of the Middle Creek area, Gallatin County, Montana [M.S. thesis]: Bozeman, Montana, Department of Earth Sciences, Montana State University.

Yaalon, D.H., and Ganor, E., 1973, The influence of dust on soils during the Quaternary: Soil Science, v. 116, no. 3, p. 146–155, doi:10.1097/00010694-197309000-00003.

Yassoglou, N.J., Nobeli, C., and Vrahamis, S.C., 1969, A study of some biosequences and lithosequences in the zone of brown forest soils in northern Greece: Morphological, physical, and chemical properties: Soil Science Society of America Journal, v. 33, p. 291–296, doi:10.2136/sssaj1969.03615995003300020035x.

Zdanowicz, C.M., Zielinski, G.A., and Germani, M.S., 1999, Mount Mazama eruption: Calendrical age verified and atmospheric impact assessed: Geology, v. 27, p. 621–624, doi:10.1130/0091-7613(1999)027<0621:MMECAV>2.3.CO;2.

Manuscript Accepted by the Society 27 February 2014

Printed in the USA

The Geological Society of America
Field Guide 37
2014

Regional setting and deposit geology of the Golden Sunlight Mine: An example of responsible resource extraction

Nancy Oyer*
Barrick Gold Corporation, 453 MT Hwy 2 E, Whitehall, Montana 59759, USA

John Childs*
Childs Geoscience Inc., 1700 West Koch Street, Suite 6, Bozeman, Montana 59715, USA

J. Brian Mahoney*
Department of Geology, University of Wisconsin, Eau Claire, Wisconsin 54702, USA

ABSTRACT

The Barrick Golden Sunlight Mine (GSM) in Whitehall, Montana, is an industry leader in safe, responsible resource extraction. With more than 3 million ounces of gold poured since 1983, and current proven and probable reserves of 318,000 ounces of gold, GSM is the largest gold producer in Montana. The gold-silver deposit is localized in a hydrothermal breccia pipe related to Late Cretaceous latite porphyry magmatism hosted by the Mesoproterozoic Belt Supergroup, and is influenced by younger cross-cutting faults and fracture systems. The deposit has been mined by both underground and open pit methods, and the current open pit operation was recently permitted for expansion. The mill and tailings operations practice efficient and environmentally responsible resource recovery by processing ore from historical tailings and dumps from around the state in addition to ore from the Golden Sunlight property. This trip will explore the complex geologic and tectonic controls on mineralization and review how GSM has addressed the technical challenges of mining, milling, and reclamation.

INTRODUCTION

The Golden Sunlight Mine (GSM), operated by Barrick Gold Corporation, is located in Jefferson County, Montana, 55 km east of Butte and 95 km northwest of Bozeman (Fig. 1). Gold and silver mineralization at GSM occurs within and around a hydrothermal breccia pipe hosted in the lower Mesoproterozoic Belt Supergroup. The hydrothermal breccia is related to Late Cretaceous latite porphyry magmatism, and influenced by younger cross-cutting faults and fracture systems. The average grade of the deposit, currently mined by conventional open pit methods, is 0.06 ounces per ton gold. From 2002 to 2009, two phases of underground open-stope operations took advantage of higher grade (0.1–0.3 ounce per ton) gold. The deposit is milled and processed on-site by conventional technology, including carbon-in-pulp, carbon column, and sand tailing retreatment.

*E-mails: noyer@barrick.com; jfchildsgeo@msn.com; mahonej@uwec.edu.

Oyer, N., Childs, J., and Mahoney, J.B., 2014, Regional setting and deposit geology of the Golden Sunlight Mine: An example of responsible resource extraction, *in* Shaw, C.A., and Tikoff, B., eds., Exploring the Northern Rocky Mountains: Geological Society of America Field Guide 37, p. 115–144, doi:10.1130/2014.0037(06).

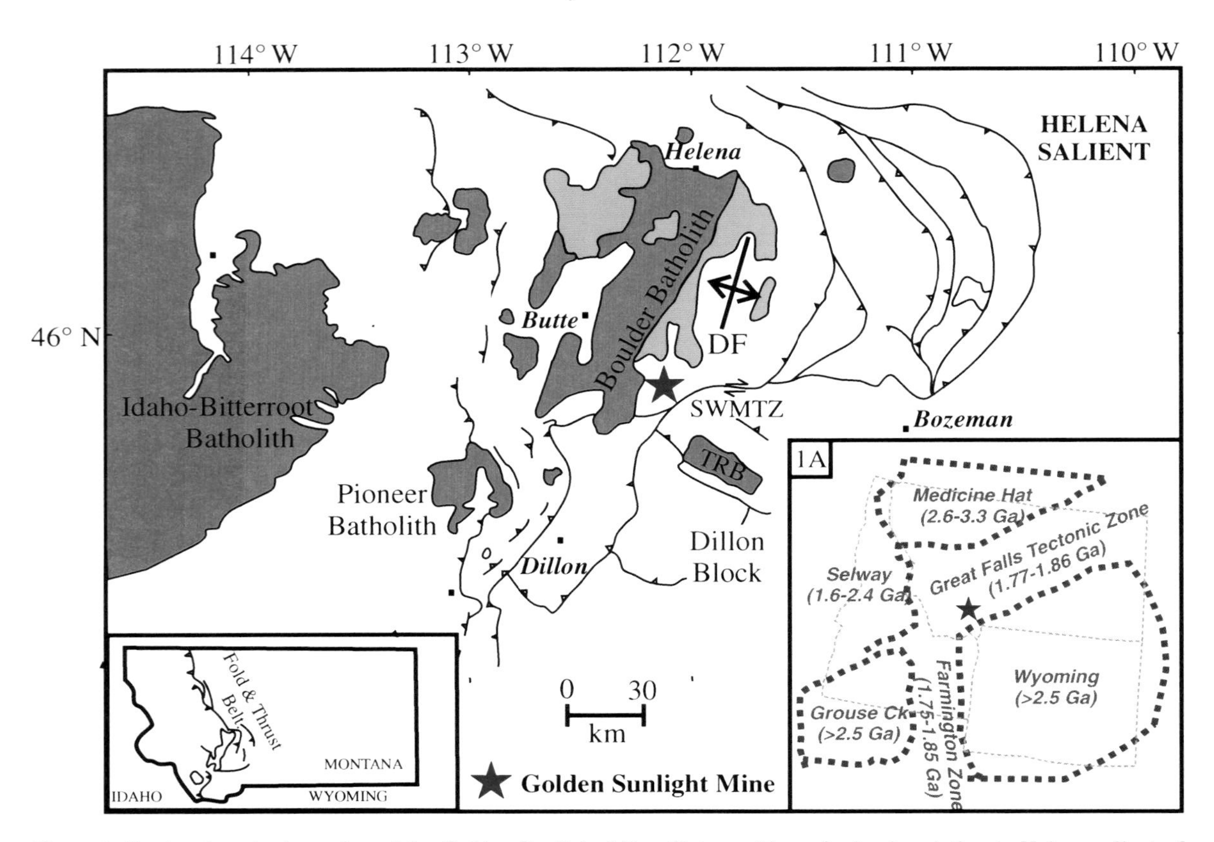

Figure 1. Regional geologic setting of the Golden Sunlight Mine. Note position of mine in relation to Helena salient of the Montana fold-and-thrust belt and the Boulder batholith (red) and Elkhorn Mountains Volcanics (green). SWMTZ—southwest Montana transverse zone; TRB—Tobacco Root batholith; DF—Devil's Fence anticlinorium. Inset 1A shows mine location in relation to Archean cratons and Proterozoic basement terranes (Lageson et al., 2001; Foster et al., 2006).

Regional Geology

The Golden Sunlight Mine is located in southwest Montana, which contains some of the most spectacular and complicated geology in the North American Cordillera (Figs. 1, 2). The geologic setting of the mine is complex, as several major geologic provinces, including the Great Falls tectonic zone, the Belt basin, the Montana fold and thrust belt, the Boulder batholith and the Basin and Range structural province overlap in the region (Figs. 1, 2). Economic mineralization in the mine centers around a Late Cretaceous hydrothermal breccia system emplaced within Mesoproterozoic Belt Supergroup strata (Fig. 3), but this setting is the culmination of a long and protracted geologic history.

The crystalline basement of southwest Montana consists of the Great Falls tectonic zone, a northeast-trending zone of structurally deformed metamorphic rocks that separates the Wyoming craton (> 2.5 Ga crust) from the Medicine Hat block of the Hearne craton (2.6–3.3 Ga crust) (Fig. 1A; Mueller et al., 2002). The Great Falls Tectonic Zone records episodic 1.77–1.86 Ga magmatism and metamorphism related to the amalgamation of the Wyoming and Hearne cratons (Fig. 1A) and the resultant crustal suture zone may have exerted significant influence on the development of younger tectonic features (O'Neill and Lopez, 1985; Mueller et al., 2002; Foster et al., 2006). Archean rocks of the Wyoming craton and Paleoproterozoic rocks of associated arc terranes in southwest Montana are referred to as the Dillon block, which is exposed in the Tobacco Root Mountains south of GSM (Figs. 1–3).

Economic mineralization at GSM is hosted by the LaHood and Greyson Formations of the Mesoproterozoic Belt Supergroup (Figs. 3, 4). These strata were deposited in the Belt basin, a large, intracratonic extensional basin that accumulated an immense thickness (15–20 km) of primarily fine-grained clastic sediment between 1.40 and 1.47 Ga (Ross and Villeneuve, 2003). The Golden Sunlight Mine is situated at the southern end of the Belt basin, which unconformably overlies the Great Falls tectonic zone (Fig. 1A). The southern margin of the Belt basin was tectonically active, as suggested by coarse conglomerates of the LaHood Formation, which record syndepositional tectonism along the ancestral east-trending Willow Creek normal fault (also known as the Perry Line) (Schmidt and Garihan, 1986). Episodic motion along the Willow Creek fault uplifted Archean and Paleoproterozoic crystalline rocks of the Dillon block, which shed sediment to the north into the southern Belt basin. The LaHood Formation is at least 1200 m thick, and interfingers with fine-grained formations of the Lower Belt Supergroup (Newland and

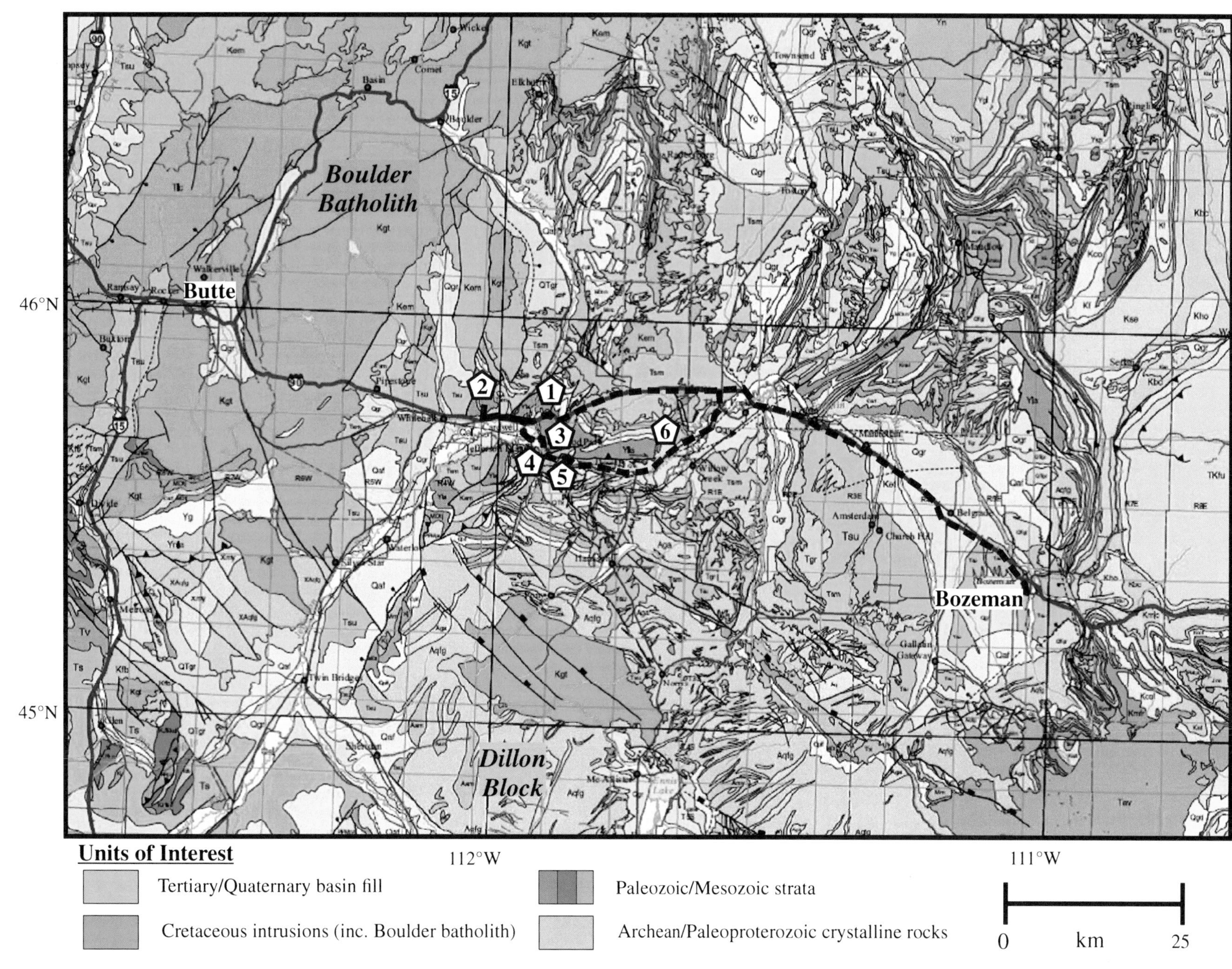

Figure 2. *Geologic Map of Montana* (scale 1:500,000), showing field trip route and location of Golden Sunlight Mine relative to major geologic features in southwest Montana. (Map modified from Vuke et al., 2007.)

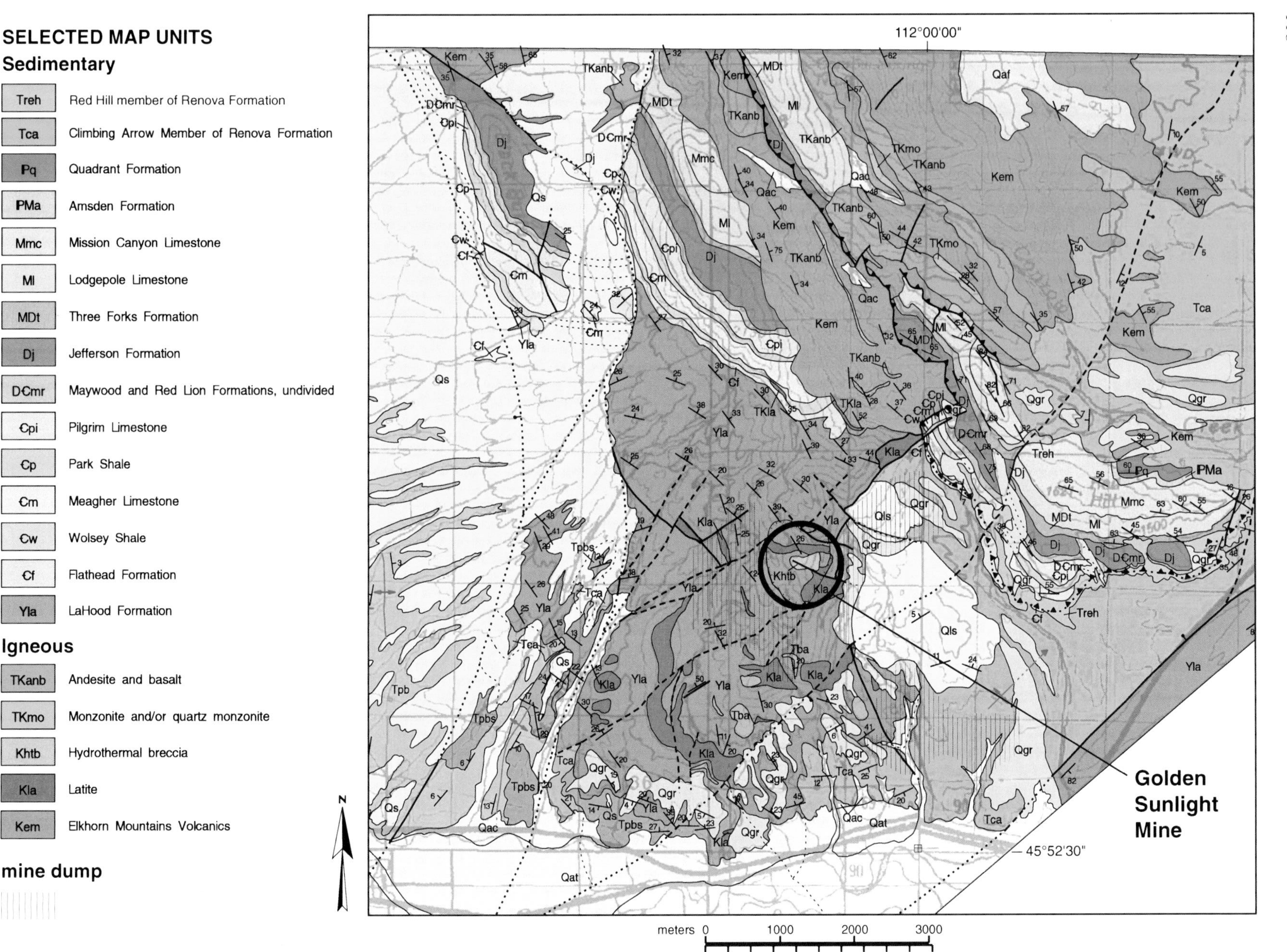

Figure 3. Geologic map of the Golden Sunlight Mine west of the south Boulder River valley. Note location of mine in northeast-dipping structural panel that exposes Mesoproterozoic LaHood Formation and cross-cutting Cretaceous latite. (Map modified from Vuke et al., 2004.)

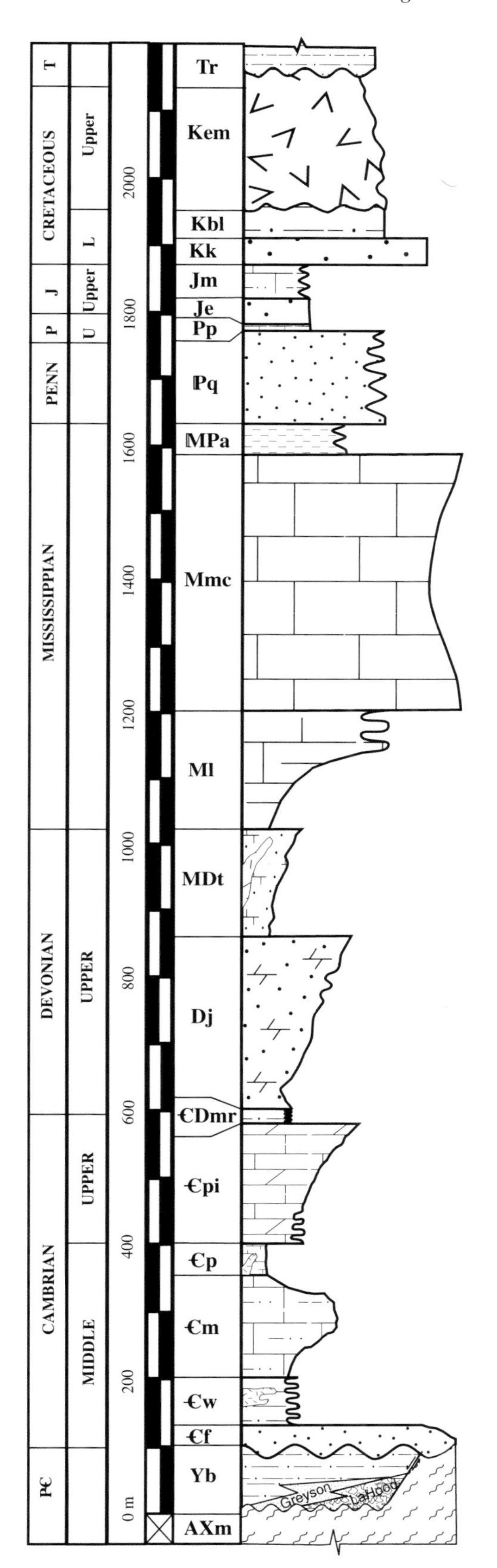

<u>Renova Fm. (Tr; ~0-300 m)</u>: Conglomerate, sandstone, siltstone and shale.

<u>Elkhorn Mountains Volcanics (Kem; > 1000 m)</u>: Andesitic to rhyolitic flows, breccias, lapilli tuff, tuff and intercalated sediments.

<u>Black Leaf Fm. (Kbl; ~35 m)</u>: Sandstone, siltstone and shale.

<u>Kootenai Fm. (Kk; 120 m)</u>: Chert lithic feldspathic arenite.

<u>Morrison Fm. (Jm; 50 m)</u>: Quartz arenite, thin bedded siltstone, shale and sandy micrite.

<u>Ellis Gp. (Je; 40 m)</u>: Thin bedded, calcareously lithic arenite.

<u>Phosphoria Fm. (Pp; 2-5 m)</u>: Black-dark gray bedded chert, cherty limestone, and oolitic phosphatic shale.

<u>Quadrant Fm. (Pq; 120 m)</u>: Quartz arenite.

<u>Amsden Fm. (MPa; 50 m)</u>: Quartz arenite, siltstone, shale and silty micrite.

<u>Mission Canyon Fm. (Mmc; 275 m)</u>: Wacke to packstone, locally rudstone.

<u>Lodgepole Fm. (Ml; 125 m)</u>: Turbiditic micrite to packstone.

<u>Three Forks Fm. (MDt; 160 m)</u>: Calcareous micaceous shale to silty micrite.

<u>Jefferson Fm. (Dj; 250 m)</u>: Fossiliferous sandy dolomitic wackestone with thin sand laminations.

<u>Maywood and Red Lion Fm. (€Dmr; 60 m)</u>: Quartz arenite, siltstone and shale.

<u>Pilgrim Fm. (€pi; 150 m)</u>: Lower portion consists of micrite; upper portion consists of bioturbated dolomitic wackestone.

<u>Park Fm. (€p; 50 m)</u>: Micaceous shale.

<u>Meagher Fm. (€m; 20 m)</u>: Fossiliferous sandy micrite to packstone.

<u>Wolsey Fm. (€w; 65 m)</u>: Micaceous sandy shale/siltstone.

<u>Flathead Fm. (€f; 15-30 m)</u>: Subfeldspathic arenite to quartz pebble conglomerate.

<u>Mesoproterozoic Belt Supergroup (Yb; > 1000 m)</u>: Siltstone and shale interbedded with micaceous subfeldspathic arenite. Contains LaHood Fm. boulder conglomerate in south.

<u>Archean/Proterozoic (AXm)</u>: Gneiss and schist.

Figure 4. Regional stratigraphic column of Archean to Tertiary rocks in southwest Montana.

Greyson Formations) to the north (Ross and Villeneuve, 2003). These formations may represent a progradational sequence of basin plain, submarine fan, slope, and shelf facies (Foster et al., 1999; Foster and Chadwick, 1999). Sedimentary facies and grain size within Belt Supergroup strata may have exerted a control on economic mineralization.

The Belt Supergroup is overlain by lower Paleozoic strata immediately north of the mine, including the Cambrian Flathead, Wolsey, Meagher, Park, and Pilgrim Formations (Fig. 4). Basal Cambrian strata are traditionally interpreted as representing deposition on a quiescent passive margin, but the recent recognition of Neoproterozoic to Cambrian alkalic magmatism, Belt Supergroup uplift, and syndepositional boulder conglomerate suggests the continental margin was tectonically active in Neoproterozoic to Cambrian time (Lund et al., 2010; McDonald et al., 2012; Mahoney et al., 2013). Lower Paleozoic strata are overlain by up to 2000 m of mid- to late Paleozoic shallow marine strata and Mesozoic marine to continental sedimentary and volcanic rocks.

Early to Late Cretaceous tectonic and magmatic activity plays an important role in economic mineralization. The Montana Disturbed Belt is a northwesterly trending zone of closely spaced west-dipping thrust faults and associated east-vergent folds that developed in Proterozoic, Paleozoic, and Mesozoic strata during Late Jurassic to Late Cretaceous contractional deformation of the Sevier orogeny in Montana. The Helena salient is a pronounced eastward bulge within the Montana Disturbed Belt that extends from north of Helena to Whitehall, Montana (Figs. 1, 2). The southern boundary of the eastward convex Helena salient is an east-west–trending transfer zone termed the "southwest Montana transverse zone" that is spatially associated with the northern end of the Dillon block and is subparallel to the trace of the ancestral Willow Creek fault (Figs. 1, 3; Schmidt and O'Neill, 1982). The Golden Sunlight Mine is immediately adjacent to this Late Cretaceous transfer zone.

The close spatial and temporal relationship between the Late Cretaceous fold-and-thrust belt of the Helena salient of the Montana Disturbed Belt, plutonic rocks of the Boulder batholith, and coeval volcanic rocks of the Elkhorn Mountains Volcanics has long been recognized (e.g., Klepper et al., 1957; Robinson et al., 1968). The Boulder magmatic system is entirely contained within the Helena salient, suggesting a close relationship between magmatism and structural deformation (Lageson et al., 2001; Ihinger et al., 2011). Magmatism associated with the Boulder batholith is pre-, syn- and post-deformational with respect to thin-skinned folding and thrusting of the Sevier orogenic system. Hundreds of satellite dikes, sills, and stocks intrude folded Proterozoic, Paleozoic, and Mesozoic strata east of the Boulder batholith, including both folded sills and stocks that clearly cross-cut folded strata.

The Golden Sunlight Mine is located ~15 km east of the eastern margin of the composite Late Cretaceous Boulder batholith, and ~1.5 km south of the comagmatic Elkhorn Mountains Volcanics. Satellite dikes, stocks, and sills are widespread east of the mine site, with the Dougherty Mountain stock, ~8 km to the east, a prime example. These satellite intrusions are roughly coeval with the Boulder batholith, the Elkhorn Mountains Volcanics, and the latite intrusions at the mine, suggesting a comagmatic origin (Ihinger et al., 2011).

Thin-skinned deformation of the Sevier orogeny transitioned into Laramide tectonism in Late Cretaceous to Paleogene time, when the frontal fold-and-thrust belt impinged on major intraforeland basin basement-cored uplifts, leading to a broken foreland and major synorogenic sedimentation (Fig. 2; Schmidt et al., 1988; Schwartz and DeCelles, 1988). The southwest Montana transverse zone acted as a major east-trending structural boundary that separated thin-skinned thrusts in the Helena salient to the north from basement-involved thick-skinned thrusts and reverse faults to the south. This partitioning was accompanied by ~15–25 km of right-lateral slip and up to 10 km of reverse dip slip motion during Late Cretaceous to Paleogene time (Schmidt et al., 1988). Transverse motion on the southwest Montana transverse zone may be responsible for the development of important structures that control mineralization in the Golden Sunlight deposit.

Extension in the northern portion of the Basin and Range structural province led to the development of a series of north-northeast–trending half-graben and graben structures in Eocene to Oligocene time (Lageson, 1989; Ruppel, 1993). Silicic to mafic volcanism occurred locally (Foster, 1987; Luedke, 1994; Dudas et al., 2010). North- and northeast-trending high angle normal faults truncate both Sevier and Laramide structural features, as well as rocks of the Boulder magmatic system, and the resulting intermontane valleys contain thick Tertiary and Quaternary basin fill.

Mine Area Geology

The Golden Sunlight Mine is located 8 km northeast of Whitehall, Montana, at the southern end of a north-trending, fault-bounded mountain range known as Bull Mountain. This range forms the west limb of a north-plunging synclinorium that exposes Mesoproterozoic to Cretaceous strata (Figs. 2, 3) (Chadwick, 1996). The Mesoproterozoic Belt Supergroup, the oldest exposed rocks in the area, includes weakly metamorphosed interbedded quartzite-siltite-argillite of the LaHood Formation, overlain by siltite-argillite of the Greyson Formation (Figs. 3, 4). The Mesoproterozoic rocks are overlain by Cambrian to Pennsylvanian sedimentary strata and the Upper Cretaceous Elkhorn Mountains Volcanics to the north (Figs. 3, 4). Late Cretaceous hypabyssal latite porphyry and basaltic andesite sills and dikes intrude all older rocks in the area, and are particularly abundant within Belt Supergroup strata.

Economic gold and silver mineralization occurs within and around a northeast-trending, west-southwest–plunging cylindrical (~210-m-diameter) hydrothermal breccia pipe (Figs. 5, 6). The Mineral Hill pit was designed and developed to optimize mining this breccia pipe (Fig. 7). The clasts in the breccia pipe are composed of latite porphyry and Mesoproterozoic LaHood and Greyson sedimentary rocks, while the matrix is composed of smaller rock fragments and hydrothermal gangue and ore minerals. Clasts are typically angular to subangular, and range in size from coarse sand to boulder. The breccia has been subdivided by texture, clast composition, clast size, and matrix composition, as beautifully documented in the historic, hand-drawn images of Chadwick (1992) (Fig. 5). The breccia pipe and associated mineralized faults, veins, and joints cut both the Proterozoic rocks and the latite. Post-mineralization lamprophyre dikes and sills cross-cut both the latite dikes and the breccia pipe (DeWitt et al., 1996). The orebody is truncated on the east by a range-front, east-dipping normal fault.

Approximately 1.5 km north of GSM in the Bonnie exploration area, Mesoproterozoic strata are overlain by the Cambrian Flathead, Wolsey, and Meagher Formations (Figs. 3, 7) (Chadwick, 1996). Gold mineralization occurs locally in the Mesoproterozoic strata, Flathead Quartzite, and Wolsey Shale (David Odt, 2013, personal commun.).

Tertiary (Oligocene through Pliocene) sedimentary strata fill the valleys flanking Bull Mountain. A large Quaternary alluvial fan caps Tertiary sedimentary strata west of GSM, and numerous landslide, or debris flow deposits, flank the eastern range front (Fig. 3). Locally these debris flows, such as the East Area deposit mined in 2011, contain economic deposits of gold. Pediment terrace gravels cap much of the Jefferson Valley east of the mine and are dissected by Holocene tributaries of the Jefferson River.

STRUCTURAL CONTROLS

The Golden Sunlight Mine lies at the juncture of several important crustal-scale structural features. The mine is located within the Great Falls tectonic zone, is proximal to the ancestral Willow Creek fault, which forms the boundary between the Archean Dillon block and the Belt basin, and lies along the southwest Montana transverse zone, which forms the southern margin of the Helena salient (Figs. 1, 2). The overlapping nature of these structural features suggests there is a fundamental crustal discontinuity that has localized crustal deformation through time. For example, the Willow Creek fault was probably a growth fault along the southern margin of the Belt basin in Mesoproterozoic time, which was reactivated as a major tear fault translating large-scale thrust movement to the north into right lateral-oblique slip along the transverse zone during Cretaceous deformation (Schmidt and O'Neill, 1982).

Foster and Chadwick (1990), who based their interpretation on a description of the structural control of the Carlin Trend in Nevada (Dean et al., 1990), suggest that crustal weakness associated with the Great Falls tectonic zone, expressed by steeply dipping northeast-trending structures, provided a pathway for brecciation and gold mineralization at GSM. The highest grade mineralization commonly occurs along the northeast-trending structures, which include faults, veins, joints, and the breccia pipe itself. A prime example is the Sunlight vein zone, an en echelon series of steeply dipping, tabular gold-enriched quartz-sericite-pyrite zones targeted by underground miners from the late 1800s to the early 1900s (David Odt, 2013, personal commun.). One ore shoot was reported to contain 33,500 tons grading 1.0 ounce per ton, and, prior to 1951, was stoped over a 175 m vertical distance (Fess Foster, 2005, personal commun.). Permitting has been approved to mine the South Area (Fig. 7), a satellite deposit that includes the Sunlight vein zone.

Northwest-trending structures are less mineralized. The east-northeast and northwest structures appear to be synthetic and antithetic structures, respectively, that formed in response to dextral movement along the northeast-trending faults. The northeast-striking, west-southwest (35°)–plunging breccia pipe is bound locally by two steeply dipping, northeast-trending fault zones, the Lone Eagle and the Fenner-Grey fault zones, mirroring the regional northeast trend of the Great Falls tectonic zone (Figs. 6A, 6C). The breccia is thought to have formed in a transtensional zone between these structures. The position of both latite and Proterozoic clasts identified in the upper breccia pipe indicates the pipe was a relatively passive feature that initially fractured in place, and then collapsed (Foster, 1991; Chadwick, 1992; Foster and Childs, 1993) (Figs. 5, 6). This is illustrated by Chadwick's "Sag Zone" (1992; black dots, Fig. 5), where at the margins of the pipe, latite clasts in the breccia occur at the same elevation as the latite sill adjacent to the breccia. In contrast, toward the middle of the breccia, latite clasts occur at progressively lower (up to 100 m) elevations. The same is true for Proterozoic quartzite clasts. The observation of gravitational collapse within the breccia pipe suggests the pipe was originally emplaced vertically, and has been subsequently rotated.

The top of the breccia pipe was offset ~450 m to the east along the Tertiary Corridor fault (Figs. 5, 6B), which itself is offset by reverse fault reactivation along the Lone Eagle fault zone (Figs. 6C). If one assumes the breccia pipe was originally vertical, then this would rotate the Corridor fault to a west-dipping structure with reverse motion.

MINERALIZATION AND ALTERATION

Mineralization at GSM is concentrated in a 210-m-diameter breccia pipe in the Mineral Hill pit, with three smaller, satellite deposits (2BOP, South Area, and North Area) flanking the pipe (Fig. 7). In the Mineral Hill pit, ~70% of the gold occurs in the breccia, with the remaining gold in the adjacent sediments and latite. Gold and gold tellurides occur as 1–10 micron grains encapsulated in pyrite (Fig. 8; Spry and Thieben, 1997). Mineralization is both disseminated and structurally controlled (see Table 1 for a list of economic minerals at GSM). Disseminated mineralization (typically less than 0.1 ounce per ton gold) occurs as sulfide in the breccia matrix, the latite, and in the Proterozoic host rocks. Structurally controlled mineralization (which can be greater than 0.1 ounce per ton gold) occurs as sulfide along faults and sulfide filled veins and joints. The sulfide is predominantly pyrite (up to 20%). The main accessory sulfides include chalcopyrite, covellite, and bornite (Foster, 1991; Spry and Thieben, 1997). Native silver is present but typically occurs as 1–10 micron grains of silver tellurides encapsulated within grains

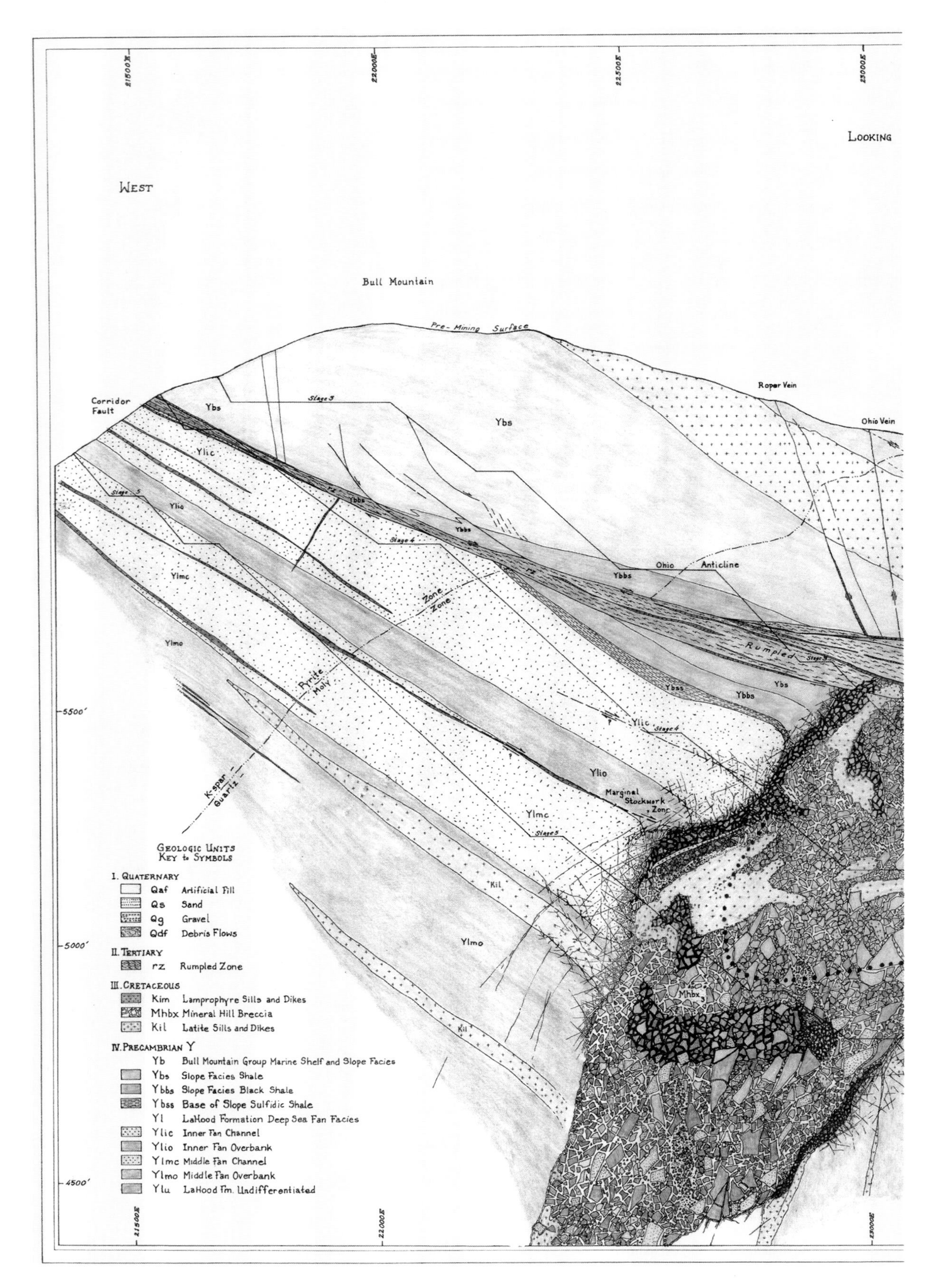

Figure 5. Hand-drawn geologic cross section through the Mineral Hill Breccia at 26,000 North by Tom Chadwick (1992).

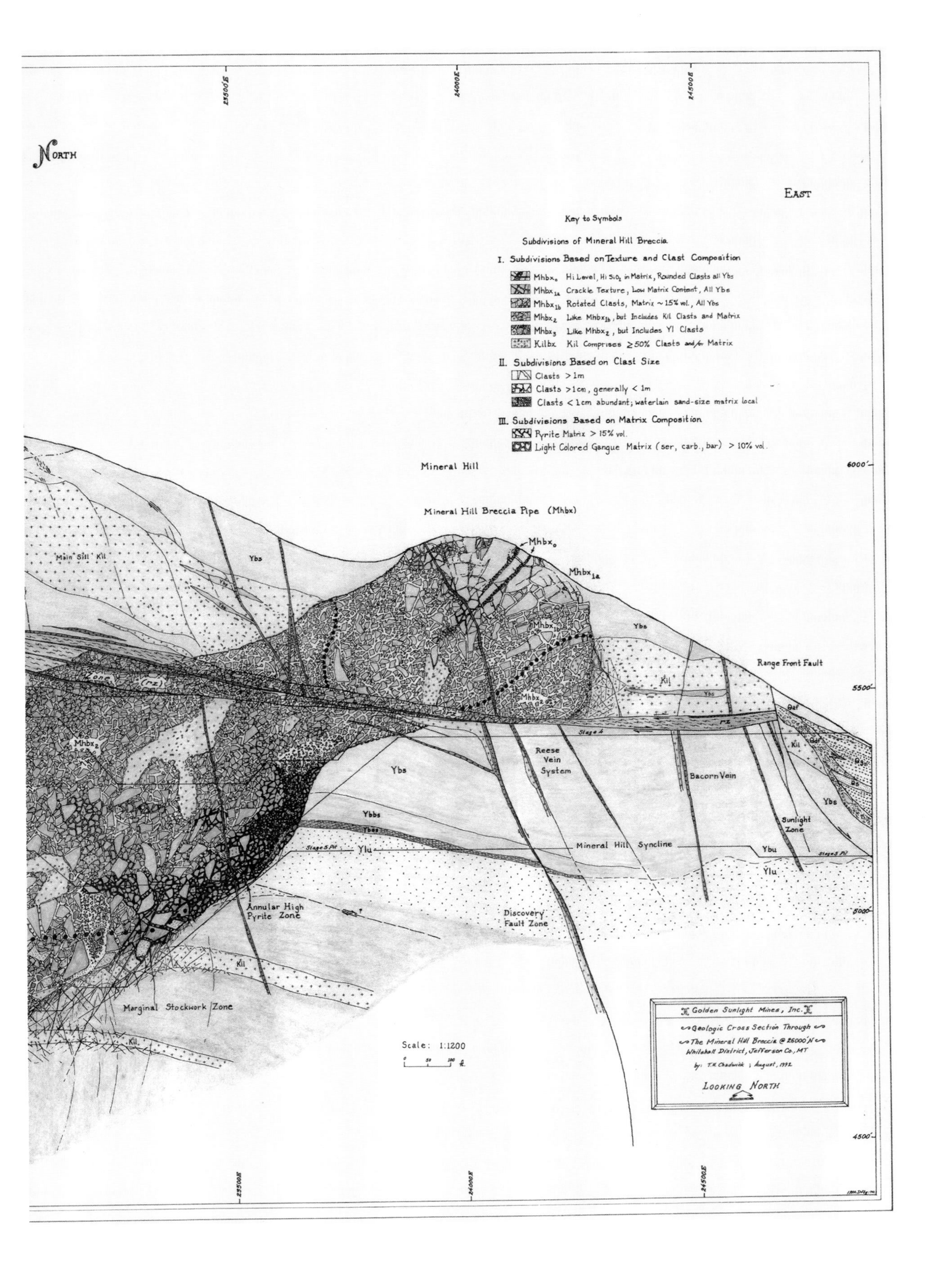
23500E
24000E
24500E
North
East
Key to Symbols
Subdivisions of Mineral Hill Breccia
I. Subdivisions Based on Texture and Clast Composition
Mhbx0 Hi Level, Hi SiO2 in Matrix, Rounded Clasts all Ybs
Mhbx1a Crackle Texture, Low Matrix Content, All Ybs
Mhbx1b Rotated Clasts, Matrix ~15% vol., All Ybs
Mhbx2 Like Mhbx1b, but Includes Kil Clasts and Matrix
Mhbx3 Like Mhbx2, but Includes Yl Clasts
Kilbx Kil Comprises ≥50% Clasts and/or Matrix
II. Subdivisions Based on Clast Size
Clasts >1m
Clasts >1cm, generally < 1m
Clasts < 1cm abundant; waterlain sand-size matrix local
III. Subdivisions Based on Matrix Composition
Pyrite Matrix > 15% vol.
Light Colored Gangue Matrix (ser, carb., bar) > 10% vol.
Mineral Hill
6000'
Mineral Hill Breccia Pipe (Mhbx)
Mhbx0
Mhbx1a
Main Sill Kil
Ybs
Mhbx1b
Range Front Fault
Kil
5500'
Mhbx2
Reese Vein System
Ybs
Bacorn Vein
Ybbs
Ylu
Mineral Hill Syncline
Ybu
Ylu
Sunlight Zone
Annular High Pyrite Zone
Discovery Fault Zone
5000'
Kil
Marginal Stockwork Zone
Scale: 1:1200
Golden Sunlight Mines, Inc.
Geologic Cross Section Through
The Mineral Hill Breccia @ 26000'N
Whitehall District, Jefferson Co., MT
by: T.H. Chadwick ; August, 1992
Looking North
4500'
23500E
24000E
24500E

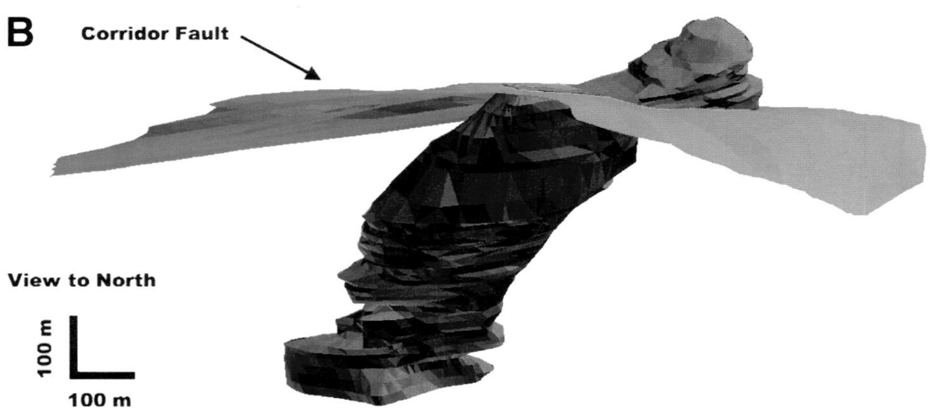

Figure 6. Two-dimensional representation of the current three-dimensional mine model illustrating the orientation of the primary mineralized breccia pipe and associated structures. (A) Mine model superimposed on oblique aerial photograph of the Golden Sunlight Mine in 2012. Note high-angle northeast-trending transverse faults flanking the southwest-plunging breccia pipe. Also note top-to-the-east truncation of the breccia pipe along the Corridor fault. (B) Cross-sectional view of the mineralized breccia pipe. View to north. Note the plunge to the southwest, and the top-to-the-east displacement of the breccia pipe along the Corridor fault. (C) View to the west of the main pit, 3 December 2013. Note the top of the breccia pipe and its orientation with respect to the high-angle, northeast-trending Lone Eagle and Fenner and Grey fault zones. Note also that the Corridor fault, which displaces the upper portion of the breccia zone, is offset by the high-angle transverse Lone Eagle fault.

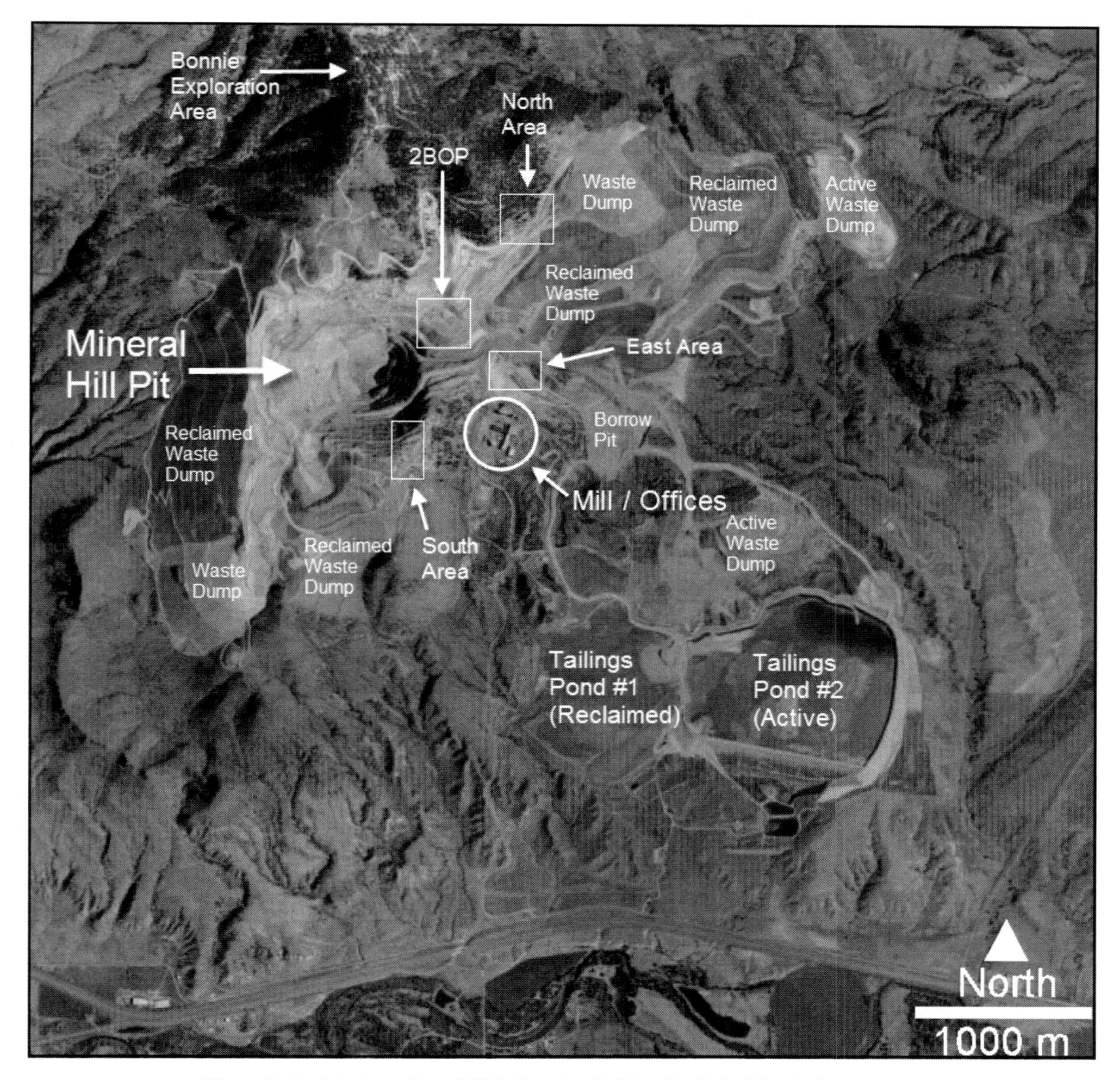

Figure 7. Aerial photo, June 2013, showing Golden Sunlight Mine infrastructure.

of pyrite (Spry and Thieben, 1997). Free gold occurs interstitially as 1–100 micron-sized particles. Gold tellurides include calaverite and buckhornite, and gold-silver tellurides include hessite, krennerite, sylvanite, and petzite (Spry and Thieben, 1997). Sericite, kaolinite, quartz, and barite occur as gangue with lesser dolomite, magnesite, dickite, orthoclase, and albite (Foster 1991; Gary Wyss, 2011, personal commun.; Peter Whittaker, 2012, personal commun.).

Brecciation decreases upward in the pipe, and the top of the breccia contained a weakly brecciated, strongly silicified cap. Outside of the breccia pipe and at the margins, the alteration is dominated by sericite, often with visible fuchsite. Argillic alteration is present to a lesser extent, and is generally marginal to the more dominant phyllic (quartz-sericite-pyrite) zone. Localized potassic alteration and molybdenite mineralization in the latite occur in the deeper levels of the breccia pipe, suggesting the pipe formed in an epithermal environment above an alkaline molybdenum porphyry system at depth (Foster, 1991; Foster and Childs, 1993).

Initial fluid inclusion and salinity analyses determined an average temperature of formation of 190 °C and a salinity of less than 1 weight percent NaCl, supporting the interpretation of a shallow epithermal system, with magmatic ore-forming fluids (Porter and Ripley, 1985). However, subsequent fluid inclusion and isotopic analyses show a wide range of temperatures and salinities, indicating a gradual continuum of multiple hydrothermal events. These events include periods of more vigorous boiling activity (up to 400 °C), as well as hydrothermal activity at

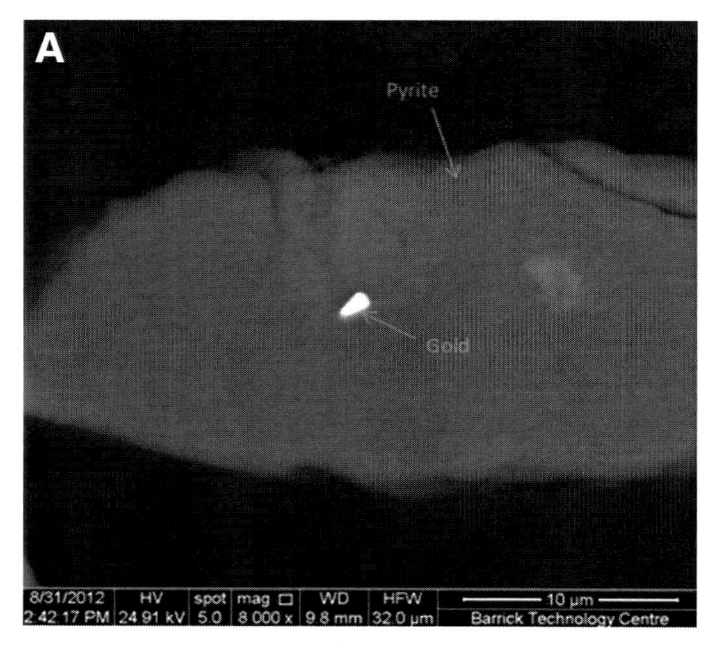

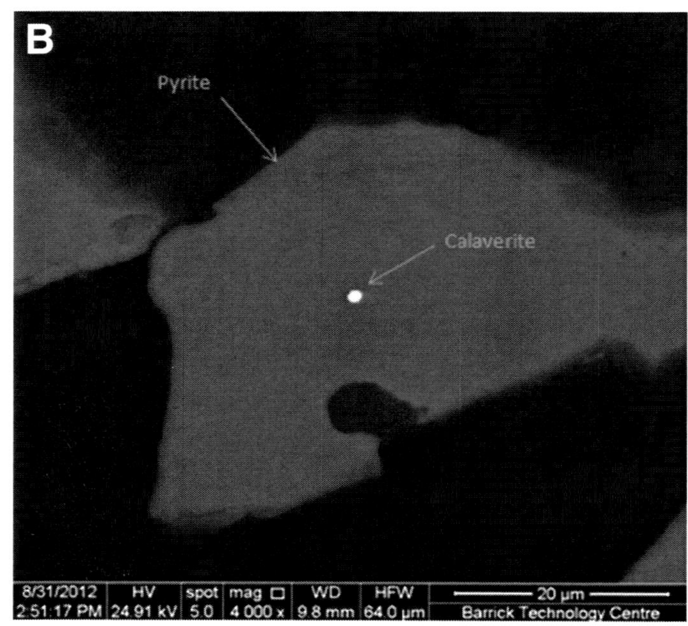

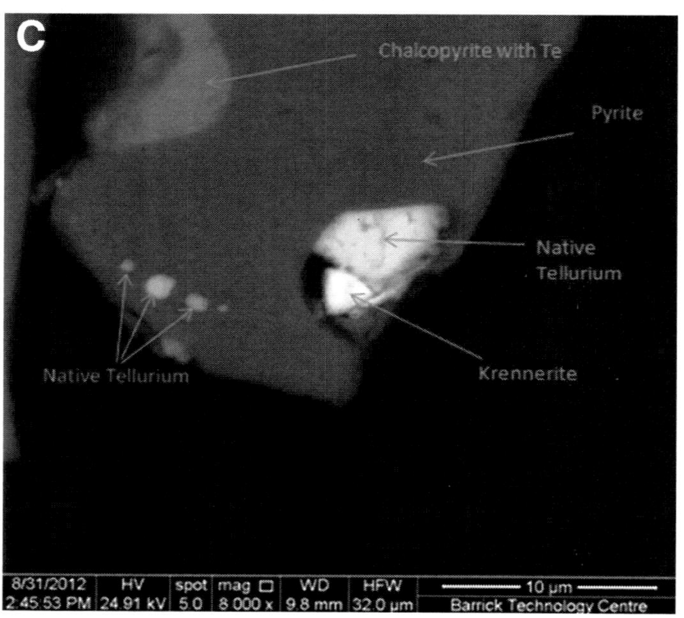

Figure 8. Photomicrographs of mineral textures within the Golden Sunlight breccia pipe: (A) photomicrograph of gold grain encapsulated in pyrite grain; (B) photomicrograph of calaverite grain encapsulated in pyrite grain; (C) photomicrograph of krennerite grain associated with tellurium, locked in pyrite. Source: Peter Lind, Erik de Jong, and Peter Whittaker, 2012, Barrick Technology Centre, Mineral Liberation Analyses, Barrick Gold Corporation internal report.

temperatures as low as 100 °C (Paredes, 1994). Higher salinities and temperatures are associated with quartz-pyrite-molybdenite stages of mineralization, while lower temperatures are associated with gold-silver mineralization (Paredes, 1994). Fluid inclusion analyses, hydrogen and oxygen isotope studies, and mass balance calculations all suggest gold ore-forming fluids were probably mixed magmatic-meteoric waters that interacted with a combination of sedimentary and igneous rocks, at low water/rock ratios, rather than being exclusively magmatic or meteoric (Paredes, 1994). Paredes (1994) suggests that early, magmatically derived quartz-pyrite-molybdenite ore-forming fluids were hot, saline, CO_2 bearing and intermittently boiling, and were followed by later, cooler, less saline gold ore–forming fluids that had a probable meteoric water component. The source of the gold in the system is controversial, with some workers suggesting remobilization of synsedimentary mineralization within the Belt Supergroup rocks as the primary source (Foster, 1991; Foster and Childs, 1993), and others arguing that the metals were derived from the Late Cretaceous alkalic magmatic system (Paredes, 1994).

The most likely mechanism for the generation of the Mineral Hill breccia pipe is brecciation caused by exsolution of a fluid-vapor phase from an alkali magma either during decompression (Porter and Ripley, 1985), or during magma withdrawal as the

TABLE 1. ECONOMIC MINERALS AT GOLDEN SUNLIGHT MINE

Minerals	Chemical composition
Gold and silver minerals	
Calaverite	$AuTe_2$
Electrum	AuAg
Gold	Au
Hessite	Ag_2Te
Jalpaite	Ag_3CuS_2
Krennerite	$(Au,Ag)Te_2$
Petzite	Ag_3AuTe_2
Silver	Ag
Sylvanite	$AgAuTe_4$
Tellurides	
Altaite	PbTe
Coloradoite	HgTe
Melonite	$NiTe_2$
Tellurium	Te
Tellurobismuthite	Be_2Te_3
Copper sulfides	
Bornite	Cu_5FeS_4
Chalcocite	Cu_2S
Chalcopyrite	$CuFeS_2$
Covellite	CuS
Tennantite	Cu_3AsS_3
Copper oxides	
Chrysocolla	$Cu_2H_2(Si_2O_5)(OH)_4$
Copper_Zn	$Cu_{0.8}Zn_{0.2}$
Cuprite	Cu_2O
Other sulfides	
Arsenopyrite	FeAsS
Galena	PbS
Sphalerite	ZnS

Sources: Spry et al. (1996, 1997); Spry and Thieben, (1997); Gary Wyss, The Center for Advanced Mineral & Metallurgical Processing (CAMP) (2011, Barrick internal reports); Peter Whittaker, Logan Jameson, and Barrick Technology Centre, 2012, 2013, Barrick internal reports. Methods used: reflected light petrography, scanning electron microscopy and/or energy dispersive x-ray spectroscopy (SEM/EDX), and mineral liberation analysis (MLA).

alkali magma cooled (Paredes, 1994). There is evidence to suggest that the breccia pipe initially fractured in place, incorporating adjacent wall rocks at the margin, and then collapsed (Foster, 1991; Chadwick, 1992). Paredes (1994) suggests that overlying wall rock caved into a slightly molten pluton. The mixing of meteoric water with magmatic vapor as a potential explosive breccia-forming process is not proven, but is suggested by the work of Porter and Ripley (1985), Spry and Thieben (1997), and Paredes (1994).

The mineralized breccia pipe was originally identified in surficial exposures at ~1785 m above sea level. The pipe plunges ~35° west-southwest, and over 610 vertical m of breccia have been drilled. However, both drill hole and underground mining data demonstrate that no gold has been found below an elevation of ~1325 m. The absence of gold in the lower portion of the breccia pipe suggests a vertical temperature gradient existed during mineralization, supporting the theory that a boiling zone, or low temperature epithermal setting existed, above which gold was deposited as the system cooled (Fess Foster, 2005, personal commun.).

Gold-silver mineralization in the 2BOP, North Area, and South Area satellite deposits outside of the Mineral Hill breccia pipe contains anomalous, but subeconomic copper, lead, zinc, molybdenum, telluride, and bismuth, as documented by trace elemental analysis of ore samples (Fig. 7; David Odt, 2013, personal commun.). These deposits are considered to be genetically related to the breccia system, and all mineralization is considered to be coeval with latite intrusion in Late Cretaceous time.

DEPOSIT GEOCHRONOLOGY

Mineralization at GSM is clearly genetically linked to Late Cretaceous magmatism, as latite emplacement and mineralization are complexly integrated. Latite at GSM occurs as discrete dikes and sills within the country rock, as clasts in the breccia pipe, and as an intrusive matrix in the breccia, suggesting that brecciation and mineralization were syngenetic with respect to latite magmatism (Foster, 1991). Therefore, accurate geochronologic analysis of the latite provides a maximum age constraint on mineralization.

Uranium-lead geochronologic analysis has been hampered by complex zircon systematics related to inheritance of Archean and Proterozoic lead associated with probable xenocrystic zircon from Belt Supergroup sedimentary rocks (DeWitt et al., 1996). The application of laser-ablation multicollector-inductively coupled plasma-mass spectrometry (LA-MC-ICPMS) utilizing a 30 micron spot provides the required spatial resolution to evaluate the zonation of xenocrystic and magmatic zircon, and permits differentiation of zircon core and rim domains. Analyzing zircon zonation permits delineation of various tectonic events recorded in the crust.

A geochronologic sample of a latite sill immediately adjacent to the breccia pipe was collected in 2012 from the 5700 bench level on the north wall of the Mineral Hill pit. The sill is mineralized with abundant pyrite, and is cut by an unmineralized lamprophyre dike. Zircon was extracted from the sample by conventional separation techniques and mounted in epoxy along with the Sri Lanka zircon standard (565 Ma), and imaged by back-scatter electron (BSE) microscopy prior to isotopic analysis. U-Pb analysis was conducted by LA-MC-ICPMS at the Arizona LaserChron Center, following procedures described by

Gehrels et al. (2008). Laser-ablation analyses were conducted on zircon cores and rims visible in BSE images. Age uncertainties were typically less than ± 1% at a 1σ level. The $^{206}Pb/^{207}Pb$ age was used for grains older than 900 Ma, and the $^{206}Pb/^{238}U$ age was used for younger grains. Analyses > 10% discordant were discarded. U-Pb data are given in Table 2.

A total of 21 U-Pb analyses from latite zircon passed the discordance filter (Table 2). The majority of the grains are Archean to Proterozoic in age ($n = 15$), with two distinct subpopulations of Mesoproterozoic ($n = 2$) and Late Cretaceous ($n = 4$) age (Fig. 9). The Archean to Proterozoic age grains appear systematically discordant, with younger Proterozoic zircon more systematically discordant than Archean grains. In particular, Proterozoic grains between 2600 and 1850 Ma form a well-developed chord with a lower intercept of ca. 900 Ma (Fig. 9). The Mesoproterozoic subpopulation was derived from rim analyses, and, while only comprising two grains, was concordant at 1660 ± 85 Ma. The Late Cretaceous subpopulation is also derived from rim analyses, and is concordant at 84.4 ± 2.1 Ma (Fig. 9).

Archean and Proterozoic zircon from the latite are interpreted as xenocrystic zircon representing detrital zircon from the Belt Supergroup, which were incorporated into the latite through crustal assimilation during magma generation. Note the strong correspondence between zircon derived from the latite with the detrital zircon population of the LaHood Formation southeast of the GSM (Fig. 10). Mesoproterozoic ages are interpreted to represent overgrowths developed during regional magmatism and metamorphism associated with development of the Great Falls tectonic zone (Mueller et al., 2002). The age of this subpopulation (1660 ± 85 Ma; n = 2) is slightly younger than the 1.77–1.86 Ga tectonic zone, but the timing seems too close to be a coincidence. The Late Cretaceous ages represent overgrowths developed during latite magmatism, and therefore represent the true age of the latite, providing a maximum age on magmatism and mineralization at GSM.

A minimum age constraint on mineralization is provided by cross-cutting, non-mineralized lamprophyre dikes. Numerous lamprophyre sills and dikes cut the gold-mineralized breccia pipe at GSM and Mesoproterozoic sedimentary rocks in the surrounding area. Lamprophyre dikes are unmineralized, and clearly cross-cut mineralized latite, quartzite, and vein systems. DeWitt et al. (1996) report a ^{40}Ar-^{39}Ar plateau date of 76.9 ± 0.5 Ma from biotite phenocrysts in the lamprophyre, which represents either a crystallization age or an uplift age. Either of these interpretations provides a minimum age constraint on mineralization at GSM.

The Late Cretaceous age of the latite sill at GSM is consistent with new geochronology from magmatic rocks in proximity to the GSM, including the southeastern portion of the Boulder batholith, satellite plutons, and Elkhorn Mountains Volcanics (Ihinger et al., 2011). The closest phases of the Boulder batholith, including the Rader Creek pluton (82.4 ± 0.5 Ma; ~20 km SW of GSM) and the Butte quartz monzonite east of Pipestone Pass (77.9 ± 1.3 Ma; ~23 km W of GSM), are either coeval or slightly younger than the latite at GSM (Ihinger et al., 2011). The Doherty Mountain stock, the closest satellite pluton of the Boulder magmatic system (~8 km E of GSM), yields a U-Pb zircon age of 79.3 ± 0.6 Ma. Folded sills adjacent to the Doherty Mountain stock yield Ar/Ar ages of ca. 80 Ma (Harlan et al., 2008). The tuff of Hadley Gulch, a rhyolitic welded tuff near the base of the Elkhorn Mountains Volcanics at the north end of Bull Mountain, yields a 79.2 ± 0.5 Ma age. Regionally, the age of the Elkhorn Mountains Volcanics ranges from ca. 78–85 Ma, while the age of plutonic rocks of the Boulder batholith ranges from ca. 75–81 Ma (Ihinger et al., 2011). These new geochronologic data suggest that latite intrusion and mineralization at GSM occurred early during the voluminous Late Cretaceous magmatic episode in southwest Montana. The widespread distribution of both lower Belt Supergroup rocks and Late Cretaceous magmatic rocks within and adjacent to the southwest Montana transverse zone suggests that the conditions responsible for mineralization at GSM are not unique within the region.

Milling

The chief technical challenge in milling at GSM is liberating micron-sized gold grains encapsulated in pyrite. Although rare free gold does occur, most of the 1–10 micron gold grains are encapsulated in 20–50 micron pyrite grains. Figure 8A shows a 1.2 micron × 0.7 micron native gold grain locked within a 30 micron pyrite grain; Figure 8B shows a 1.2 micron grain of a gold telluride, calaverite, locked in the center of a 50 micron grain of pyrite; and Figure 8C shows a 2 micron krennerite (gold-silver telluride) grain associated with tellurium and locked in a 20 micron pyrite grain. Unless the gold or gold-telluride grain encapsulated in the pyrite is exposed, or liberated, the mill will not be able to recover the gold. The average recovery of gold at the Golden Sunlight mill is 80 percent.

The Golden Sunlight milling process begins with conventional two-stage crushing. The mill circuit is shown in Figure 11. Primary grinding is completed in a rod and ball mill circuit where wet grinding reduces the ore to 65 percent < 100 mesh, or 150 microns. Ninety percent of the gold recovered in the mill is leached at that grind size. The grind circuit product from the rod and ball mill is thickened with flocculant, and the overflow with free gold is pumped to the carbon columns, where the gold is adsorbed on to activated carbon made from burnt coconut shells. The underflow, or ground ore slurry, is mixed with sodium cyanide, lime, and compressed air, and agitated in leach tanks to keep the solids in suspension during the process. The leaching process places gold into solution as a gold-cyanide compound, with gold recovered by carbon adsorption in a carbon-in-pulp (CIP) circuit. The sand fraction from the leach tailings is separated and goes to the sand tailing retreatment (STR) circuit, where the pyrite-rich, gold-bearing sulfides are separated from the sand by gravity concentration in spiral-shaped launders. The gold-bearing, pyrite-rich concentrate is then reduced in the regrind mill to ~80 percent < 400 mesh, or 37 microns, which liberates more of the fine gold,

TABLE 2. U-Pb GEOCHRONOLOGIC ANALYSES

						Isotope ratios					Apparent ages (Ma)								
Analysis	U (ppm)	$^{206}Pb/^{204}Pb$	U/Th	$^{206}Pb^*/^{207}Pb^*$	± (%)	$^{207}Pb^*/^{235}U^*$	± (%)	$^{206}Pb^*/^{238}U$	± (%)	Error corr.	$^{206}Pb^*/^{238}U^*$	± (Ma)	$^{207}Pb^*/^{235}U$	± (Ma)	$^{206}Pb^*/^{207}Pb^*$	± (Ma)	Best age (Ma)	± (Ma)	Conc. (%)
01JBM12-16	855	30855	0.9	20.3468	4.6	0.0872	5.1	0.0129	2.1	0.42	82.4	1.8	84.9	4.1	154.9	107.3	82.4	1.8	NA
01JBM12-30	306	21120	2.2	22.1247	16.2	0.0819	16.5	0.0131	3.2	0.20	84.2	2.7	79.9	12.7	–44.9	396.7	84.2	2.7	NA
01JBM12-25	513	31818	1.2	21.1534	4.1	0.0866	4.8	0.0133	2.6	0.53	85.1	2.2	84.4	3.9	63.1	98.0	85.1	2.2	NA
01JBM12-9	445	52997	1.6	22.0129	6.4	0.0862	7.0	0.0138	2.9	0.42	88.1	2.6	83.9	5.6	–32.6	154.4	88.1	2.6	NA
01JBM12-29	403	1140737	12.5	9.6648	1.3	4.2096	1.8	0.2951	1.2	0.70	1666.8	18.2	1675.9	14.5	1687.2	23.1	1687.2	23.1	98.8
01JBM12-28	194	348479	2.0	9.5417	0.4	4.1077	1.1	0.2843	1.0	0.92	1612.8	14.6	1655.8	9.1	1710.8	8.0	1710.8	8.0	94.3
01JBM12-19	318	817981	13.4	7.3882	0.7	6.6207	1.2	0.3548	1.0	0.80	1957.3	16.7	2062.1	11.0	2168.5	13.0	2168.5	13.0	90.3
01JBM12-34	163	252597	0.9	7.2685	0.8	6.0140	2.6	0.3170	2.5	0.95	1775.2	39.1	1977.9	23.0	2197.0	13.9	2197.0	13.9	80.8
01JBM12-27	92	367508	1.2	6.9691	1.8	6.6950	2.7	0.3384	2.0	0.76	1878.9	33.4	2072.0	23.8	2269.8	30.2	2269.8	30.2	82.8
01JBM12-2	57	108462	0.9	6.6755	1.2	8.1856	2.3	0.3963	2.0	0.86	2152.0	36.7	2251.8	21.1	2343.6	20.4	2343.6	20.4	91.8
01JBM12-11	278	675307	2.9	6.6280	3.3	7.5690	4.2	0.3638	2.6	0.62	2000.4	44.2	2181.2	37.3	2355.8	55.8	2355.8	55.8	84.9
01JBM12-4	404	831406	2.5	6.5575	0.9	8.8135	2.2	0.4192	2.0	0.91	2256.7	37.3	2318.9	19.7	2374.1	15.5	2374.1	15.5	95.1
01JBM12-13	34	84766	0.4	6.0990	1.2	10.4738	2.7	0.4633	2.5	0.90	2454.1	50.6	2477.6	25.5	2496.9	19.9	2496.9	19.9	98.3
01JBM12-21	1041	307684	28.9	6.0447	0.1	10.4591	1.3	0.4585	1.3	0.99	2433.1	25.5	2476.3	11.7	2512.0	2.2	2512.0	2.2	96.9
01JBM12-17	70	225420	2.3	5.8683	2.2	10.9060	3.4	0.4642	2.6	0.77	2457.9	53.5	2515.1	31.6	2561.6	36.3	2561.6	36.3	96.0
01JBM12-14	717	1264859	1.3	5.6425	1.3	11.4635	2.1	0.4691	1.7	0.80	2479.7	35.4	2561.6	20.0	2627.1	21.1	2627.1	21.1	94.4
01JBM12-5	362	428910	1.0	5.5021	0.3	12.4660	1.5	0.4975	1.5	0.98	2602.8	32.0	2640.2	14.3	2668.9	4.7	2668.9	4.7	97.5
01JBM12-22	805	1634110	1.8	4.8611	3.0	13.6671	4.0	0.4818	2.6	0.65	2535.3	54.1	2726.9	37.5	2872.1	48.8	2872.1	48.8	88.3
01JBM12-1	418	1112620	10.8	4.6063	0.6	16.8025	1.7	0.5613	1.6	0.93	2872.2	36.8	2923.6	16.4	2959.2	10.0	2959.2	10.0	97.1
01JBM12-23	282	611801	38.6	4.1535	2.3	19.6403	3.0	0.5916	2.0	0.67	2996.1	48.1	3073.8	29.2	3125.0	35.9	3125.0	35.9	95.9
01JBM12-10	355	588490	2.7	3.9662	1.8	16.9978	2.6	0.4890	1.9	0.73	2566.1	40.1	2934.7	24.9	3198.2	28.1	3198.2	28.1	80.2

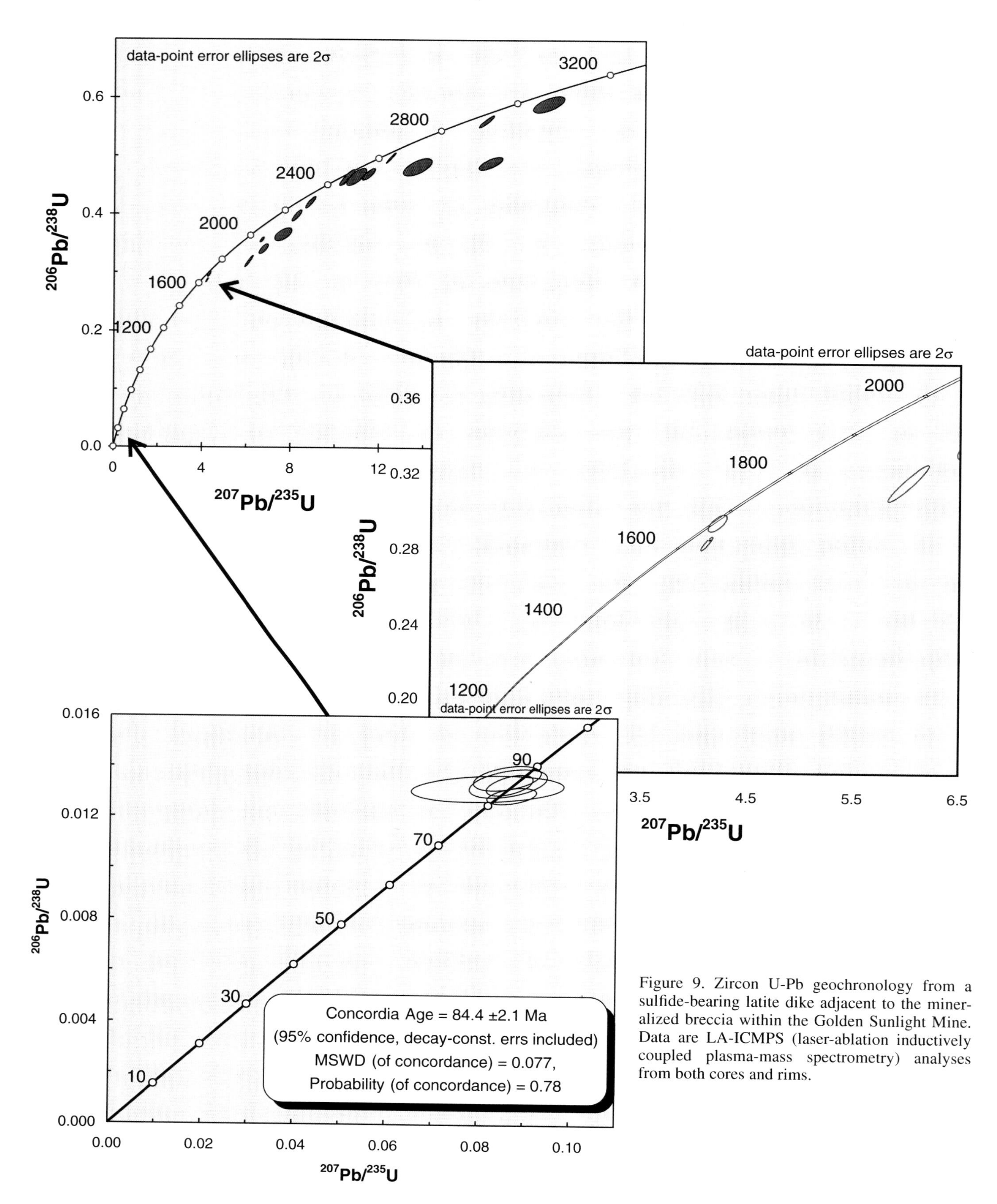

Figure 9. Zircon U-Pb geochronology from a sulfide-bearing latite dike adjacent to the mineralized breccia within the Golden Sunlight Mine. Data are LA-ICMPS (laser-ablation inductively coupled plasma-mass spectrometry) analyses from both cores and rims.

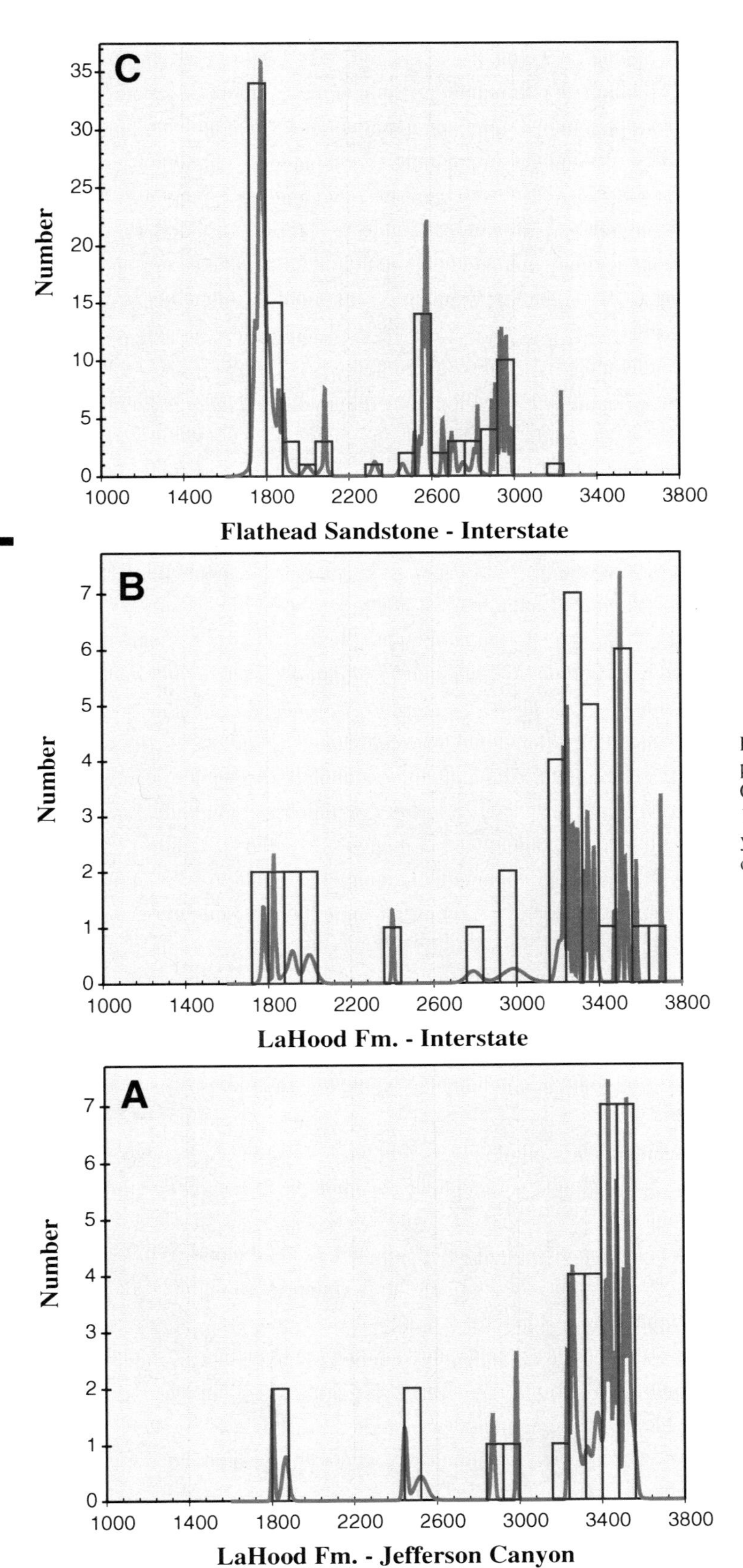

Figure 10. Detrital zircon probability plots of Belt Supergroup strata and the Cambrian Flathead sandstone (Stop 1). The *x* axis shows age in millions of years and the *y* axis shows the number of grains.

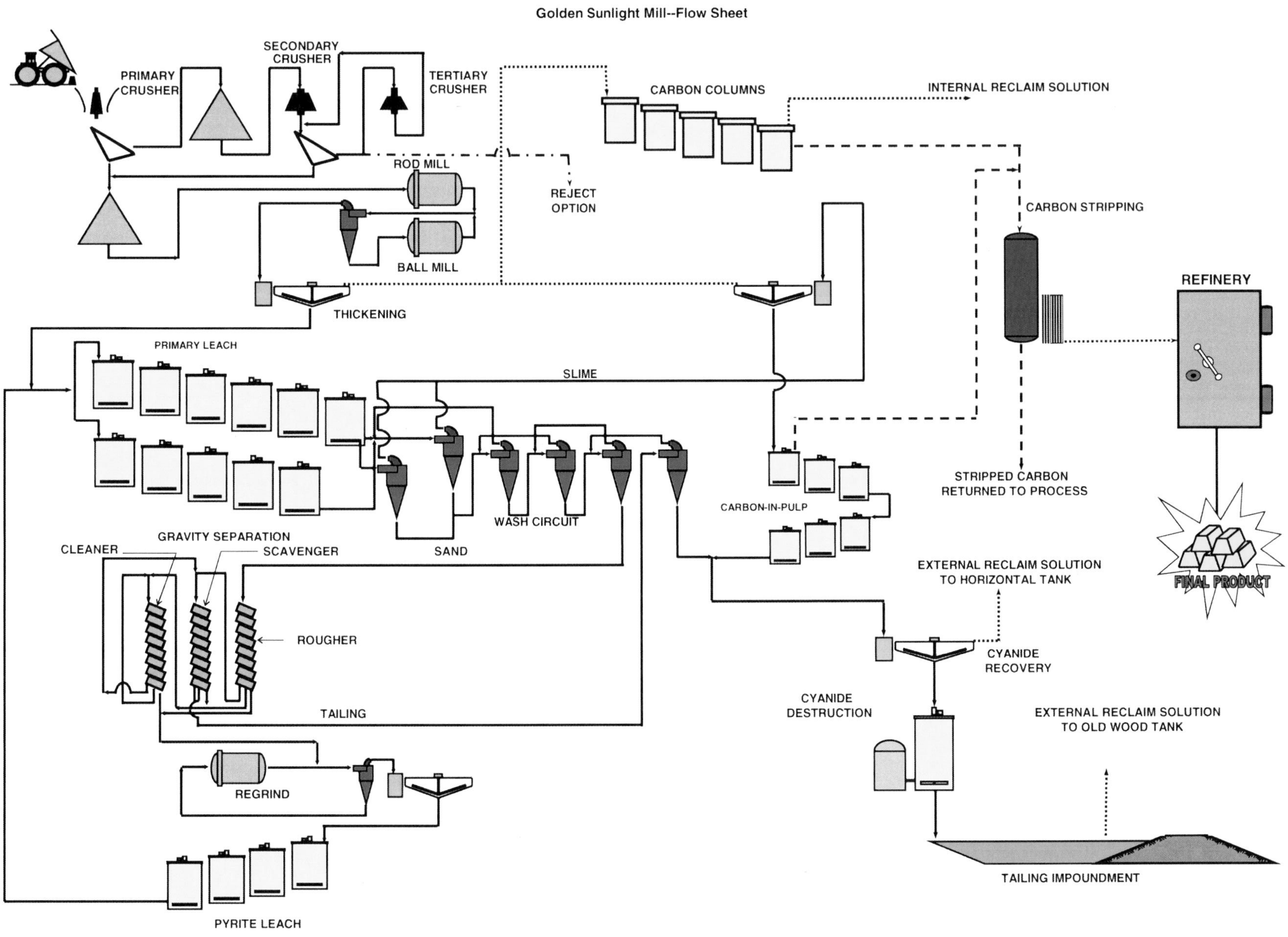

Golden Sunlight Mill--Flow Sheet

PRIMARY CRUSHER
SECONDARY CRUSHER
TERTIARY CRUSHER
CARBON COLUMNS
INTERNAL RECLAIM SOLUTION
ROD MILL
REJECT OPTION
BALL MILL
CARBON STRIPPING
REFINERY
THICKENING
PRIMARY LEACH
SLIME
WASH CIRCUIT
CARBON-IN-PULP
STRIPPED CARBON RETURNED TO PROCESS
FINAL PRODUCT
GRAVITY SEPARATION
CLEANER
SCAVENGER
SAND
EXTERNAL RECLAIM SOLUTION TO HORIZONTAL TANK
ROUGHER
CYANIDE RECOVERY
CYANIDE DESTRUCTION
TAILING
EXTERNAL RECLAIM SOLUTION TO OLD WOOD TANK
REGRIND
TAILING IMPOUNDMENT
PYRITE LEACH

Figure 11. Mill flow sheet.

allowing that additional 10 percent of gold still encapsulated in pyrite to be leached and recovered.

Gold is removed from the carbon in both the carbon columns and CIP circuit in pressure-stripping vessels, and returned to solution for electrowinning onto steel wool cathodes. The gold-laden steel wool is smelted and poured into bars that vary in composition, with an average of 75 percent gold, 8 percent silver, and 13 percent copper and other metal impurities.

Tailing products from both the gravity and the CIP circuit are recombined and sent to a cyanide recovery thickener. Cyanide in the tailings slurry is treated in a cyanide destruction process that utilizes a combination of ammonium bisulfite and hydrogen peroxide to remove ~99.9 percent of the cyanide in the tailings.

The potential to increase recovery by installing an additional regrind mill does exist. However, due to the cost of power as well as that of grinding media, grinding the entire ore stream to 80 percent < 40 microns is not economically feasible at the current grade of GSM deposits. Recent lab work, however, indicates that grinding only the pyrite concentrate to 5 microns could significantly increase the overall mill recovery. While grinding equipment to produce such a fine grind was not available when the STR circuit was originally installed, such equipment is available today. The Golden Sunlight Mine is further investigating the benefits of finer pyrite concentrate grinding with the help of the Barrick Technology Centre in Vancouver, British Columbia.

Reclamation

After blasting, the rock is segregated into waste rock and ore. Waste rock is trucked to a waste dump area that has been pre-stripped of topsoil that is stockpiled for later use. Upon completion, the dump is recontoured to a natural configuration, and the stockpiled topsoil is placed over the dump in a uniform thickness to create a water balance cap. The topsoil is subsequently amended with compost, and reseeded with native grasses and shrubs. Ore is trucked to the mill for removal of gold, and, after processing, the tailings are transported by pipe to the tailings impoundment facility in the form of a 50 percent slurry (half water and half ground rock). Cyanide in the tailings slurry is treated in a cyanide destruction process that removes ~99.9% of the cyanide in the tailings. The Golden Sunlight Mine earned the "International Cyanide Management Code For the Manufacture, Transport, and Use of Cyanide in the Production of Gold" (Code) certification in 2011. The Code is a voluntary industry program that focuses exclusively on the safe management of cyanide by gold mining companies, and companies involved with the production and transport of cyanide to gold mining companies. The Code was developed by a multi-stakeholder steering committee under the guidance of the United Nations Environmental Program and the then-International Council on Metals and the Environment. To receive certification, companies must have their entire operations audited by an independent third party. Operations meeting the Code's strict requirements are able to use a unique trademark symbol, identifying the company as a certified operation. Audit results are made public, to keep stakeholders informed. Per the Code website (www.cyanidecode.org/about-cyanide-code), "The objective of the Code is to improve the management of cyanide used in gold mining and to assist in the protection of human health and the reduction of environmental impacts." The Code is managed by the International Cyanide Management Institute.

Instead of waiting until the end of mine life to begin reclamation, GSM conducts reclamation concurrently with mining. To date, roughly half of the ~970 ha (2400 acres) of mining-related land disturbance (waste rock disposal sites, tailings, and other areas) has already been reclaimed (Fig. 7) concurrent with active mining operations, at a cost exceeding $12 million. The Golden Sunlight Mine has posted more than $98 million in financial assurance with the state of Montana and the U.S. Bureau of Land Management (BLM) for reclamation and closure.

The mine engages all stakeholders—employees, area citizens, community leaders, local businesses, and government—in closure related decision making. The mine established a model relationship through the Community Transition Advisory Committee (CTAC). CTAC's membership is a diverse cross section of people from the local community and Jefferson County. CTAC's goal is to sustain mining operations in the area and mitigate economic hardship throughout the mine life cycle, including closure, through diversification of the local economy. Barrick Gold Corporation strives to maintain a high degree of transparency and an excellent reputation not only with citizens and legislators in Jefferson and adjacent counties, but also with the BLM, the Montana Department of Environmental Quality, and the U.S. Forest Service.

Golden Sunlight's reclamation goals are to return mine site land to a condition capable of supporting industry, recreation, superior wildlife habitat, and grazing. Wildlife habitat and the mine's environmental policy extend well beyond the boundaries of the mine area. The Golden Sunlight Mine's land holdings include 215 ha (530 acres) of wetland adjacent to the Jefferson River, 260 ha (640 acres) of elk-calving ground on the Bull Mountains, and the 1425 ha (3520 acre) Candlestick Ranch property along the Boulder River. The Golden Sunlight Mine has opened much of its lands to the public for recreational hunting and fishing, and supports habitat conservation organizations.

Third Party Ore

Montana's long history of mining, much of which pre-dates modern mining and environmental regulation, left a collection of legacy tailings impoundments and waste rock piles that require cleanup. With taxpayers responsible for future cleanup costs, GSM found a creative solution. Many historic tailings impoundments and waste rock piles contain not only acid generating sulfides, but also gold that can be extracted profitably at current prices. Golden Sunlight offered to reprocess and store this material in its modern facilities, removing the need for taxpayer-funded cleanups. This project encouraged businesses to obtain permits to remove the contaminated material, collect it, truck it to the mine, and sell it. The material is processed at Golden Sunlight's mill and the new tailings are stored in Golden Sunlight's tailings impoundment.

Since 2010, tailings and waste rock have been removed from dozens of historic mines. Golden Sunlight has received ~508,000 tons of tailings and has paid out approximately $35 million to local businesses collecting and transporting the material, and has recovered more than 34,000 ounces of gold. In addition to creating jobs throughout the state of Montana, the third party ore program accomplishes what would have cost taxpayers millions of dollars in cleanup costs. For this reason, the Bureau of Land Management (BLM), U.S. Forest Service, and the Montana Department of Environmental Quality view this program as a win-win situation. To promote responsible management of this program, GSM does not process any material from sites found to be in nonconformance with their permits, and will shut down any operator until agency compliance is met, encouraging responsible mining through a very close and transparent relationship with the regulatory community. In 2012, GSM won the prestigious national BLM environmental award for hard rock mining, in large part to recognize this project.

ROAD LOG

We travel west from Bozeman, Montana, to the Golden Sunlight Mine (GSM) on I-90 and back to Bozeman via Montana Highway 2 through Jefferson Canyon and then on I-90. The field trip route is shown on a portion of the *Geologic Map of Montana* (Fig. 2). The western part of the route is covered by the Butte South 30′ × 60′ Quadrangle (McDonald et al., 2012) while the eastern part is covered by the Bozeman 30′ × 60′ Quadrangle (Vuke et al., 2014). Please note that the Datum for all UTM locations for the field stops is WGS84.

Cumulative distance (mi)	*Incremental distance (mi)*	*Cumulative distance (km)*	*Directions*
0.0	0.0	0.0	Pull out of the Strand Union Building parking area at Montana State University, turn right on West Grant Street, and proceed to stop sign.
0.2	0.2	0.3	Turn left on South 11th Ave. and proceed one block.
0.3	0.1	0.5	Turn right on West Lincoln Ave. and proceed to stop sign.
0.8	0.5	1.3	Turn right on South 19th Ave. and proceed north across West Main Street. As we travel North on 19th Ave., the Bridger Range appears ahead and to the right (Fig. 2). Tight fold hinges defined by resistant Paleozoic limestone units are clearly visible. The foothills of the range are underlain by Archean gneiss and schist. The range is a basement-cored Laramide uplift, which has been modified by Tertiary extension along west-dipping Basin and Range normal faults that have down-dropped the hanging wall block to form the Gallatin Valley (Lageson, 1989). The range comprises an east-verging overturned anticlinorium broken by numerous Laramide reverse faults. Ross Pass is a prominent saddle in the middle of the Bridger Mountains that lies immediately south of the prominent Ross Peak. This pass can be seen behind us as we travel west on I-90. The pass marks a major structural boundary between fold-and-thrust–style deformation involving Proterozoic Belt Supergroup rocks to the north and basement cored Laramide-style deformation to the south (Lageson et al., 1983). We will examine a similar structural transition zone later in the field trip when we traverse the Jefferson Canyon.
4.5	3.7	7.2	Continue north on North 19th Ave., cross I-90 on overpass, and turn left onto I-90. Proceed west toward Butte. As we travel west, notice the excellent views of the mountain ranges that border the Gallatin Valley (Fig. 2). The high country of the Gallatin Range to the south is underlain by thick accumulations of basalt and basaltic andesite of the Eocene Absaroka Volcanics. The northern Madison Range and the Spanish Peaks are visible to the southwest. The trace of a major normal fault is visible along the range front at the southern margin of the Gallatin Valley.
9.5	5.0	15.2	Large gravel quarries on the right are developed in ancient river gravels in the floodplain of the Gallatin River.
11.3	1.8	18.1	Belgrade Exit
14.2–15.8	2.9	22.7–25.28	Subdivisions on the left and right are a result of rapid population growth in the Gallatin Valley.

Cum. distance (mi)	*Increm. distance (mi)*	*Cum. distance (km)*	*Directions*
			The low Horseshoe Hills lie to the north (Fig. 2). These hills are underlain by northeast-trending tight to open folds and thrusts of the Montana fold-and-thrust belt (Verrall, 1955; Lageson, 1992). We will traverse numerous thrusts and upright to overturned, east-verging folds as we proceed to Cottonwood Pass west of Three Forks.
17.4	3.2	27.8	Cross Gallatin River. Inferred Central Park normal fault crosses the highway just to the west of Gallatin River.
21.0	3.6	33.6	Manhattan Exit
23.0	2.0	36.8	Horseshoe Hills are on the right across the Gallatin River.
25.0–26.4	2.0	40.0–42.2	Paleozoic strata are visible to the right across Gallatin River.
26.4	1.4	42.2	Logan Exit The prominent cliffs across the Gallatin River at Logan are the type locality for the Mississippian Madison Limestone (Sando and Dutro, 1974). The Paleozoic strata here, including the Madison Limestone, form the western limb of a tight anticline in the hanging wall of the Nixon thrust (Vuke et al., 2014; Lageson, 1992).
27.3	0.9	43.7	Start descent from Quaternary alluvial deposits onto the floodplain of the Madison River (Fig. 2), which is underlain by Miocene tuffaceous siltstone, sandstone, and conglomerate (Vuke et al., 2014).
29.8	2.5	47.7	Imbricate faults and thrusts in Paleozoic strata underlie the low hills to the north.
31.1	1.3	49.8	Cross Madison River.
31.4	0.3	50.2	Three Forks Exit Holcim Inc. operates a cement plant on the Missouri River three miles north of the highway here. The mine is in Paleozoic carbonate rocks in the Trident syncline, which is bounded by the Green thrust on the east and the Trident thrust on the west. The cement plant was built in 1910. Currently, limestone is mined near the plant, and other ingredients for the cement are mined elsewhere in southwestern Montana and transported to the facility (Mike Mullaney, 2013, personal commun.). The Holcim operation has 74 employees, some of whom are third-generation employees at Trident. Holcim is an international company and operates 12 cement plants in the U.S. with 1800 employees.
32.5	1.1	52.0	Cross Jefferson River.
35.0	2.5	56.0	Pass Exit 274 for Montana Highway 287 and Helena. Wheat Montana, a family owned farm, mill, and bakery north of the interstate, is a great place for coffee and a muffin. Continue west on I-90.
37.6	2.6	60.2	On the left, northwest-dipping Paleozoic sedimentary rocks are exposed in the hanging wall of the imbricate Lombard thrust. We will examine thrusts at the east end of these exposures during our last stop of the day (Fig. 2).
38.3	0.7	61.3	Jefferson County line
38.7–39.5	0.4	61.9–63.2	Low roadcuts in Eocene or Cretaceous andesite porphyry are visible.
42.2	3.5	67.5	Milligan Canyon Exit
45.0	2.8	72.0	Complexly folded Paleozoic rocks underlie the wooded hills ahead on both the right and left.
48.4	3.4	77.4	Cottonwood Pass
49.8	1.4	79.7	Start traverse through a spectacular folded section of Paleozoic rocks in outcrops and roadcuts on both sides of the highway (Fig. 2). On the north side of the highway, steeply dipping Paleozoic rocks extend to the northeast up the west side of Cottonwood Canyon, forming the east limb of a tight northeast-plunging anticline. Note abrupt changes in dip direction due to folding.

Cum. distance (mi)	*Increm. distance (mi)*	*Cum. distance (km)*	*Directions*
50.5	0.7	80.8	Approximate contact between the Paleozoic section and the underlying Proterozoic Belt rocks to the west.
50.7–51.0	0.2	81.1–81.6	**STOP 1: Sulfide-Bearing Shale and Sandstone of the LaHood Formation (UTM 0430253 E; 5081185 N)** *Please use extreme caution and wear your safety vests at this stop.* Stay on the shoulder and avoid entering the traffic lanes on this dangerous corner. This stop has excellent roadcut exposures of the thin to thickly bedded Mesoproterozoic LaHood Formation shale, siltstone, and pebbly sandstone (Foster and Chadwick, 1999; Foster et al., 1999). These rocks strike N70E and dip ~70° to the north. Several "bundles" of thickening upward sandstone beds can be seen in this outcrop, suggesting that these rocks were deposited in an outer fan environment. The beds display scour features and graded bedding and are cut by numerous faults. Approximately 1/3 of the way from the east end of the outcrop, a fold hinge occurs between relatively planar beds above and below it. This is interpreted as a possible syn-sedimentary slump structure. A prominent channel is visible immediately above the folded strata. Detrital zircon data from both the LaHood Formation and overlying Cambrian Flathead Sandstone from the roadcut ~800 m east are shown in Figure 10. Note the strong Archean component in the LaHood Formation, consistent with derivation of sediment from the Archean Dillon block to the south. The Flathead Sandstone has a much different detrital zircon signature, dominated by a major 1780 Ma peak of unknown provenance. The lack of Archean grains relative to the LaHood Formation suggests the Dillon block was not an active source in Cambrian time. Perhaps the most interesting feature of this outcrop is the strong alteration of shale beds. Supergene bleaching and limonite staining of these beds occurred as a result of oxidation and acid leaching of fine-grained syn-sedimentary sulfides. The shale beds were deposited in an anoxic environment, and are black where unoxidized. See the Golden Sunlight Mine straight ahead across the Boulder River Valley (Fig. 3). The low hills in the Boulder River Valley are underlain by Mesoproterozoic Belt Supergroup sedimentary rocks and Tertiary valley-filling sediment. The foothills below the mine are underlain by deformed Paleozoic sedimentary rocks. The Golden Sunlight orebody occurs within a Late Cretaceous breccia pipe hosted by LaHood Formation of the Belt Supergroup, which will be discussed during our mine visit. Belt Supergroup rocks in the mine area are overlain by Paleozoic sedimentary rocks to the north, which are in turn overlain further to the north by the Cretaceous Elkhorn Mountains Volcanics (Fig. 4). The Elkhorn Mountains Volcanics are the cogenetic extrusive equivalents of the Boulder batholith exposed on the skyline to the west.
51.5	0.8	82.4	On the right, a buttress outcrop of Devonian Jefferson Dolomite is exposed in the thickened hinge of a northeast-plunging syncline, which lies adjacent to the overturned anticline that we just traversed. This synclinal hinge is visible behind us for the next few miles as we descend into the valley of the Boulder River.
53.4	1.9	85.4	Cross Boulder River.
53.7	0.3	85.9	Take the Cardwell–Boulder Exit.
53.9	0.2	86.2	Turn right onto the frontage road (Montana Highway 2) and follow it west.
54.3	0.4	86.9	The east limb of the Cardwell syncline is defined by a prominent outcrop of the Cambrian Flathead Formation. The west limb of this syncline is marked

Cum. distance (mi)	*Increm. distance (mi)*	*Cum. distance (km)*	*Directions*
			by an east-dipping exposure of the same quartzite ~120 m to the west. The Flathead is underlain by iron-stained Mesoproterozoic Belt Supergroup feldspathic sandstone and siltstone.
54.4	0.1	87	West limb of the Cardwell syncline.
55.2	0.8	88.3	Flashing yellow light at the intersection with Montana Highway 69 leading north to Boulder. Do not turn. Continue straight ahead on Montana Highway 2. At 1:30, note the V-shaped valley reclaimed by GSM. The low hill at 12:30 is underlain by a latite sill, similar to the latite that hosts some of the ore in and adjacent to the breccia pipe orebody. Breccia pipes similar to the pipe at GSM typically occur in clusters rather than as a single pipe. The prominent hill just left (south) of the V-shaped dump is underlain by a latite intrusive complex and small breccia pipes, which are similar to the mineralized pipe that hosts the main Golden Sunlight orebody ~1.5 km to the north. No significant mineralization has been discovered in the southern intrusive center, even though its composition and breccia bodies are otherwise similar to the main mineralized pipe to the north. The north end of the Tobacco Root Mountains is visible to the south (left). Gold mineralization in this area occurs in veins such as those worked at the Mayflower Mine, which is located in the sparsely wooded foothills.
56.7	1.5	90.7	Turn right onto gravel road leading to GSM. The Golden Sunlight sign is easy to miss from the highway unless you are looking for it.
56.9	0.2	91	Mine access road swings sharply to the left. The water tank at 12:00 and the street grid on the right are part of the Sunlight Business Park, which has been created by the mine in coordination with Jefferson (County) Local Development Corporation. The land for this business park was donated by the mine, and the development is designed to attract businesses that will provide jobs when the mine closes.
57.8	0.9	92.5	Reclaimed tailings pond is visible on the right and the southern latite intrusive complex is exposed in steep slopes to the left (Figs. 3, 7).
58.3–58.4	0.5	93.3–93.4	Outcrops along the mine access road include bright-red debris flows that have been deposited east of a range-front fault, located at the base of the steep slopes west of the access road. Figure 7 shows the location of the East Area, a debris flow that contained sufficient gold to be mined in 2011.
58.7	0.4	93.9	Access road bends to the left. Mine dumps on upper slope straight ahead are from three levels of mining on the Golden Sunlight vein prior to initiation of the open pit operation in the 1980s. Approximately 30,000 tons at 1 ounce of gold per ton were mined from the Golden Sunlight vein (Fess Foster, 2005, personal commun.).
58.8	0.1	94.1	**STOP 2: Golden Sunlight Mine (UTM 0421455 E; 5083553 N)** Turn right into the visitor parking lot of GSM. We will meet mine personnel at the gate house up the stairway on the north edge of the parking lot. Please stay with the field trip group at the vehicles while GSM personnel organize the mine tour. Refer to the Mine Area Geology, Structural Controls, Mineralization and Alteration, and Deposit Geochronology sections of this chapter for details regarding the mine geology. The mine tour will provide a comprehensive overview of the geology of the deposit, ore processing facility, reclamation process, and the third party ore program. Stops planned in the mine include: 1. Overview of the Mineral Hill open pit; 2. Examination and sampling of ore stockpile; 3. Primary crusher;

Cum. distance (mi)	Increm. distance (mi)	Cum. distance (km)	Directions
			4. Mill; and 5. Reclamation and third party ore. Please note that the stops within the mine are subject to modification depending on ongoing mine operations and safety factors. When we leave the mine, we will retrace our route on Montana Highway 2 east to where the frontage road passes under I-90 at Cardwell. Rather than getting on I-90, we will go under the freeway and turn left on the Jefferson Canyon Road.
58.8	0.0	94.1	Leave visitor lot at GSM.
60.8	2.0	97.3	Turn left onto Frontage Road.
62.4	1.6	99.8	Intersection with Boulder Road; stay straight on Highway 2.
63.7	1.3	101.9	Pass under I-90.
63.8	0.1	102.1	Immediately south of I-90, intersection with road to Lewis and Clark Caverns and Yellowstone Park (Highway 2, Jefferson Canyon road). ***Please reset mileage.*** This allows people interested in running the Jefferson Canyon portion of the field trip in the future to do so easily without starting from GSM.
0.0	0.0	0.0	Turn left on Lewis and Clark Caverns Road (Montana Highway 2), and continue east.
0.5	0.5	0.8	Cross Boulder River roughly where the highway joins the Jefferson River.
1.7	1.2	2.7	Exposures on left of Mesoproterozoic Belt Supergroup have been interpreted as progradational basin plain, submarine fan, slope, and shelf sedimentation deposited north of the ancestral Willow Creek fault (Foster et al., 1999; Foster and Chadwick, 1999). The range front here is bounded on the west by the north-striking Starretts Ditch fault. This structure is a late Cenozoic Basin and Range fault that separates Mesoproterozoic LaHood Formation on the east from Eocene to Oligocene basin-filling sediments of the Renova Formation on the west (Kuenzi and Fields, 1971; Vuke et al., 2014). This young fault is responsible for uplift of the block that we will be driving though as we continue east. This recent uplift led to incision of the Jefferson River and development of the Jefferson Canyon, which stranded distinctive tilted paleo-river gravels, termed the "Ballard Gravels." These gravels are as much as 450 m (1500 feet) vertically above the present river level (Schmidt et al., 1987).
2.2	0.5	3.5	LaHood Bar and Steakhouse. This local geologists' watering hole is famous for warm hospitality, great food, and well-lubricated geologic discussions. They serve some of the best steaks in southwest Montana at a reasonable price. Depending on timing, we may stop here for lunch. Shadan LaHood, a Lebanese immigrant, built a lodge here in the 1920s. It was used as a Civilian Conservation Corp. facility during the Great Depression. The burnt remnant of the lodge is visible immediately south of the restaurant.
2.3	0.1	3.7	Enter Jefferson Canyon. The road follows the northeast bank of Jefferson River. This canyon was first recorded on 26 July 1805 by William Clark of the Lewis and Clark expedition. The canyon saw little traffic until 1881. The Jefferson Canyon provides a unique transect of the southwest Montana transverse zone. This zone is a long-lived structural feature that has been active from at least the Mesoproterozoic through the Cenozoic (Schmidt and O'Neill, 1982; Schmidt and Garihan, 1986; Schmidt et al., 1988). During the Mesoproterozoic, the Willow Creek fault, which is roughly coincident with the southwest Montana transverse zone in the Jefferson Canyon area, formed the east-west–trending southern margin of the Belt basin. The coarse sediment of the LaHood Formation was shed off the uplifted Archean

Cum. distance (mi)	*Increm. distance (mi)*	*Cum. distance (km)*	*Directions*
			highlands immediately to the south of the Willow Creek fault, and deposited along the southern margin of the Belt basin. These coarse-grained rocks are found as far east as the Bridger Range north of Bozeman, and as far west as the Highland Mountains south of Butte (Fig. 2). Jefferson Canyon provides some of the best exposures of the LaHood Formation as well as structural features of the southwest Montana transverse zone. The transverse zone forms the structural transition from the thin-skinned deformation of the Sevier fold-and-thrust belt to the north, to the Laramide deformation of the Rocky Mountain foreland, characterized by steep reverse faults and uplifted Archean crystalline blocks to the south (Schmidt and O'Neill, 1982). The zone was proposed as a lateral ramp in the fold-and-thrust belt that developed along the Willow Creek fault where the thin-skinned structures to the north terminated against Archean basement exposures to the south (Schmidt and Garihan, 1986; Schmidt et al., 1987, 1988). The southwest Montana transverse zone forms the southern margin of the Helena tectonic salient, which is an eastward tectonic bulge in the fold-and-thrust belt. The generally east-northeast faults of the transverse zone are typically oblique reverse faults with a dextral sense of slip. Structures in the Rocky Mountain foreland south of the transverse zone are dominated by northwest-trending faults and folds that involve Archean basement rocks (Foster and Childs, 1993). Numerous previous field trips provide detailed descriptions of the geology of the Jefferson Canyon. These include but are not limited to Johns et al. (1981), Lageson and Montagne (1981), Hawley et al. (1982), Lageson et al. (1983), Schmidt et al. (1987, 1989), and Lewis (1988). Mapping of exposures within and surrounding the Whitehall valley led to the interpretation of the Mesoproterozoic LaHood, Newland, and Greyson Formations as products of prograding basin plain, submarine fan, slope, and shelf sedimentation along the fault-controlled southern margin of the Belt basin (Foster et al., 1999; Foster and Chadwick, 1999). Relatively uncommon blocks and slabs of limestone within the LaHood Formation were interpreted as possible olistoliths shed from a carbonate shelf facies developed in shallow water on the uplifted block south of the Willow Creek fault. McTeague and Schmitt (2003) suggest that some of the LaHood Formation in the Jefferson Canyon area is made up of subaerial alluvial fan deposits.
2.9	0.6	4.6	**STOP 3: Exposure of LaHood Sandstone and Conglomerate (UTM 0428806 E; 5077271 N)** At a road sign indicating a bend to the left, pull off to the left and park. *Use extreme caution near the highway here.* A 6-m-wide near vertical coarse conglomerate layer in coarse-grained sandstone of the LaHood Formation is well exposed. Subangular to subrounded clasts of various Archean lithologies up to 2 m in diameter are present in three boulder conglomerate layers. Graded beds are well developed here with tops to the west. The type section of the LaHood Formation (Alexander, 1955) is in a gully to east. A fault zone offsets the conglomerate at the west end of the outcrop.
3.1–3.7	0.2	5.0–5.9	Note excellent highway outcrops of dark-gray, coarse-grained sandstones and coarse conglomerate of the Proterozoic LaHood Formation. Zircon provenance studies by Frost and Winston (1987), Ross et al. (1992), and Ross and Villeneuve (2003) indicate that these coarse sediments were derived locally from Archean rocks of the Tobacco Root Mountains immediately to the south (Fig. 10).

Cum. distance (mi)	*Increm. distance (mi)*	*Cum. distance (km)*	*Directions*
3.8	0.1	6.1	**STOP 4: View Stop at Folded Paleozoic and Mesozoic Section (UTM 0429260 E; 5075797 N)** Looking west across the Jefferson River, a section of rocks ranging from the Mississippian Mission Canyon Limestone through the Cretaceous Kootenai Formation dip northwest, forming the southeast limb of a northeast-plunging, overturned syncline (Fig. 12). The overturned north limb of the syncline is cut off by a splay of the right-reverse Cave fault, which places overturned Cambrian Wolsey Shale and limestone of the Meagher Formation and the underlying Proterozoic LaHood Formation on the north against Jurassic and Permian rocks to the south (Schmidt et al., 1987). The Cambrian Flathead Formation (quartzite) is missing from the sequence. The east-northeast–trending Cave fault and the overturned syncline are representative of the deformation in the southwest Montana transverse zone where it impinges on the Montana fold-and-thrust belt to the north (Schmidt et al., 1987).
3.9	0.1	6.2	Note the good highway exposures of LaHood conglomerate.
4.1	0.2	6.6	Small outcrops of Paleozoic limestone are visible on the left and massive cliffs of generally northwest-dipping Mission Canyon limestone and other Paleozoic units are cut by thrusts on the right across river.
4.7	0.6	7.5	**STOP 5: Limespur Quarry (UTM 0430360 E; 5075605 N)** Pull off to the left. The open cut and large cavern 12 m above the road are the remnants of a limestone quarry developed along a specific bedding horizon within the northwest-dipping Mission Canyon Formation. The portal below the main cut ~3 m above the highway was a probable haulage level for rock mined via underground stoping. The limestone was mined starting in 1899 or 1900 for use as a flux in smelters in the Butte District thirty miles to the west (Blake, 1953). The quarry was in operation until 1935. To the east, steeply dipping Proterozoic through Mississippian strata are exposed beneath the Jefferson Canyon fault, which has dextral oblique movement and has placed the Mission Canyon Formation on top of the early Paleozoic section. Schmidt et al. (1987) estimate that net displacement on the Jefferson Canyon fault is ~9 km where the fault is exposed on the road to the Lewis and Clark Caverns. If time permits and people are interested, we will traverse east from the Limespur Quarry to examine this structurally complex thrust zone.
5.0	0.3	8.0	Note folds in limestones to the right across the river. Extensive historic placer workings (dating to 1889) are present well above the Jefferson River. These

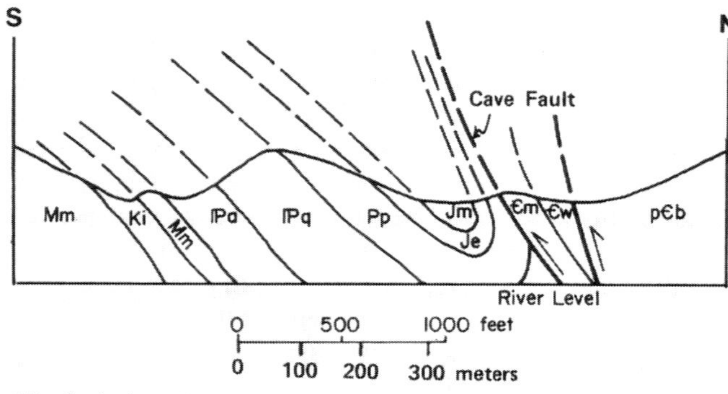

Figure 12. Jefferson Canyon area cross section and photograph from Schmidt et al. (1987) (Stop 4).

Cum. distance (mi)	*Increm. distance (mi)*	*Cum. distance (km)*	*Directions*
			paleo-placers were worked using water conveyed via a tunnel and flume from the South Boulder River west of Jefferson Canyon (Blake, 1953).
5.8	0.8	9.3	Coarse sandstone and conglomerate of the LaHood Formation are visible on the left.
6.0–6.2	0.2	9.6–9.9	Intensely folded and faulted Paleozoic and Mesozoic rocks are cut by a northeast-trending thrust.
6.3	0.3	10.1	Note probable landslide deposits that occupy the tributary valley on the south side of the Jefferson River.
7.4	1.1	11.8	Lewis and Clark Caverns State Park on the left. President Theodore Roosevelt designated this area as a National Monument in 1908. The land was given to the state of Montana in 1932 and became Montana's first state park. These caverns are worth a visit if you have a chance. Schmidt et al. (1987) suggest that the caverns formed due to increasing gradient of the Jefferson River during basin and range uplift, resulting in groundwater flow along bedding plane faults and a fractured fold hinge in the Mississippian Madison Limestone.
9.2	1.8	14.7	Outcrops on the left are coarse lithic tuff of the Cretaceous Elkhorn Mountains Volcanics. The volcanic rocks are cut by north-trending thrusts and are overlain by Mesoproterozoic Belt rocks in the hanging wall of the Jefferson Canyon fault ~1.5 km to the north.
9.6	0.4	15.4	Jefferson River veers to the south away from the highway and enters a canyon cut through a faulted northeast-trending anticline in Paleozoic and Mesozoic strata. This deformation is within the southwest Montana transverse zone south of the Jefferson Canyon fault.
9.8	0.2	15.7	Outcrop on the left exposes the faulted Elkhorn Mountains Volcanics.
11.0	1.2	17.6	Here and at the intersection with Montana Highway 287 just ahead, the Sappington Talc Mill can be seen ~2 km to the south across the Jefferson River. The Sappington Mill is one of two mills operated by Imerys Talc. The second mill is located in the town of Three Forks and will be visible from the highway once we start traveling east on I-90 later in the trip.
11.2	0.2	17.9	Site of "Rockin' the Rivers" yearly rock concert and the proverbial "Bridge for Sale."
11.8	0.6	18.9	Excellent view of the Bridger Range straight ahead.
12.7	0.9	20.3	Intersection with Montana Highway 287. Turn left. The Sappington Talc Mill operated by Imerys Talc is again visible ~3 km to the southwest.
14.8	2.1	23.7	View ahead exposes the Paleozoic section in the hanging wall of the Lombard imbricate thrust.
15.4	0.6	24.6	Road on the right leads to Willow Creek, Montana. Stay on Highway 287. Continue straight ahead.
17.9	2.5	28.6	Broadwater County line
18.0	0.1	28.8	Milligan Canyon Road leads to the left, stay on Highway 287.
18.5	0.5	29.6	Buttress outcrops of Cambrian Meagher Limestone and Pilgrim Limestone are exposed in the hanging wall of the Lombard thrust.
19.3	0.8	30.9	**STOP 6: Highway Thrust (UTM 0450614 E; 5079019 N)** *At pullout: park on the right and use extreme caution when crossing the highway.* Thrust slices of white to gray-weathering glauconitic quartzite unit of the Cambrian Wolsey Shale overlie strongly hematite-stained, coarse-grained Belt sandstones of the LaHood Formation (Fig. 13). The Wolsey Shale is contorted and tectonically thinned due to thrusting. The Cambrian Meagher Limestone is in fault contact with the underlying Wolsey, and basal beds of the Meagher are missing (Robinson, 1963).

Figure 13. Structural interpretation of the Highway thrust fault. Image from Google Earth (Stop 6).

Cum. distance (mi)	*Increm. distance (mi)*	*Cum. distance (km)*	*Directions*
20.1	0.8	32.2	Roadcut in massive limestone of the Meagher Formation (?).
20.3	0.2	32.5	Strongly brecciated, silicified, and sheared Meagher Formation or Pilgrim Limestone are cut by gouge zones along thrust faults.
21.4–21.7	1.1	34.2–34.7	Road outcrops on left of strongly sheared Meagher Limestone and Wolsey Shale.
22.6	1.2	36.2	Intersection with road leading 5 km to the east to Three Forks. Site of the Three Forks Talc Mill operated by Imerys Talc. Continue straight north on Highway 287.
23.9	1.3	38.2	Turn right onto on-ramp for I-90 heading east toward Bozeman.
26.7	2.8	42.7	Cross Gallatin River.
27.3	0.6	43.7	Three Forks Exit
28.0	0.7	44.8	Cross Madison River.
32.3	4.3	51.7	Logan Exit
37.9	5.6	60.6	Manhattan Exit
41.8	3.9	66.9	Cross Gallatin River.
46.8	5.0	74.9	Belgrade Exit
54.4	7.6	87.0	Take 19th Ave. Exit.
54.9	0.5	87.8	Turn right on South 19th Ave.
56.3	1.4	90.1	Continue south across West Oak St.
57.3	1.0	91.7	Continue south across West Main St.
57.9	0.6	92.6	Continue south across West College St.
58.4	0.5	93.4	Turn left on West Lincoln St.
58.9	0.5	94.2	Turn left on South 11th Ave.
59.0	0.1	94.4	Turn right on West Grant St.
59.2	0.2	94.7	Turn left into Strand Union Building Parking Area at MSU. *End of field trip.*

ACKNOWLEDGMENTS

Special thanks to our principal reviewers: Katie McDonald, associate research geologist at the Montana Bureau of Mines and Geology (MBMG), and Fess Foster, consulting geologist and former director of Geology and Environmental Affairs at GSM. Thanks to Susan Vuke, associate research geologist at MBMG, for critical input and review of both the road log and manuscript. Thanks to Barrick Golden Sunlight Mine's Chief Geologist David Odt, Chief Metallurgist Tom Van Norman, Mill Superintendent Rick Jordan, and Environmental Superintendent Mark Thompson for critical input and review of the manuscript. Sandy Underwood, Aaron Norby, and especially Helen Lynn at Childs Geoscience Inc., provided help with figures, editing, and layout of the road log route.

REFERENCES CITED

Alexander, R.G., Jr., 1955, The Geology of the Whitehall Area, Montana: Billings, Montana, Yellowstone-Bighorn Research Association Contribution 195, 111 p.

Blake, O.D., 1953, Stratigraphy and structure in the Three Forks area, Field trip no. 1, *in* Blake, O.D., ed., Guidebook of Field Excursions: Geological Society of America Rocky Mountain Section, Sixth Annual Meeting: Butte, Montana School of Mines, p. 19–30.

Chadwick, T., 1992, Geologic Cross Section Looking North through the Mineral Hill Breccia at 26,000 Feet North, Whitehall District, Jefferson County, Montana: Golden Sunlight Mines, Inc., 1 sheet, scale 1:12,000.

Chadwick, T., 1996, Geology of the Whitehall District, Jefferson County, Montana (Revised): Golden Sunlight Mines, Inc., scale 1:12,000.

Dean, D.A., Benedetto, K.M.F., and Durgin, D.C., 1990, Southern extension of the Carlin Trend: Influence of structure and stratigraphy on gold deposition, Field Trip 4, *in* Buffa, R.H., and Coyner, A.R., eds., Geology and Ore Deposits of the Great Basin, Volume 1: Reno, Nevada, Geological Society of Nevada, p. 81–93.

DeWitt, E., Foord, E., Zartman, R.E., Pearson, R.C., and Foster, F., 1996, Chronology of Late Cretaceous Igneous and Hydrothermal Events at the Golden Sunlight Gold-Silver Breccia Pipe, Southwestern Montana: U.S. Geological Survey Bulletin 2166, 49 p.

Dudas, F.O., Ispolatov, V.O., Harlan, S.S., and Snee, L.W., 2010, ^{40}Ar/^{39}Ar geochronology and geochemical reconnaissance of the Eocene Lowland Creek volcanic field, west-central Montana: Journal of Geology, v. 118, no. 3, p. 295–304.

Foster, D.A., Mueller, P.A., Mogk, D.W., Wooden, J.L., and Vogl, J.J., 2006, Proterozoic evolution of the western margin of the Wyoming Craton; implications for the tectonic and magmatic evolution of the Northern Rocky Mountains: Canadian Journal of Earth Sciences, v. 43, p. 1601–1619, doi:10.1139/e06-052.

Foster, F., 1987, Epithermal precious-metal systems associated with an Eocene cauldron: Lowland Creek Volcanic Field, southwestern Montana, *in* Berg, R., and Breuninger, R., eds., Guidebook of the Helena Area, West-Central Montana: Butte, Montana Bureau of Mines and Geology Special Publication 95, p. 53–54.

Foster, F., 1991, Geology and general overview of the Golden Sunlight Mine, Jefferson County, Montana, *in* The Association of Exploration Geochemists 15th International Geochemical Exploration Symposium Field Trip Guidebook to Mineral Deposits of Montana: Amsterdam, Elsevier Science Publishers, p. 26–36.

Foster, F., and Chadwick, T., 1990, Relationship of the Golden Sunlight Mine to the Great Falls Tectonic Zone, *in* Moye, F.J., ed., Geology and Ore Deposits of the Trans-Challis Fault System/Great Falls Tectonic Zone: Salmon, Idaho, Tobacco Root Geological Society 15th Annual Field Conference Guidebook, p. 77–81.

Foster, F., and Chadwick, T., 1999, Some observations regarding formation correlations and regional paleogeography of the southern Helena Embayment: Belt Symposium III Abstracts, Montana Bureau of Mines and Geology Open-File Report 381, p. 20–22.

Foster, F., and Childs, J.F., 1993, An overview of significant lode gold systems in Montana, and their regional geologic setting: Exploration and Mining Geology, v. 2, no. 3, p. 217–244.

Foster, F., Chadwick, T., and Nilsen, T.H., 1999, Paleodepositional setting and synsedimentary mineralization in Belt Supergroup rocks of the Whitehall, Montana area: Belt Symposium III Abstracts: Montana Bureau of Mines and Geology Open-File Report 381, p. 23–25.

Frost, C.D., and Winston, D., 1987, Nd isotope systematics of coarse- and fine-grained sediments: Examples from the Middle Proterozoic Belt-Purcell Supergroup: The Journal of Geology, v. 95, p. 309–327, doi:10.1086/629132.

Gehrels, G., Valecia, V., and Ruiz, J., 2008, Enhanced precision, accuracy, efficiency, and spatial resolution of U-Pb ages by laser ablation-multi-collector-inductively coupled plasma-mass spectrometry: Geochemistry, Geophysics, Geosystems: Technical Brief, v. 9, no. 3, p. 1–13.

Harlan, S.S., Geissman, J.W., Whisner, S.C., and Schmidt, C.J., 2008, Paleomagnetism and geochronology of sills of the Doherty Mountain area, southwestern Montana; implications for the timing of fold-and-thrust belt deformation and vertical-axis rotations along the southern margin of the Helena salient: Geological Society of America Bulletin, v. 120, p. 1091–1104, doi:10.1130/B26313.1.

Hawley, D., Bonnet-Nicolaysen, A., and Coppinger, W., 1982, Stratigraphy, Depositional Environments, and Paleotectonics of the LaHood Formation: Guidebook for Field Trip Held in Conjunction with the 35th Annual Meeting of the Rocky Mountain Section of the Geological Society of America: Bozeman, Montana, Department of Earth Science, Montana State University, 20 p.

Ihinger, P., Mahoney, J.B., Johnson, B.R., Kohel, C., Guy, A.K., Kimbrough, D.L., and Friedman, R.M., 2011, Late Cretaceous magmatism in southwest Montana: The Boulder batholith and Elkhorn Mountains Volcanics: Geological Society of America Abstracts with Programs, v. 43, no. 5, p. 647.

Johns, W.M., Berg, R.B., and Dresser, H.W., 1981, First day geologic road log Part 4, Three Forks to Twin Bridges via U.S. Highways 10 and 287 and State Highway 287, *in* Tucker, T.E., ed., Field Conference & Symposium Guidebook to Southwest Montana: Montana Geological Society, p. 388–392.

Klepper, M.R., Weeks, R.A., and Ruppel, E.T., 1957, Geology of the Southern Elkhorn Mountains, Jefferson and Broadwater Counties, Montana: U.S. Geological Survey Professional Paper 292, 82 p.

Kuenzi, W.D., and Fields, R.W., 1971, Tertiary stratigraphy structure and geologic history, Jefferson basin, Montana: Geological Society of America Bulletin, v. 85, p. 1563–1580.

Lageson, D.R., 1989, Reactivation of a Proterozoic continental margin, Bridger Range, southwestern Montana, *in* French, D.E., and Grabb, R.F., eds., Geologic Resources of Montana: Montana Geological Society Field Conference Guidebook, p. 279–298.

Lageson, D.R., 1992, Structural analysis of the Horseshoe Hills transverse fold-thrust zone, Gallatin County, Montana: A preliminary report, *in* Guidebook for the Red Lodge-Beartooth Mountains-Stillwater area, Tobacco Root Geological Society Seventeenth Annual Field Conference: Northwest Geology, v. 20/21, p. 117–124.

Lageson, D.R., and Montagne, J., 1981, Road log from Dillon to Three Forks, Montana, *in* Tucker, T.E., ed., Field Conference and Symposium Guidebook to Southwest Montana: Montana Geological Society, p. 399–406.

Lageson, D.R., Schmidt, C.J., Dresser, H.W., Welker, M., Berg, R.B., and James, H.L., 1983, Road log No. 1, Bozeman to Helena via Battle Ridge Pass, White Sulphur Springs, Townsend and Toston, *in* Smith, D.L., ed., Guidebook of the Fold and Thrust Belt, West Central Montana: Montana Bureau of Mines and Geology Special Publication 86, p. 1–32.

Lageson, D.R., Schmitt, J.G., Horton, B.K., Kalakay, T.J., and Burton, B.R., 2001, Influence of Late Cretaceous magmatism on the Sevier orogenic wedge, western Montana: Geology, v. 29, no. 8, p. 723–726, doi:10.1130/0091-7613(2001)029<0723:IOLCMO>2.0.CO;2.

Lewis, S.E., 1988, Field guide to mesoscopic features in the LaHood Formation, Jefferson Canyon area, southwest Montana, *in* Lewis, S.E., and Berg, R.B., eds., Precambrian and Mesozoic Plate Margins: Montana, Idaho and Wyoming with Field Guides, 8th International Conference on Basement Tectonics: Montana Bureau of Mines and Geology Special Publication 96, p. 155–158.

Luedke, R.G., 1994, Map Showing Distribution, Composition and Age of Early and Middle Cenozoic Volcanic Centers in Idaho, Montana, West-Central

South Dakota and Wyoming: U.S. Geological Survey Map I-2291-C, 2 plates, scale 1:1,000,000.
Lund, K., Aleinikoff, J.N., Evans, K.V., du Bray, E.A., Dewitt, E.H., and Unruh, D.M., 2010, SHRIMP U-Pb dating of recurrent Cryogenian and Late Cambrian–Early Ordovician alkalic magmatism in central Idaho: Implications for Rodinian rift tectonics: Geological Society of America Bulletin, v. 122, p. 430–453, doi:10.1130/B26565.1.
Mahoney, J.B., Link, P.K., Todt, M.K., Taylor, S.S., and Balgord, E., 2013, Neoproterozoic to Cambrian passive margin? Evidence for active tectonism during the Sauk transgression in central Idaho and Montana: Geological Society of America Abstracts with Programs, v. 45, no. 7, p. 166.
McDonald, C., Elliott, C.G., Vuke, S.M., Lonn, J.D., and Berg, R.B., 2012, Geologic Map of the Butte South 30′ × 60′ Quadrangle, Southwestern Montana: Montana Bureau of Mines and Geology Open-File Report 622, scale 1:100,000, 1 sheet.
McTeague, M.S., and Schmitt, J.G., 2003, Facies analysis of the LaHood Formation, a Proterozoic fan-delta deposit: Northwest Geology, v. 32, p. 214.
Mueller, P., Heatherington, A., Kelley, D., Wooden, J., and Mogk, D., 2002, Paleoproterozoic crust within the Great Falls tectonic zone: Implications for the assembly of southern Laurentia: Geology, v. 30, p. 127–130, doi:10.1130/0091-7613(2002)030<0127:PCWTGF>2.0.CO;2.
O'Neill, J.M., and Lopez, D.A., 1985, Character and regional significance of the Great Falls Tectonic Zone, East-Central Idaho and West-Central Montana: American Association of Petroleum Geologists Bulletin, v. 69, p. 437–447.
Paredes, M., 1994, A fluid inclusion, stable isotope and multi-element study of the Golden Sunlight Deposit, Montana [M.S. thesis]: Ames, Iowa State University, 174 p.
Porter, E.W., and Ripley, E., 1985, Petrologic and stable isotope study of the gold-bearing breccia pipe at the Golden Sunlight deposit, Montana: Economic Geology and the Bulletin of the Society of Economic Geologists, v. 80, p. 1689–1706, doi:10.2113/gsecongeo.80.6.1689.
Robinson, G.D., 1963, Geology of the Three Forks Quadrangle, Montana: U.S. Geological Survey Professional Paper 370, 140 p.
Robinson, G.D., Klepper, C.G., and Obradovich, J.D., 1968, Overlapping plutonism, volcanism, and tectonism in the Boulder Batholith region, western Montana, *in* Coates, R.R., Hay, R.L., and Anderson, C.A., eds., Studies in Volcanology: Geological Society of America Memoir 116, p. 557–576.
Ross, G.M., and Villeneuve, M., 2003, Provenance of the Mesoproterozoic (1.45 Ga) Belt basin (western North America): Another piece in the pre-Rodinia paleogeographic puzzle: Geological Society of America Bulletin, v. 115, p. 1191–1217, doi:10.1130/B25209.1.
Ross, G.M., Parrish, R.R., and Winston, D., 1992, Provenance and U-Pb geochronology of the Mesoproterozoic Belt Supergroup (northwestern United States): Implications for age of deposition and pre-Panthalassa plate reconstructions: Earth and Planetary Science Letters, v. 113, p. 57–76, doi:10.1016/0012-821X(92)90211-D.
Ruppel, E.T., 1993, Cenozoic Tectonic Evolution of Southwest Montana and East-Central Idaho: U.S. Geological Survey Professional Paper 1224, 24 p.
Sando, W.J., and Dutro, J.T., 1974, Type Sections of the Madison Group (Mississippian) and Its Subdivisions in Montana: U.S. Geological Survey Professional Paper 842, 22 p.
Schmidt, C.J., and Garihan, J.M., 1986, Middle Proterozoic and Laramide tectonic activity along the southern margin of the Belt basin, *in* Roberts, S., ed., Belt Supergroup: A Guide to the Proterozoic Rocks of Western Montana and Adjacent Areas: Montana Bureau of Mines and Geology Special Publication 94, p. 217–235.
Schmidt, C.J., and O'Neill, J.M., 1982, Structural evolution of the Southwest Montana Transverse Zone, *in* Powers, R.W., ed., Geologic Studies of the Cordilleran Thrust Belt—1982: Denver, Rocky Mountain Association of Geologists, v. 1, p. 193–218.
Schmidt, C.J., Aram, R., and Hawley, D., 1987, The Jefferson River Canyon area, southwestern Montana, *in* Beus, S.S., ed., Rocky Mountain Section: Boulder, Colorado, Geology of North America, Geological Society of America Centennial Field Guide, v. 2, p. 63–68.
Schmidt, C.J., O'Neill, J.M., and Brandon, W.C., 1988, Influence of Rocky Mountain foreland uplifts on the development of the frontal fold and thrust belt, southwest Montana, *in* Schmidt, C.J., and Perry, W.J., Jr., eds., Interaction of the Rocky Mountain Foreland and the Cordilleran Thrust Belt: Geological Society of America Memoir 171, p. 171–201.
Schmidt, C.J., Genovese, P., and Foster, F., 1989, Road log—Structure and economic geology along the transverse thrust zone of southwestern Montana, *in* French, D.E., and Grabb, R.F., eds., Geologic Resources of Montana: Billings, Montana Geological Society Field Conference Guidebook, Centennial Edition, p. 482–501.
Schwartz, R.K. and DeCelles, P.G., 1988, Cordilleran Foreland Basin evolution in response to interactive Cretaceous thrusting and foreland partitioning, southwestern Montana, *in* Schmidt, C.J. and Perry, W.J., Jr., eds., Interaction of the Rocky Mountain Foreland and the Cordilleran Thrust Belt: Geological Society of America Memoir 171, p. 489–514.
Spry, P.G., and Thieben, S.E., 1997, A Preliminary Study of Gold-Bearing Minerals in Concentrate from the Western Part of the Mineral Hill Breccia Pipe, Golden Sunlight Deposit: Internal Report to Golden Sunlight Mines, Inc., 8 p.
Spry, P.G., Paredes, M.M., Foster, F., Truckle, J., and Chadwick, T., 1996, A genetic link between gold-silver telluride and porphyry molybdenum mineralization at the Golden Sunlight Deposit, Whitehall, Montana: Fluid inclusion and stable isotope studies: Economic Geology and the Bulletin of the Society of Economic Geologists, v. 91, p. 507–526, doi:10.2113/gsecongeo.91.3.507.
Spry, P.G., Foster, F., Truckle, J., and Chadwick, T.H., 1997, The mineralogy of the Golden Sunlight gold-silver telluride deposit, Montana, USA: Mineralogy and Petrology, v. 59, p. 143–164, doi:10.1007/BF01161857.
Verrall, P., 1955, Geology of the Horseshoe Hills area, Montana [Ph.D. dissertation]: New Jersey, Princeton, Princeton University, 260 p.
Vuke, S.M., Coppinger, W.W., and Cox, B.E., 2004, Geologic Map of the Cenozoic Deposits of the Upper Jefferson Valley: Montana Bureau of Mines and Geology Open File Report 505, 38 p., scale 1:50,000, 1 sheet.
Vuke, S.M., Porter, K.W., Lonn, J.D., and Lopez, D.A., 2007, Geologic Map of Montana: Montana Bureau of Mines and Geology Geologic Map 62, 73 p., 2 sheets, scale 1:500,000.
Vuke, S.M., Lonn, J.D., Berg, R.B., and Schmidt, C.J., 2014, Geologic Map of the Bozeman 30′ × 60′ Quadrangle, Southwestern Montana: Montana Bureau of Mines and Geology Open-File Report 647, 41 p., 1 sheet, scale 1:100,000.

Manuscript Accepted by the Society 7 March 2014

Printed in the USA

The Geological Society of America
Field Guide 37
2014

Polyphase collapse of the Cordilleran hinterland: The Anaconda metamorphic core complex of western Montana— The Snoke symposium field trip

Thomas J. Kalakay*
Department of Geology, Rocky Mountain College, 1511 Poly, Billings, Montana 59102, USA

David A. Foster*
Department of Geological Sciences, University of Florida, Gainesville, Florida 32611, USA

Jeffrey D. Lonn*
Montana Bureau of Mines and Geology, 1300 West Park Street, Butte, Montana 59701, USA

ABSTRACT

The Anaconda and Bitterroot metamorphic core complexes are located in western Montana, along the eastern edge of the Cordilleran hinterland. This multi-tiered extensional terrain contains exceptional exposures that collectively exhibit a crustal cross section through orogenic continental crust (i.e., middle through upper crust). The core complex footwall rocks consist of Late Cretaceous arc-related plutons and Eocene granitic plutons intruded into deformed and metamorphosed Midproterozoic Belt Supergroup and Paleozoic to Cretaceous shelf-platform strata. Late Cretaceous shear zones and folds dominate footwall structure, representing significant thinning of the stratigraphic section. Eocene detachments, mylonites, and plutonic suites distinctly overprint the Late Cretaceous structures. A stark example of this Eocene overprint is the Anaconda detachment, which resulted in eastward translation of the Late Cretaceous, arc-related Boulder batholith. This field trip will cover a transect through the Anaconda core complex from the Philipsburg valley to Butte, Montana. Field trip participants will examine key locations that clarify the distinction between the timing and structural style of Late Cretaceous crustal thickening and/or collapse features versus those related to Eocene core complex development.

*E-mails: kalakayt@rocky.edu; dafoster@ufl.edu; jlonn@mtech.edu.

Kalakay, T.J., Foster, D.A., and Lonn, J.D., 2014, Polyphase collapse of the Cordilleran hinterland: The Anaconda metamorphic core complex of western Montana—The Snoke symposium field trip, *in* Shaw, C.A., and Tikoff, B., eds., Exploring the Northern Rocky Mountains: Geological Society of America Field Guide 37, p. 145–159, doi:10.1130/2014.0037(07). For permission to copy, contact editing@geosociety.org.

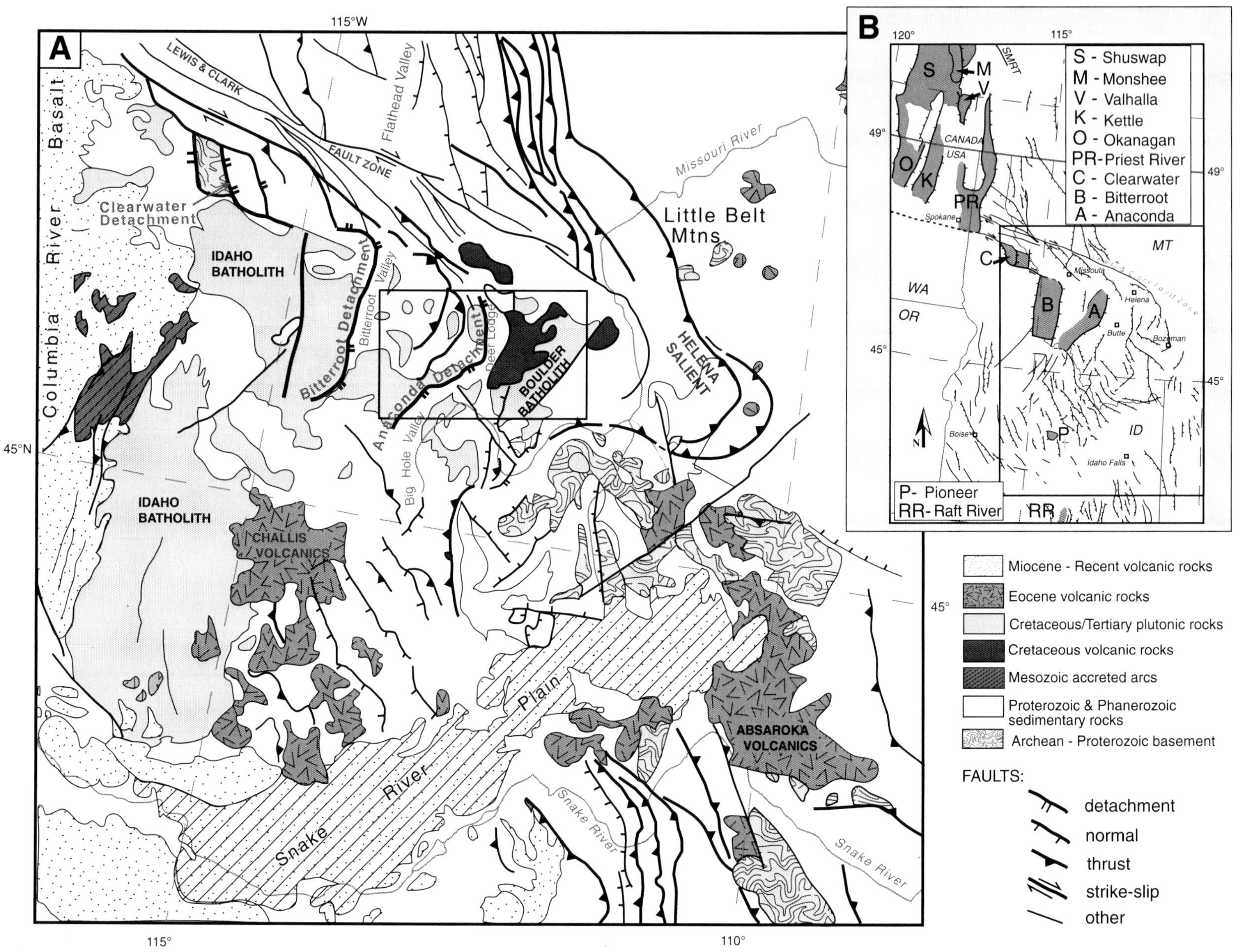

Figure 1. (A) Tectonic map of the northern U.S. Rocky Mountains from Foster et al. (2010), showing major Phanerozoic structures and tectonic elements in the vicinity of the Anaconda detachment, which bounds the Anaconda metamorphic core complex. The box shows the area of the map in Figure 2. (B) Inset map showing a larger area depicting the location of the Anaconda metamorphic core complex and other Eocene metamorphic core complexes. The core complexes are shaded gray with names corresponding to the abbreviations listed in the key. The box shows the area of the map in A. ID—Idaho; MT—Montana; OR—Oregon; WA—Washington.

INTRODUCTION

Continental extension was widespread throughout northwestern North America from the latest Cretaceous to mid-Tertiary time (Armstrong, 1978; Coney, 1980; Hodges and Walker, 1992; Foster and Fanning, 1997; Wells, 1997; Foster et al., 2007; Fig. 1). North of the Snake River plain and into the Canadian Cordillera, intense extension and exhumation of middle-crustal rocks occurred during the Eocene (Parrish et al., 1988; Foster and Fanning, 1997; Doughty and Price, 2000; Foster et al., 2007; Vogl et al., 2012). In Montana, low-angle detachment faulting related to this event is well documented in the Eocene Bitterroot metamorphic core complex (Hyndman, 1980; Garmezy, 1983; House and Hodges, 1994; Foster and Fanning, 1997; Foster et al., 2001; Foster and Raza, 2002), where it predates the formation of present basins and ranges and was coeval with regional basaltic to rhyolitic magmatism (e.g., the Challis, Lowland Creek, and Absaroka Volcanic fields of Smedes, 1962, and Smedes and Prostka, 1972, respectively). Until recently, the magnitude of Eocene extension in areas east of the Bitterroot core complex was not well understood. Regional extensional deformation is recorded by an east-dipping, normal-sense mylonite zone along the eastern front of the Flint Creek and Anaconda Ranges (O'Neill, et al., 2002; Kalakay et al., 2003; Foster et al., 2007, 2010). This top-to-the-east fault juxtaposes a footwall sequence of plutonic and high-grade metamorphic rocks with a hanging wall succession of unmetamorphosed sediments, landslide breccias, and volcanic rocks (Fig. 2). From our mapping, the sinuous fault zone can be traced for more than 100 km.

New questions raised by the discovery of the Anaconda detachment call for a better understanding of the geometry, magnitude, and timing of crustal extension throughout western Montana. This field excursion crosses a structural transect that is critical for understanding the complex deformational history of the Anaconda metamorphic core complex. With exceptional exposure this transect illuminates several key elements, including those related to Late Cretaceous and Eocene deformation.

Geologic Setting

Emmons and Calkins (1913), Csejtey (1962), Desmarais (1983), Heise (1983), Buckley (1990), Wallace et al. (1992), and Lewis (1998) conducted geologic mapping in the Anaconda and Flint Creek Ranges, which consist primarily of metamorphosed Belt quartzite, argillite, and pelite (Emmons and Calkins, 1913; Desmarais, 1983; Heise, 1983; Wallace et al., 1992; Lewis, 1998). In places, metamorphosed sedimentary units consisting of middle Cambrian to Cretaceous rocks are also exposed (Fig. 3). All units have been intruded by two distinct generations of plutons, dikes, and sills. Most are either Late Cretaceous or early–middle Eocene in age (Grice, 2006; Foster et al., 2010).

Within the core of the Anaconda and Flint Creek Ranges, rocks showing metamorphic grades as high as upper amphibolite facies are common. Kyanite-bearing metapelites are typically overprinted by lower pressure high-temperature assemblages indicating a polyphase metamorphic history (Grice, 2006). A general increase in metamorphic grade toward the east indicates there was significant tectonic exhumation associated with movement along the Anaconda detachment. This period of extension was superimposed on hinterland structures of the Sevier orogenic belt, which developed during Late Cretaceous contraction and metamorphism.

Despite intense geologic investigation, many aspects of the structure and stratigraphy of the Anaconda and Flint Creek Ranges have been difficult to explain. Regional geologic features that pose particularly difficult problems are (1) east- and west-verging isoclinal and often recumbent folds, (2) older on younger fault relationships, and (3) apparent stratigraphic thinning of units along the eastern margin of the ranges.

Wallace et al. (1992) proposed an elaborate stacking sequence of hinterland thrusting and folding to explain these features. However, this model is faced by a number of difficulties. Most notably, there is no recognized input from crustal extension. In this chapter, we discuss evidence that extensional tectonics did play a significant role and that typical contraction models cannot adequately explain the complex geology of the Anaconda and Flint Creek Ranges.

Anaconda Core Complex Overview

The Anaconda metamorphic core complex (O'Neill et al., 2004; Foster et al., 2007, 2010) is comprised of three structural-metamorphic domains: (1) a metamorphic-plutonic footwall exposed in the Anaconda and Flint Creek Ranges, (2) a low-grade hanging wall exposed along the western edge of and within the Deer Lodge Valley (Fig. 2), and (3) a brittle-plastic detachment fault system exposed along the eastern flanks of the Anaconda and Flint Creek Ranges (Figs. 2, 3).

Footwall rocks of the Anaconda complex are made up of Late Cretaceous to Eocene granitic plutons intruded into metamorphosed Mesoproterozoic Belt Supergroup and Middle Cambrian to Cretaceous shelf-platform strata (Figs. 2, 3) (Emmons and Calkins, 1913, 1915; Desmarais, 1983; Heise, 1983; Wallace et al., 1992; Lonn et al., 2003; Grice, 2006; Foster et al., 2010). In the Flint Creek Range, footwall rocks are comprised of granodiorite to granite plutons of the Late Cretaceous Mount Powell batholith, Royal stock, and Lost Creek stock. These plutons intruded deformed and metamorphosed Middle Cambrian to Cretaceous strata, and in a few areas metamorphosed Belt strata (Emmons and Calkins, 1913, 1915; Allen, 1966; Hyndman et al., 1982; Lonn et al., 2003; O'Neill et al., 2004). In the Anaconda Range, the footwall is largely Late Cretaceous diorite to granodiorite and early to middle Eocene granitic plutons, which intruded deformed Belt Supergroup and metamorphosed Middle Cambrian strata (Fig. 3) (Desmarais, 1983; Wallace et al., 1992; Lonn et al., 2003; O'Neill et al., 2004; Foster et al., 2010). Upper amphibolite facies metamorphism and nappe-style folding (e.g., Lake of the Isle shear zone in Fig. 3) of the Belt and Cambrian strata occurred in Late Cretaceous time with peak metamorphic temperatures (>700 °C)

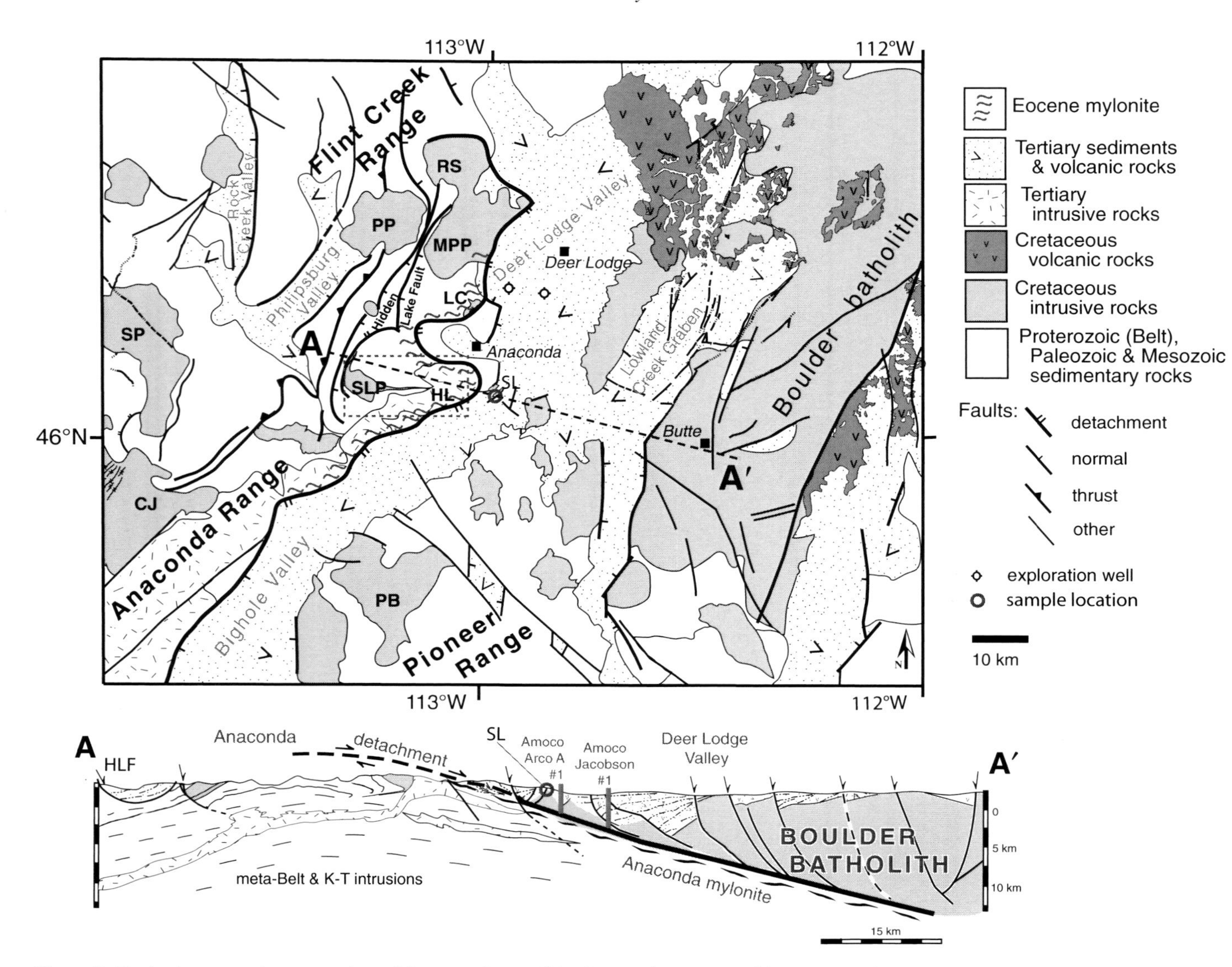

Figure 2. Geologic map and cross section of the central part of the Anaconda metamorphic core complex, Boulder batholith, and adjacent regions (modified from Foster et al., 2010). The map was compiled from Emmons and Calkins (1913), Lonn et al. (2003), O'Neill et al. (2004), and Foster et al. (2007). Cretaceous intrusive rock at location denoted by the circle labeled SL is from coarse-grained granodiorite interpreted to be the detached top of the Storm Lake pluton in the footwall. Abbreviations: SP—Sapphire pluton; CJ—Chief Joseph pluton; SLP—Storm Lake pluton; PB—Pioneer batholith; PP—Philipsburg batholith; MPP—Mount Powell pluton; RS—Royal stock; HL—Hearst Lake plutonic suite; LC—Lost Creek stock. The dashed box shows the area of Figure 3.

accompanying intrusion of granodioritic plutons at ca. 75 Ma, based on U-Pb zircon data (Grice, 2006; Figs. 4, 5). Cretaceous metamorphism and deformation took place at pressures of 4.6–6.0 kbar based on metamorphic thermobarometry of garnet bearing metapelitic rocks (Grice, 2006; Fig. 6).

The hanging wall of the Anaconda core complex is made up of an array of asymmetric fault-bound basins containing unmetamorphosed Tertiary clastic, volcaniclastic, and volcanic strata. These strata are exposed in the Deer Lodge Valley and preserved in a reentrant between the Anaconda and Flint Creek Ranges (O'Neill et al., 2004; Foster et al., 2007, 2010) (Fig. 3). The stratigraphically lowest rocks in these basins are moderately west tilted (~50–60°), poorly sorted, and poorly consolidated conglomerates, sandstones, breccias, and mega-breccias (Kalakay et al., 2003; O'Neill et al., 2004). These strata grade upwards into progressively less tilted (~0–25°) volcanic lava flows, volcanic tuffs, and volcaniclastic deposits of the Eocene Lowland Creek volcanic field (ca. 53–49 Ma; Dudás et al., 2010). The upward decrease in the tilt of these basin fill strata indicates deposition synchronous with extension.

Metamorphic and plutonic rocks of the footwall are juxtaposed with the hanging wall rocks along a gently east-dipping, low-angle brittle-plastic detachment system, which shows top-to-the-east-southeast displacement (Emmons and Calkins, 1913;

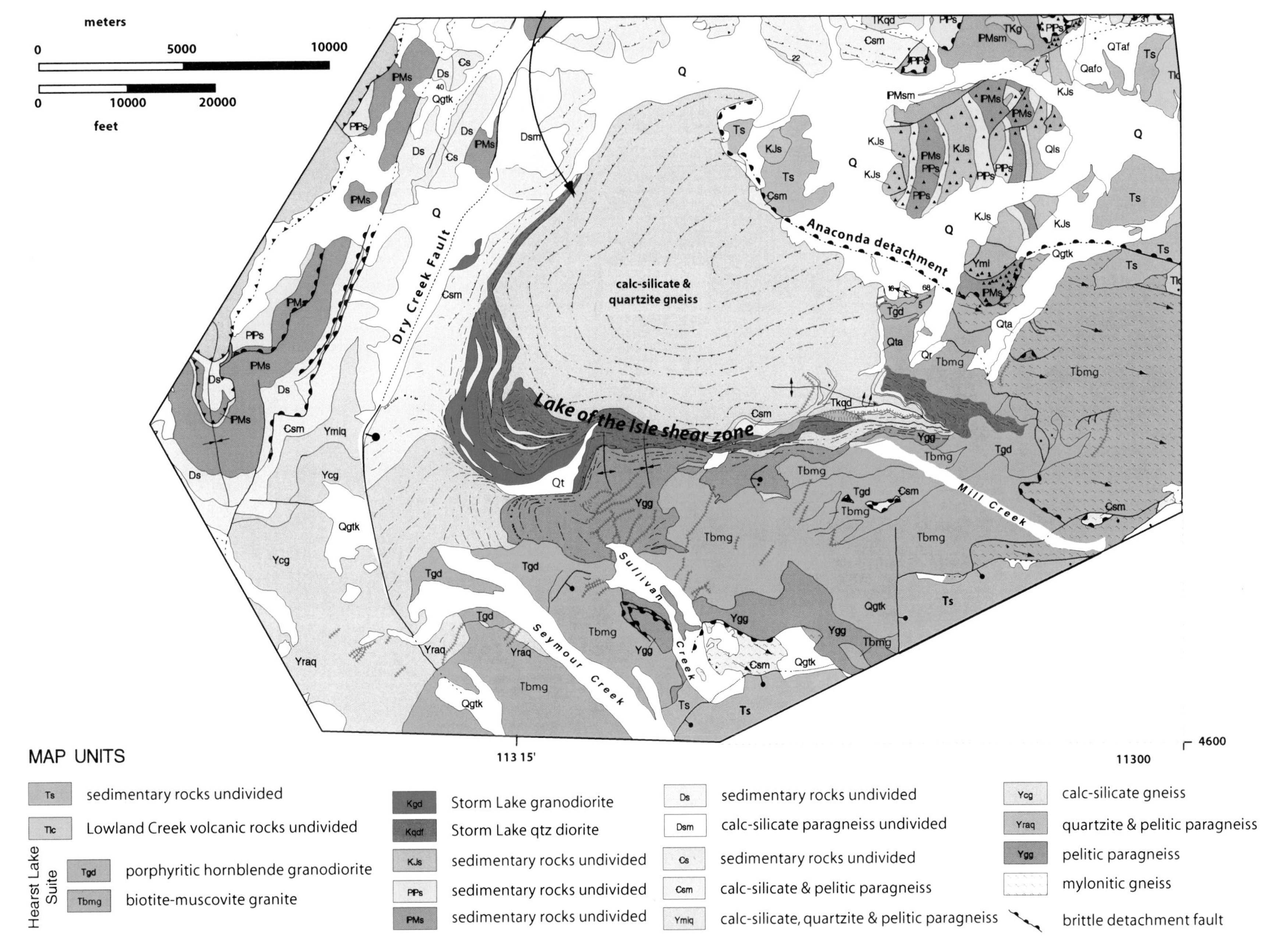

Figure 3. Geologic map of Anaconda and Mill Creek area in the northern Anaconda Range (modified from Lonn et al., 2003), showing relationships between Cretaceous and Eocene structures and plutonic rocks. The two colors for the Late Cretaceous (75 Ma; Grice, 2006) Storm Lake stock are for the granodiorite (gd) and quartz diorite (qd) compositions. The Hearst Lake suite plutons are Eocene with ages of 53 Ma for biotite granodiorite (bg) and 47 Ma for two-mica granite (gr) (Foster et al., 2010).

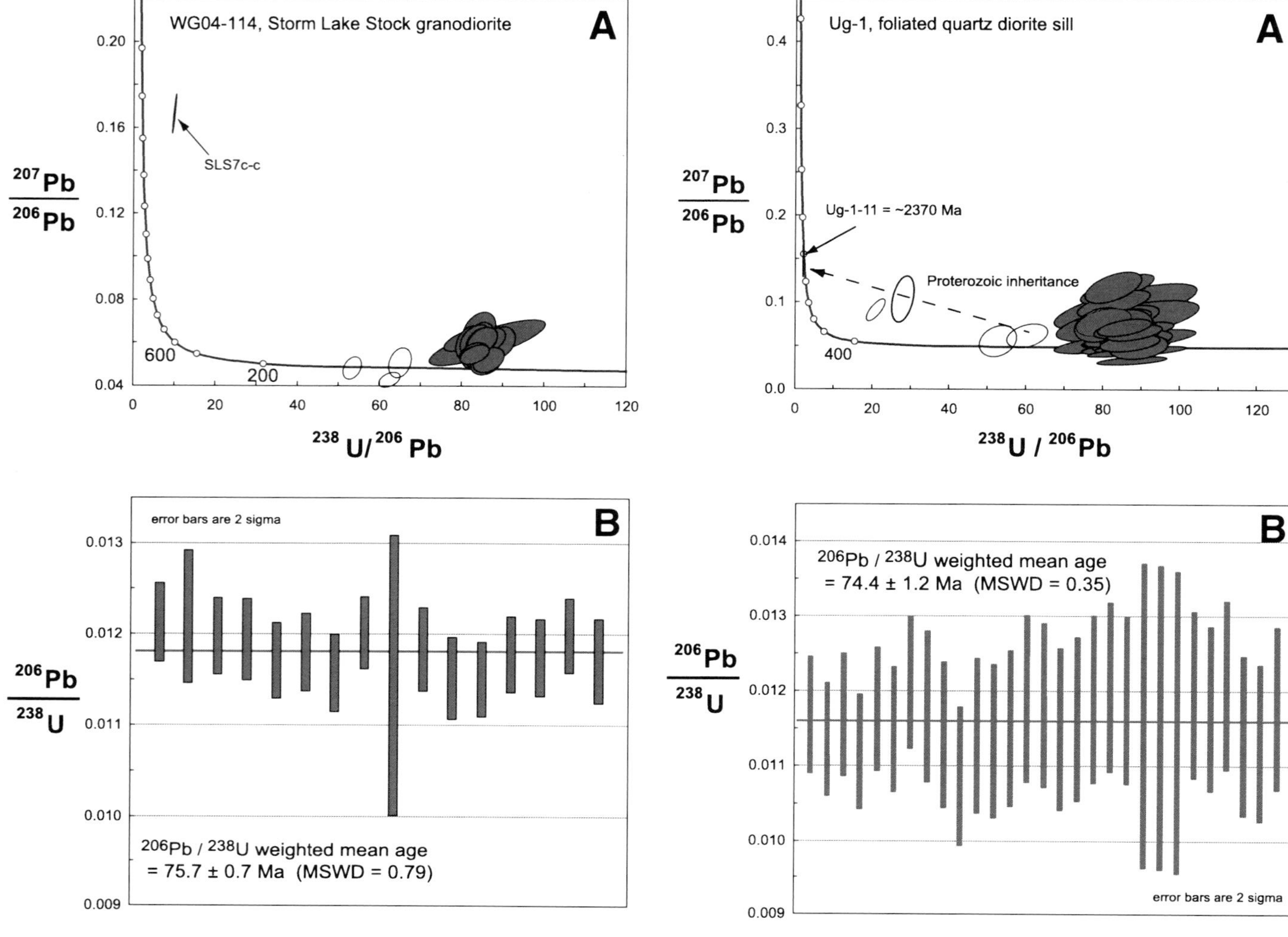

Figure 4. U-Pb zircon data from undeformed hornblende-biotite granodiorite from the Storm Lake stock. MSWD—mean square of weighted deviation.

Figure 5. U-Pb zircon data from a sheared hornblende-biotite quartz diorite sill in the Lake of the Isle shear zone. MSWD—mean square of weighted deviation.

O'Neill and Lageson, 2003; Kalakay et al., 2003; O'Neill et al., 2004; Foster et al., 2007, 2010). The Anaconda detachment has a mapped strike length of at least 100 km from the northern Flint Creek Range to the southern Anaconda Range (Kalakay et al., 2003; O'Neill et al., 2004; Foster et al., 2007; Fig. 2).

The Anaconda detachment is characterized by greenschist facies mylonite, ultramylonite, pseudotachylite, and overprinting brittle normal faults (Figs. 3, 7). Along the eastern flank of the northeastern Anaconda Range, the detachment is characterized by a 300–500-m-thick lower to middle greenschist facies mylonitic shear zone of stretched two-mica granite, biotite granite, granodiorite, and mylonitic micaeous quartzite (Emmons and Calkins, 1913; Kalakay et al., 2003; O'Neill et al., 2004; Foster et al., 2007, 2010, Grice, 2006). Fractured K-feldspar porphyroclasts in the granitoids are encased by a matrix of plastically deformed quartz. The micaeous quartzite exhibits unannealed quartz ribbons with undulatory extinction and mica fish. These metamorphic textures are indicative of deformation at temperatures lower than ~400–450 °C (Passchier and Trouw, 2005). Greenschist facies mylonites exhibit shallow-plunging, mineral-stretching lineations and kinematic indicators, which show top-to-the-east-southeast (102–110°) sense of motion (Kalakay et al., 2003; O'Neill et al., 2004; Grice, 2006). Strain in the greenschist facies mylonites is heterogeneous and distributed into 0.1–2-m-thick zones of ultramylonite alternating with ~5–15-m-thick zones of mylonite and protomylonite (Fig. 8) and some bands of pseudotachylite (Kalakay et al., 2003; Foster et al., 2007, 2010). The mylonites exposed in the northeastern Anaconda Range are cut by an array of closely spaced east-dipping brittle normal faults. Many brittle faults are listric-shaped becoming sub-horizontal with depth and parallel to ultramylonite zones in the granitoids (Fig. 7E). Slickenline striations on the brittle fault surfaces show top-to-the-east-southeast (100–110°) displacement parallel to the stretching direction in the greenschist facies mylonites (Kalakay et al., 2003).

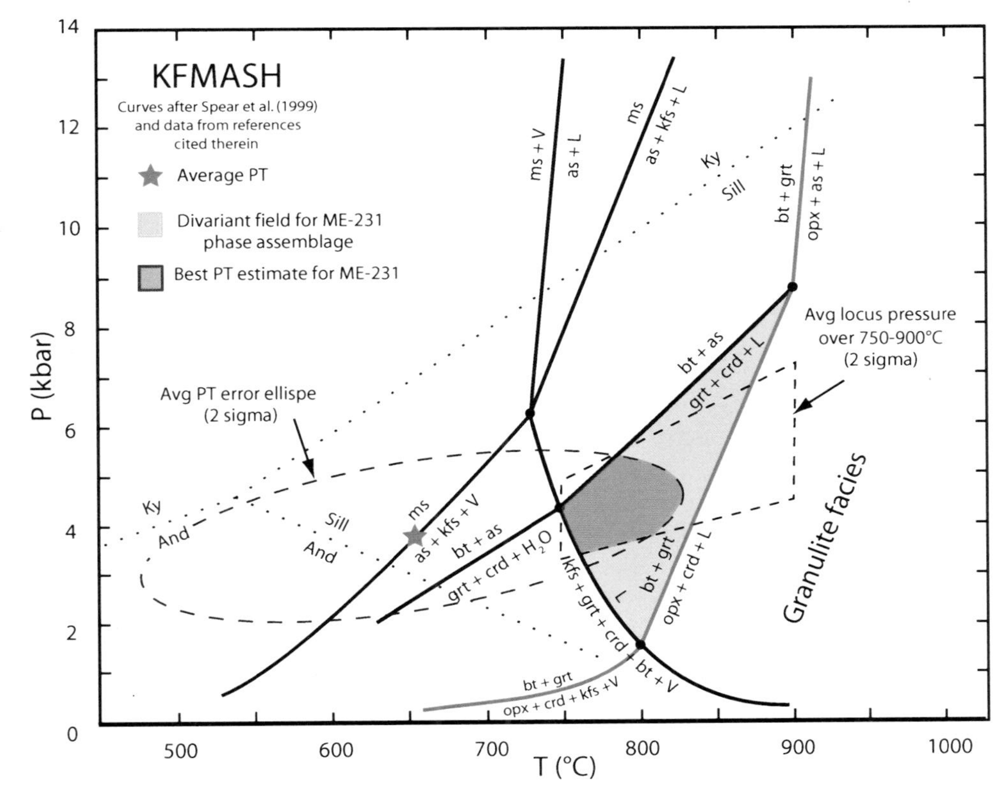

Figure 6. Summary of metamorphic assemblages and thermobarometry of a sample of migmatized metapelite from the Greyson Formation in the Lake of the Isle shear zone (Mill Creek location). Average and locus PT (pressure-temperature) estimates were calculated based on phase compositions suing THERMOCALC v. 3.21. And—andalusite; Ky—kyanite; Sill—sillimanite; ms—muscovite; as—aluminosilicate; V—vapor; bt—biotite; grt—garnet; opx—orthopyroxene; crd—corderite; kfs—K-feldspar; L—liquid.

Exposures of the brittle-plastic detachment are not continuous along strike in the Anaconda and Flint Creek Ranges because segments have been removed by erosion, cut out by younger brittle normal faults, and covered by hanging wall fault slivers or thick talus (Foster et al., 2007). Isolated exposures of the detachment along eastern flanks of the central and southern Anaconda Range are characterized by low-grade mylonitic two-mica granites and granodiorite cut by a series of east-dipping, northeast-trending brittle normal faults similar to those found in the northeastern Anaconda Range (Wallace et al., 1992). Along the eastern flanks of the Flint Creek Range, greenschist facies mylonite is found locally in the Lost Creek stock and metasedimentary rocks equivalent to the Belt and Middle Cambrian section (Allen, 1966; Lonn et al., 2003; O'Neill et al., 2004). These mylonites are cut by high-angle normal faults similar to those found in other parts of the detachment (O'Neill et al., 2004).

Along the eastern flanks of the Flint Creek and Anaconda Ranges, the detachment dips gently (~10–30°) beneath the Deer Lodge Valley. The gentle dip of the detachment is also revealed by industry exploration wells that intersected greenschist mylonites at the base of the Tertiary basin fill in the western Deer Lodge Valley at depths of ≤5 km (Fig. 2, McLeod, 1987). The downward projection of the low-angle detachment is aligned with sub-horizontal seismic reflectors beneath the Boulder batholith (Vejmelek and Smithson, 1995), suggesting that the detachment shallows with depth and continues to the east (Fig. 2; Foster et al., 2010). This uniform shallow dip along with the listric faults soling into the shear zone is consistent with the deeper parts of the Anaconda detachment originating at low angles within the brittle-plastic transition.

The detachment is not well exposed along the western margin of the Anaconda core complex. The trace of the detachment is inferred in several places by the juxtaposition of brittle faulted upper plate rocks with plastically deformed metamorphic and plutonic rocks. The western part of the detachment probably originated as a series of east-dipping listric-shaped normal faults east of a breakaway zone, which is inferred to be located east of the Georgetown thrust and is either no longer exposed or was removed by erosion (O'Neill et al., 2004; Foster et al., 2007, 2010). Upper amphibolite facies mylonite and extreme attenuation of footwall strata in the eastern part of the complex footwall (O'Neill et al., 2004) are related to Cretaceous deformation and are not Eocene fabrics (Grice et al., 2005); these Cretaceous fabrics are locally overprinted by the Eocene brittle-plastic fabrics (Fig. 3).

Timing and Rate of Eocene Extension

$^{40}Ar/^{39}Ar$ cooling ages obtained from samples collected from extensional fault blocks and exhumed metamorphic core complex footwalls may be used to determine the onset extension if the base of partial retention zone for a thermochronologic system is identified within the fault block (John and Foster, 1993; Foster and John, 1999; Stockli, 2005). The partial retention zone corresponds to an interval of crustal depths specific to each thermochronometer

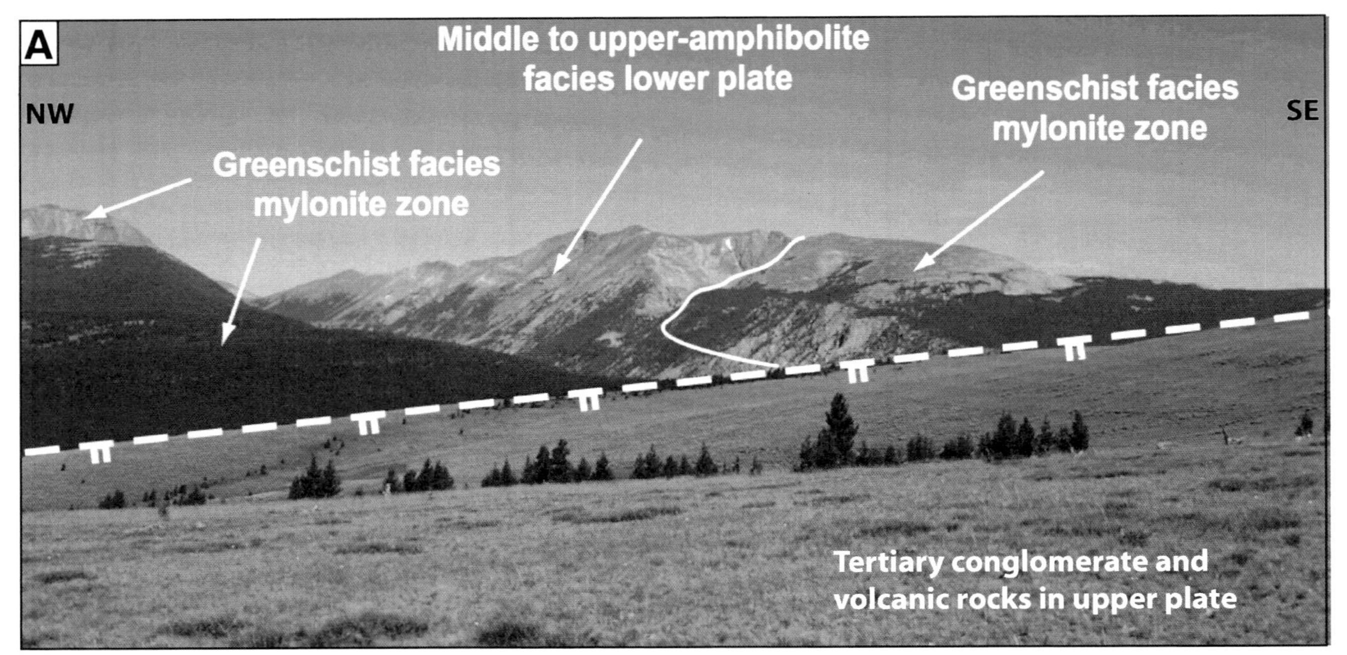
A
Middle to upper-amphibolite facies lower plate
NW
Greenschist facies mylonite zone
SE
Greenschist facies mylonite zone
Tertiary conglomerate and volcanic rocks in upper plate

B
Listric fault cutting mylonite
W
E

C
Cretaceous nappe
NW
SE

D
Extensional shear bands
W
E

E
Listric fault cutting mylonite
E
W

Figure 7. Field photos from Foster et al. (2010). (A) The footwall and hanging wall of the Anaconda metamorphic core complex looking WNW into the Mill Creek Valley. Hachured dashed line—detachment; solid line—igneous-metamorphic rock contact. (B) A listric normal fault (dashed line) cutting greenschist facies mylonite when looking north at the canyon wall of the Mill Creek Valley. (C) The north wall of the Mill Creek Valley showing a large Cretaceous nappe defined by metamorphosed Belt Supergroup and Cambrian strata. The ductile strain in the nappe occurred at upper amphibolite conditions at ca. 75 Ma. Dashed line marks upper contact of metamorphosed Hasmark Formation. (D) Late-stage semi-brittle extensional shear bands cutting greenschist facies mylonite and ultramylonite in the Mill Creek Valley (view is to the north). Thick lines mark brittle faults with shear bands (thin lines) connecting them. (E) The south side of the Clear Creek Valley, showing listric normal faults (dashed line) that cut through and sole into the greenschist facies mylonites beneath the Anaconda detachment (view to the south).

Figure 8. (A) Field photograph of greenschist facies mylonite and ultramylonite developed within Eocene granite from the eastern part of the Mill Creek Valley (orientation: SSE to the right side of the photo). (B) Photomicrograph showing mica fish that grew in Cambrian quartzite mylonite beneath the Anaconda detachment (orientation: SSE to the right side of the photo).

where progressively higher temperatures result in only partial retention of radiogenic ^{40}Ar. For biotite, the argon partial retention zone occurs between ~250–330 °C (McDougall and Harrison, 1999). At pre-extension depths deeper than the partial retention zone, temperatures are too hot for radiogenic ^{40}Ar to be retained prior to the onset of extension and cooling (e.g., John and Foster, 1993). At pre-extension depths shallower than the partial retention zone, the crust is cool enough to allow retention of radiogenic ^{40}Ar, and rocks containing K-bearing minerals record ^{40}Ar/^{39}Ar ages related to earlier cooling events. Within the partial retention zone, temperatures progressively increase, giving rise to a progression from near argon closure at shallower depths to almost complete loss of daughter isotopes prior to extension at deeper levels. K-bearing minerals residing within the partial retention zone record "mixed ages" upon cooling (e.g., Foster and John, 1999). At the onset of extension, rapid exhumation and cooling of rocks immediately below the base of the partial retention zone—the depth transition between no argon retention and partial retention—quenches minerals that previously retained no radiogenic argon, thereby recording the onset of extension (Foster and John, 1999; Stockli, 2005).

A plot of muscovite and biotite ^{40}Ar/^{39}Ar cooling ages against distance along a section in slip direction (105°) of the Anaconda detachment through the Mill Creek Valley transect is given in Figure 9A (Foster et al., 2010). This diagram was constructed by orthogonally projecting sample locations to the transect line (Fig. 9B), and includes an applied error of ± 1 km to account for projection errors and elevation (e.g., Foster and John, 1999; Brichau et al., 2005).

The mica ^{40}Ar/^{39}Ar cooling ages young to the ESE across the footwall to the Anaconda detachment (Foster et al., 2010). There is a rapid decrease from Late Cretaceous ages (≥74 Ma) to early Eocene ages (ca. 53 Ma) over a distance of ~5 km and then a much more gradual decrease to late Eocene ages further ESE (ca. 40–39 Ma). The change in the slope of mica cooling ages at ~5 km corresponds to a quenched paleoisotherm marking the base of the biotite partial retention zone, or ~330 °C (Foster and John, 1999; Stockli, 2005). Samples to the ESE of this point were at higher temperatures prior to the onset of exhumation and cooling of the lower plate. The top of the partial retention zone lies to the west of this point, but east of sample WG04-114 from the Storm Late stock, which gave a biotite cooling age of ca. 74 Ma.

The mica cooling age at the base of the Eocene partial annealing zone is between ca. 54 and 52 Ma, which we interpret to be the time that tectonic exhumation began to rapidly cool the footwall beneath the Anaconda detachment. Eocene granodiorite of the Hearst Lake suite was intruded into the eastern part of the footwall at 53 ± 1 Ma (Foster et al., 2007) and was subsequently overprinted by greenschist facies fabrics. Intrusion of the granodiorite was presumably related to the change in tectonic setting that initiated regional extension starting at 54–53 Ma (Foster et al., 2001, 2007). An early Eocene age for the onset of extension is also consistent with the age of oldest Lowland Creek volcanic rocks (ca. 53 Ma: Dudás et al., 2010; Foster et al., 2010). The

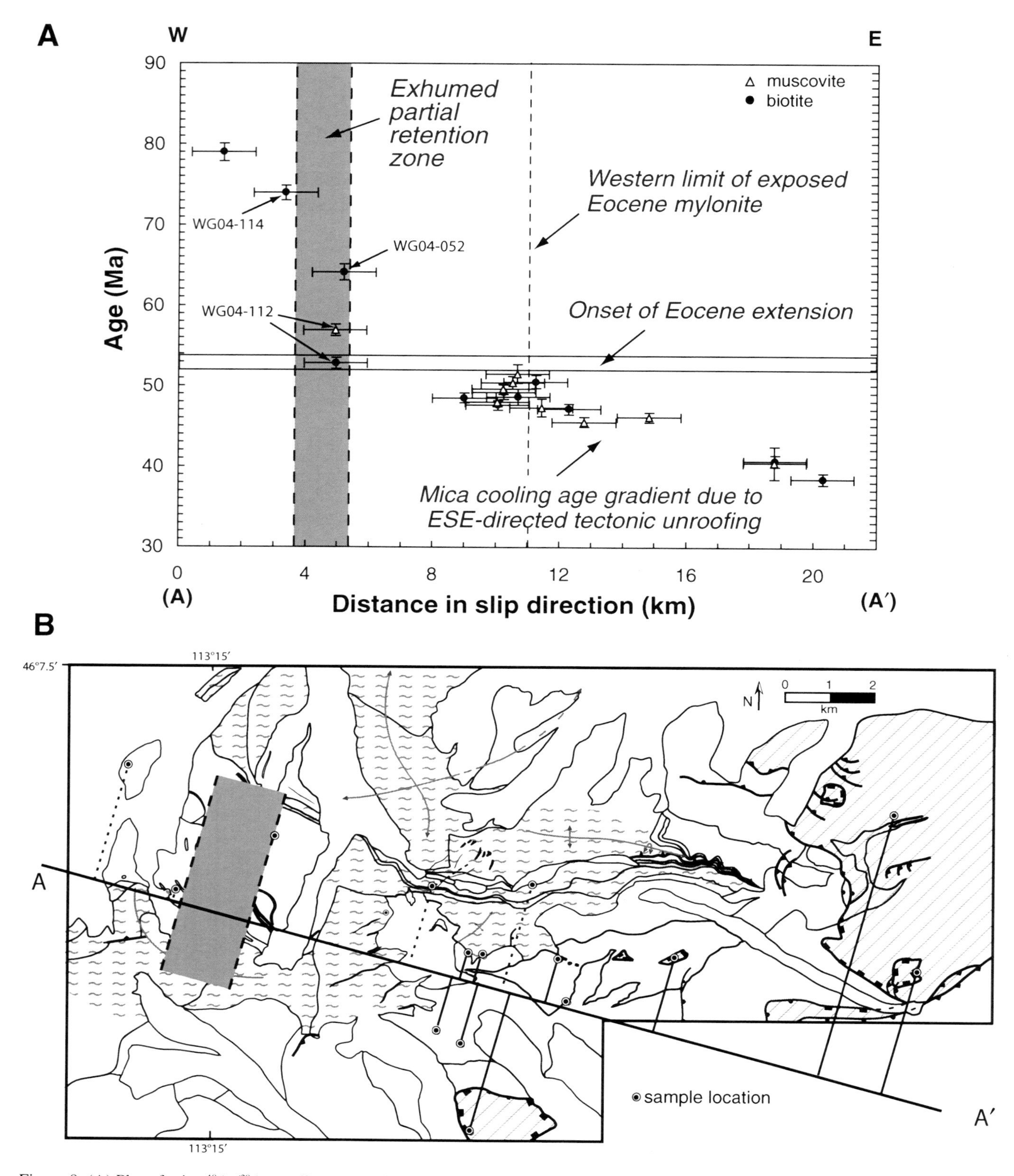

Figure 9. (A) Plot of mica $^{40}Ar/^{39}Ar$ cooling age against distance for sample locations projected to a section parallel to the slip direction of the Anaconda detachment (105°) from Foster et al. (2010). Sample locations were orthogonally projected to section A–A′ as shown in part B. The errors on the cooling ages are ± 2σ, and errors on the location are fixed at ± 1 km to account for uncertainty in projection to section A–A′ and elevation. The shaded band on upper plot (A) represents the location, or paleodepth, of the partial retention zone for mica prior to Eocene slip on the detachment and exhumation. The location of the partial retention zone is also shown for reference on the map (B).

ca. 41–39 Ma mica cooling ages obtained from the greenschist mylonite in the easternmost lower plate record when this part of the lower plate cooled through ~350 °C and was exhumed through the brittle-plastic transition. These cooling ages indicate that extension accommodated by slip on the Anaconda detachment continued at least into late Eocene time.

The inverse of the slope on the age-distance plot for muscovite and biotite $^{40}Ar/^{39}Ar$ cooling ages ≤ 51 Ma (Fig. 10) provides the basis to estimate the rate of slip on the Anaconda detachment (Foster et al., 2010). The biotite and muscovite data give extension rates of 0.93 ± 0.33 km/Ma and 0.87 ± 0.48 km/Ma (2σ), respectively, for the Eocene detachment system (see Foster and John, 1999, for method to calculate slip rate). This average rate does not account for any increase or decrease in slip rate between 51 and 39 Ma.

Magnitude of Eocene Extension

Displacement on the Anaconda detachment is constrained by reconstructing the granodiorite phase of the Storm Lake stock with a correlative Cretaceous granodiorite within a detached block to the east in the Deer Lodge Valley (see site marked "sample location" in Fig. 2). The granodiorite in both areas is unfoliated medium- to coarse-grained biotite ± hornblende granodiorite with concordant Late Cretaceous biotite $^{40}Ar/^{39}Ar$ cooling ages (Foster et al., 2010), consistent post-Cretaceous cooling histories (Fig. 6), and complementary major and trace element concentrations (Grice, 2006). Restoring the displaced top of the Storm Lake stock gives between 25 and 28 km of heave on the Anaconda detachment. This value of heave is greater than the amount indicated by the slip rate of ~0.9 km/Ma between 53 and 39 Ma (~13 km), but is consistent with this slip rate if the Anaconda detachment continued to be active until ca. 27–25 Ma as indicated by apatite fission-track data (Foster and Raza, 2002).

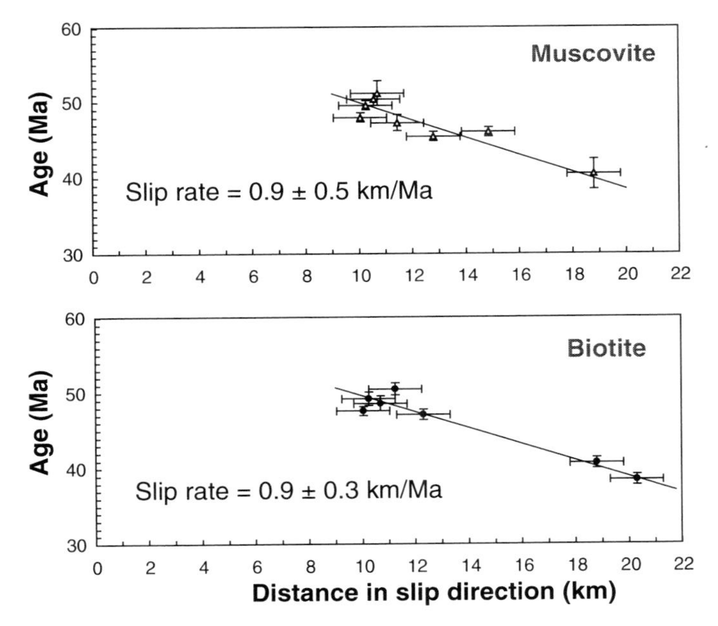

Figure 10. Calculations of the slip rate on the Anaconda detachment based on the inverse age versus distance gradient for samples that were at temperatures higher than ~330 °C and did not retain radiogenic argon before exhumation (from Foster et al., 2010).

Late Cretaceous metamorphic rocks beneath the detachment were metamorphosed at pressures of 4–6 kbar or depths of ~12–18 km, based on metamorphic thermobarmometry data (Grice, 2006). Throw on the Anaconda detachment since the Late Cretaceous, therefore, cannot be more than 12–18 km. The amount in Eocene time, however, may have been significantly less due to erosional and tectonic exhumation of the Cretaceous metamorphic and plutonic complex prior to Eocene extension.

The Anaconda detachment dips moderately east beneath the Boulder batholith (Foster et al., 2010). The Boulder batholith (Figs. 1, 2) is a composite of >15 plutons emplaced at shallow depths (<5 km) into a contemporaneous volcanic carapace (Tilling et al., 1968). Magmatism occurred between ca. 80 and 70 Ma with the most voluminous plutons, including the Butte pluton (granodiorite), intruded between ca. 75 and 70 Ma (Tilling et al., 1968; Robinson et al., 1968; Hamilton and Myers, 1974; Kalakay et al., 2001; Lund et al., 2002). The Boulder batholith hosts the Butte porphyry Cu-Mo deposit, which formed between ca. 67 and 62 Ma (Lund et al., 2002).

The Boulder batholith spans the western Helena salient (Fig. 1) with lateral thrust ramps along the northern and southern boundaries. The Lombard thrust, which transported Middle Proterozoic Belt rocks over Paleozoic and Mesozoic strata, lies immediately east of the Boulder batholith (Schmidt et al., 1990; Lageson et al., 2001). In most places, the eastern contact of the Boulder batholith is a steep west-dipping mylonite zone (Rutland et al., 1989; Kalakay et al., 2001). The Boulder batholith, therefore, is bounded by thrust faults except in the west where it is in the hanging wall of the Anaconda detachment (Fig. 2). Seismic reflection studies show a highly reflective and laminated lower crust below the batholith starting at ~12–18 km depth (Vejmelek and Smithson, 1995), which could be the down-dip projection of the Anaconda mylonite.

The 25–28 km displacement estimate for the Anaconda detachment indicates that the Boulder batholith was translated east at least that amount after emplacement (Foster et al., 2010). This reconstruction also implies that the hydrothermal systems that produced the Butte deposit were sourced to the west. Late Cretaceous–Paleocene plutons in the Flint Creek and Anaconda Ranges, including the Mt. Powell batholith, were intruded at deeper crustal levels than the Boulder batholith (Hyndman et al., 1982), and crystallized within the age range of the Butte deposits (Lund et al., 2002). Equivalent aged plutons beneath the Deer Lodge Valley could be related to mineralization at shallower levels in the Boulder batholith.

FIELD EXCURSION OUTLINE

The field trip will depart Bozeman, Montana (45°39′56″ N; 111°02′52″), and proceed west on Interstate 90. Approximately

98 miles from Bozeman, take exit 208 and merge onto MT Highway 1 North. After 5 miles on Highway 1, turn left onto Montana Highway 274, the Mill Creek road. Proceed south for ~8.5 miles on Mill Creek Road to Mill Creek pass.

STOP 1. Mill Creek Pass Viewing Area

Park in the gravel parking area. Location: 46°01′14″N; 112°59′04″W.

Looking west from this locality, there is an excellent view of the high country of the northern Anaconda Range. The high country to the west is mostly underlain by silicic intrusive rocks in contact with metamorphosed Belt Supergroup rocks. The rocks in the foreground are unmetamorphosed sedimentary and volcanic rocks. Rocks exposed at this locality are a complex mixture of conglomerate, conglomeratic sandstone, and sedimentary breccia. These strata are variably tilted and disrupted by a system of brittle normal faults. Many recent landslides also disrupt this sequence.

A thick mylonite zone forms the large cliff seen immediately across Mill Creek canyon. The zone is comprised of mylonitic two-mica granite capped by mylonitic quartzite. The quartzite is interpreted as a deformed sequence of Cambrian Flathead Sandstone (Emmons and Calkins, 1913). The small drainage immediately north of Mill Creek occupies a synformal trough that is part of the mega-mullion structure developed in the mylonitic foliation. The low eastward dip of the Anaconda mylonite zone is evident from this locality. From the cliff on the north side of Mill Creek, the trace of the mylonite zone curves beneath this field stop and turns back west to a point near the base of Short Peak, which is seen in the distance. Short Peak is capped by a mylonitic paragneiss sequence. We interpret these rocks as metamorphosed Cambrian units (i.e., Silver Hill and Flathead Sandstone) overlying uppermost Belt rocks. The sequence is strikingly similar to a less deformed metamorphic succession found to the west near Storm Lake. It is remarkably different, however, from the upper Belt through Cambrian section found at structurally lower levels in upper Mill Creek. The interesting stratigraphic and structural complications, described above, will be discussed at this stop.

At this location, we will also discuss the age and structure of the hanging wall volcanic rocks and sedimentary sequences. An example of the basal section of the hanging wall volcanic rocks can be observed in the roadcut at the south end of the parking area. $^{40}Ar/^{39}Ar$ analyses of biotite from a rhyolite unit in this outcrop gave an age of 53.7 ± 1.4 Ma (Foster et al., 2010).

From Stop 1 drive north on Mill Creek Road toward Highway 1.

STOP 2. Detached Block of the Storm Lake Granodiorite

Park in the pullout along Mill Creek Road. Location: 46°04′24″N; 112°56′23″W.

At Stop 2 we will examine an outcrop of granodiorite that is in the hanging wall of the Eocene Anaconda detachment. We interpret these rocks to be correlated with the ca. 75 Ma Storm Lake stock in the footwall, located ~15 miles to the west. The major and trace element compositions along with $^{40}Ar/^{39}Ar$ cooling ages are consistent with this granodiorite being the detached equivalent of the Storm Lake stock. Restoring this location to the part of the stock in the footwall that exhibits the same Late Cretaceous cooling ages indicates ~25–28 km of displacement for the Anaconda detachment system (Foster et al., 2010).

From Stop 2, continue north on Mill Creek Road. When you reach Montana Highway 1, turn left (west) toward Anaconda. From the intersection with Mill Creek Road, drive ~22.7 miles to Georgetown Lake Dam. Be very careful of the speed limit as you drive through Anaconda. At the dam, turn left into a pullout area.

STOP 3. Georgetown Thrust

Park in Georgetown Lake Dam parking area. Location: 46°10′58″N; 113°16′02″W.

Georgetown Lake is on the west side of the Anaconda core complex. The southern footwall structural dome of the core complex occupies the high elevation areas east-southeast of the lake where the Eocene Anaconda detachment is largely eroded away. The road east to Anaconda travels past the area of the inferred western breakaway fault system, which is now eroded away. The highway passes along a structural depression in the detachment where some exposures of the detachment can be seen, along with examples of the deformed footwall and brecciated hanging wall rocks.

The outcrop in this location is the Mesoproterozoic Helena Formation of the Belt Supergroup. The exposures are in the hanging wall of the Late Cretaceous Georgetown thrust fault, which places the Helena Formation on top of Mississippian through Cambrian sedimentary rocks. The trace of the thrust fault crosses the highway ~1.6 km (1 mile) east of this location. This thrust fault is intruded by the ca. 75–72 Ma Philipsburg batholith 8–10 km (5–6 miles) north of this location.

From Stop 3 drive east on Montana Highway 1 toward Anaconda for 6.5 miles to the entrance to Old State Highway 10.

STOP 4. Cretaceous Metamorphism and Attenuation with Overprinting Eocene Deformation

Park just off the highway at Yankee Flat overview. Location: 46°09′28″N; 113°08′39″W.

The south-facing cliff to the north of the highway exposes metamorphosed lower plate rocks of the core complex along with the Anaconda detachment. The lower part of the cliff exposes a metamorphosed and attenuated stratigraphic section of Belt quartzite through Cambrian Flathead, Silver Hill, and Hasmark Formations. Note the subtle low-angle unconformity between Proterozoic dark-colored rocks near the bottom of the outcrop and the white Cambrian Flathead sandstone. The angular unconformity in this region averages only 2–5°, and can usually only be seen from a distance. Visible above the Flathead is the Silver Hill Formation, which is only ~150 feet thick here, in contrast to 325 feet on the northwest side of the Flint Creek Range (Lonn et al., 2003). The thinness of the Silver Hill Formation here is attributed to severe internal deformation that resulted from the Late Cretaceous deformation. High-grade metamorphism occurred during the Late Cretaceous, but this is overprinted higher up in the cliff by the Eocene Anaconda detachment. Metamorphosed and highly brecciated, the Cambrian Hasmark Formation carbonate is exposed directly beneath the Anaconda detachment.

STOP 4A (Optional)

If time permits, climb up through this section, starting from a location on the north side of Highway 1. Drive 0.1 miles east on Highway 1. Turn left on Cable Road and drive ~0.5 miles. Location: 46°09′47″N; 113°08′02″W.

Meosproterozoic Belt quartzite is exposed at the base of the slope, above which are the Cambrian Flathead and Silver Hill Formations, which are mylonitic and highly attenuated compared to sections outside of the core complex. The detachment fault is located higher on the slopes, with the fractured Cambrian Hasmark Formation in the immediate footwall. Listric normal faults that cut through the Hasmark sole into the Anaconda detachment.

From Stop 4 continue east on Montana Highway 1 for 1.1 miles to Foster Creek Road. Turn left on Foster Creek Road and drive 0.4 miles.

STOP 5. Cretaceous and/or Eocene Deformation: The Northern Extension of the Lake of the Isle Shear Zone

Park near the quarry on Foster Creek Road. Location: 46°10′04″N; 113°07′28″W.

At first glance this quarry appears to expose well-bedded marble (Cambrian Hasmark Formation?), but a closer look reveals tight isoclinal folds within the layers. The axial planes appear to be parallel to the layers and some appear to be sheath folds. We attribute these folds to layer-parallel shear, but do not know whether they are the result of the Cretaceous tectonism discussed above or movement along the Eocene Anaconda detachment fault. The Anaconda detachment fault zone is at most a few hundred feet above this spot, but layer-parallel shear is also characteristic of the Late Cretaceous strain event. Eocene breccia can be seen near the top of the cliff. Deformation in this exposure is, therefore, both Late Cretaceous and Eocene in age and difficult to separate. It is possible that these rocks were excised from the lower plate of the detachment.

From Stop 5 return to Montana Highway 1 and drive southeast ~2.8 miles to the junction with Cable Road. Stop 6 is near the east side of the quarry along Blue Eyed Nellie Gulch.

STOP 6. Brecciated Eocene Units in the Hanging Wall of the Anaconda Detachment

Stop near east side of quarry. Location: 46°09′55″N; 113°03′22″W.

Here we observe several brecciated Eocene sedimentary, volcanic, and megabreccia units in the immediate hanging wall of the Anaconda detachment. O'Neill and Lageson (2003) refer to this megabreccia as the "West Valley chaos."

Return to Montana Highway 1 and drive toward Anaconda. Proceed through Anaconda to Cedar Street. Turn left on Cedar Street and proceed to the end of the road and enter the parking area.

STOP 7. Lowland Creek Volcanic Rocks and Hanging Wall Conglomerates

Park at trailhead for Upper Works historic trail, North Cedar Street, Anaconda. Location: 46°08′02″N; 112°56′49″W.

The clastic sedimentary rocks along this trail are interlayered with Eocene Lowland Creek volcanic rocks. O'Neill and Lageson (2003) interpreted the fluvial and alluvial sediments (Anaconda beds) to have been derived mainly from the upper plate of the Anaconda detachment. The conglomerates include fragments of the Missoula Group and Newland limestone along with other Paleozoic and Mesozoic units. The trail passes a series of east- and west-dipping normal faults.

From Stop 7 return to Montana Highway 1. From the intersection of Cedar Street and Highway 1 drive south 2.6 miles. Turn left on Montana Highway 48N and drive for 0.3 miles. Turn left on Galen Road and proceed 1.9 miles. Turn left on Lost Creek Road and drive to the entrance of Lost Creek State Park.

STOP 8. Late Cretaceous and Eocene Plutonism and Mylonitic Overprinting

Entrance of Lost Creek State Park on Lost Creek Road. Location: 46°11′53″N; 112°58′45″W.

On the drive into Lost Creek State Park, Lost Creek Canyon comes into view ahead. The timbered plateau that the Lost Creek Canyon cuts is capped by mylonite of the Anaconda detachment. Tan rocks near the canyon mouth are the Lost Creek granite, and lighter-colored rocks that are found further up-canyon are Belt Supergroup metasediments.

Giant boulders along the road mark the Pleistocene terminal moraine of the Lost Creek glacier. White cliffs on the right side of the road are brecciated dolomite assigned to the Cambrian Hasmark Formation between strands of the detachment fault. Just east of the entrance to Lost Creek State Park, at the mouth of Timber Gulch on the right, is the trace of the lower strand of the Anaconda detachment, beneath which are brecciated and mylonitic rocks of the Lost Creek Granite.

At the park entrance, the impressive cliffs are composed of early Tertiary Lost Creek granite (Winegar, 1971). Dark-colored rocks at the top of the right cliff are part of an older, Cretaceous gabbroic sill. Biotite from the mylonitic Lost Creek stock gave an $^{40}Ar/^{39}Ar$ age of 38.8 ± 1.6 Ma (Foster et al., 2010), which is much younger than the crystallization age of the pluton. The cooling age of this biotite is related to tectonic exhumation of the eastern edge of the footwall beneath the Anaconda detachment.

If time permits, walk east, back across the cattle guard, for 100 yards along the road to granite outcrops at the curve in the road. The sloping granite surface is actually the fault plane of the lower strand of the Eocene Anaconda detachment. The granite here contains bands of dark-colored mylonite, ultramylonite, and cataclasite, and is so brecciated that no consistent attitude for the mylonitic foliation can be found. The ductile fabrics are overprinted by brittle ones that developed as unroofing of the core complex brought these rocks up into the brittle zone. The best exposures of mylonitic fabrics are in the boulders blasted from the roadbed.

Retrace route back to I-90 and return ~115 miles back to Bozeman.

ACKNOWLEDGMENTS

This research was funded by grants from the Australian Research Council and National Science Foundation to Foster and U.S. Geological Survey EDMAP grants to Kalakay. We thank Colin Shaw and Basil Tikoff for providing helpful reviews of the manuscript.

REFERENCES CITED

Allen, J.C., 1966, Structure and petrology of the Royal Stock, Flint Creek Range, central-western Montana: Geological Society of America Bulletin, v. 77, p. 291–302, doi:10.1130/0016-7606(1966)77[291:SAPOTR]2.0.CO;2.

Armstrong, R.L., 1978, Cenozoic igneous history of the U.S. Cordillera from lat 42° to 49° N, *in* Smith, R.B., and Eaton, G.P., eds., Cenozoic Tectonics and Regional Geophysics of the Western Cordillera: Geological Society of America Memoir, 152, p. 265–282.

Brichau, S., Ring, U., Ketcham, R.A., Carter, A., Stockli, D., and Brunel, M., 2005, Constraining the long-term evolution of the slip rate for a major extensional fault system in the central Aegean, Greece, using thermochronology: Earth and Planetary Science Letters, v. 241, p. 293–306.

Buckley, S.N., 1990, Ductile extension in a Late Cretaceous fold and thrust belt, Granite County, west-central Montana [Master's thesis]: Missoula, Montana, University of Montana, 42 p.

Coney, P.J., 1980, Cordilleran Metamorphic core complexes, *in* Crittenden, M.D., Jr., Coney, P.J., and Davis, G.H., eds., Cordilleran Metamorphic Core Complexes: Geological Society of America Memoir 153, p. 7–31, doi:10.1130/MEM153-p7.

Csejtey, B., 1962, Geology of the southeast flank of the Flint Creek Range, western Montana [Ph.D. dissertation]: Princeton, New Jersey, Princeton University, 208 p.

Desmarais, N.R., 1983, Geology and geochronology of the Chief Joseph plutonic-metamorphic complex, Idaho-Montana [Ph.D. dissertation]: Seattle, Washington, University of Washington, 125 p.

Doughty, P.T., and Price, R.A., 2000, Geology of the Purcell Trench rift valley and the Sandpoint Conglomerate: Eocene en echelon normal faulting and synrift sedimentation along the eastern flank of the Priest River metamorphic complex, northern Idaho: Geological Society of America Bulletin, v. 112, p. 1356–1374, doi:10.1130/0016-7606(2000)112<1356:GOTPTR>2.0.CO;2.

Dudás, F.O., Ispolatov, V.O., Harlan, S.S., and Snee, L.W., 2010, $^{40}Ar/^{39}Ar$ geochronology and geochemical reconnaissance of the Eocene Lowland Creek volcanic field, west-central Montana: The Journal of Geology, v. 118, p. 295–304, doi:10.1086/651523.

Emmons, W.H., and Calkins, F.C., 1913, Geology and Ore Deposits of the Phillipsburg Quadrangle, Montana: U.S. Geological Survey Professional Paper 78, 271 p.

Emmons, W.H., and Calkins, F.C., 1915, Description of the Philipsburg Quadrangle, Montana: Geological Atlas Folio, Report GF-0196, 26 p.

Foster, D.A., and Fanning, C.M., 1997, Geochronology of the Idaho-Bitterroot batholith and Bitterroot metamorphic core complex: Magmatism preceding and contemporaneous with extension: Geological Society of America Bulletin, v. 109, p. 379–394, doi:10.1130/0016-7606(1997)109<0379:GOTNIB>2.3.CO;2.

Foster, D.A., and John, B.E., 1999, Quantifying tectonic exhumation in an extensional orogen with thermochronology: Examples from the southern Basin and Range Province, *in* Ring, U., Brandon, M.T., Lister, G.S., and Willet, S.D., eds., Exhumation Processes: Normal Faulting, Ductile Flow and Erosion: Geological Society of London Special Publication 154, p. 343–364.

Foster, D.A., and Raza, A., 2002, Low-temperature thermochronological record of exhumation of the Bitterroot metamorphic core complex, northern Cordilleran Orogen: Tectonophysics, v. 349, p. 23–36, doi:10.1016/S0040-1951(02)00044-6.

Foster, D.A., Schafer, C., Fanning, M.C., and Hyndman, D.W., 2001, Relationships between crustal partial melting, plutonism, and exhumation: Idaho-Bitterroot batholith: Tectonophysics, v. 342, p. 313–350, doi:10.1016/S0040-1951(01)00169-X.

Foster, D.A., Doughty, P.T., Kalakay, T.J., Fanning, C.M., Coyner, S., Grice, W.C., and Vogl, J.J., 2007, Kinematics and timing of exhumation of Eocene metamorphic core complexes along the Lewis and Clark fault zone, northern Rocky Mountains, USA, *in* Till, A., Roeske, S., Sample, J., and Foster, D.A., eds., Exhumation along Major Continental Strike-Slip Systems: Geological Society of America Special Paper 434, p. 205–229.

Foster, D.A., Grice, W.C., and Kalakay, T.J., 2010, Extension of the Anaconda metamorphic core complex: $^{40}Ar/^{39}Ar$ thermochronology with implications for Eocene tectonics of the northern Rocky Mountains and the Boulder batholith: Lithosphere, v. 2, p. 232–246, doi:10.1130/L94.1.

Garmezy, L., 1983, Geology and geochronology of the southeast border of the Bitterroot dome: Implications for the structural evolution of the mylonitic carapace [Ph.D. dissertation]: University Park, Pennsylvania State University, 276 p.

Grice, W.C., Jr., Foster, D.A., and Kalakay, T.J., 2005, Quantifying exhumation and cooling of the Eocene Anaconda metamorphic core complex, western Montana: Geological Society of America Abstracts with Programs, v. 37, no. 7, p. 230.

Hamilton, W., and Myers, W.B., 1974, Nature of the Boulder Batholith of Montana: Geological Society of America Bulletin, v. 85, p. 365–378, doi:10.1130/0016-7606(1974)85<365:NOTBBO>2.0.CO;2.

Heise, B.A., 1983, Structural geology of the Mt. Haggin area, Deer Lodge County, Montana [Master's thesis]: University of Montana, Missoula, Montana, 77 p.

Hodges, K.V., and Walker, J.D., 1992, Extension in the Cretaceous Sevier orogen, North American Cordillera: Geological Society of America Bulletin, v. 104, p. 560–569, doi:10.1130/0016-7606(1992)104<0560:EITCSO>2.3.CO;2.

House, M.A., and Hodges, K.V., 1994, Limits on the tectonic significance of rapid cooling events in extensional settings: Insights from the Bitterroot metamorphic core complex, Idaho-Montana: Geology, v. 22, p. 1007–1010, doi:10.1130/0091-7613(1994)022<1007:LOTTSO>2.3.CO;2.

Hyndman, D.W., 1980, Bitterroot dome–Sapphire tectonic block, an example of a plutonic-core gneiss-dome complex with its detached suprastructure, *in* Crittenden, M.D., Jr., Coney, P.J., and Davis, G.H., eds., Cordilleran Metamorphic Core Complexes: Geological Society of America Memoir 153, p. 427–443, doi:10.1130/MEM153-p427.

Hyndman, D.W., Silverman, A.J., Ehinger, R., Benoit, W.R., and Wold, R., 1982, The Philipsburg Batholith, Western Montana; Mineralogy, Petrology, Internal Variation and Evolution: State of Montana, Bureau of Mines and Geology Memoir 49, 37 p.

John, B.E., and Foster, D.A., 1993, Structural and thermal constraints on the initiation angle of detachment faulting in the southern Basin and Range; the Chemehuevi Mountains case study: Geological Society of America Bulletin, v. 105, p. 1091–1108, doi:10.1130/0016-7606(1993)105<1091:SATCOT>2.3.CO;2.

Kalakay, T.J., John, B.E., and Lageson, D.R., 2001, Fault-controlled pluton emplacement in the Sevier fold-and-thrust belt, SW Montana: Journal of Structural Geology, v. 23, p. 1151–1165, doi:10.1016/S0191-8141(00)00182-6.

Kalakay, T.J., Foster, D.A., and Thomas, R.A., 2003, Geometry, kinematics and timing of extension in the Anaconda extensional terrane, western Montana: Northwest Geology, v. 32, p. 42–72.

Lageson, D.R., Schmitt, J.G., Kalakay, T.J., Horton, B.K., and Burton, B.R., 2001, Influence of late Cretaceous magmatism on the Sevier orogenic wedge, western Montana: Geology, v. 29, p. 723–726, doi:10.1130/0091-7613(2001)029<0723:IOLCMO>2.0.CO;2.

Lewis, R.S., 1998, Preliminary Geologic Map of the Butte 1° × 2° quadrangle: Montana Bureau of Mines and Geology Open-File Report MBMG-363, scale 1:250,000.

Lonn, J.D., McDonald, C., Lewis, R.S., Kalakay, T.J., O'Neill, J.M., Berg, R.B., and Hargrave, P., 2003, Preliminary Geologic Map of the Philipsburg 30′ × 60′ Quadrangle, Western Montana: Montana Bureau of Mines and Geology Open File Report MBMG-483, scale 1:100,000.

Lund, K., Aleinikoff, J.N., Kunk, M.J., Unruh, D.M., Zeihen, G.D., Hodges, W.C., du Bray, E.A., and O'Neill, J.M., 2002, SHRIMP U-Pb and ^{40}Ar/^{39}Ar age constraints for relating plutonism and mineralization in the Boulder Batholith region, Montana: Economic Geology and the Bulletin of the Society of Economic Geologists, v. 97, p. 241–267, doi:10.2113/gsecongeo.97.2.241.

McDougall, I., and Harrison, T.M., 1999, Geochronology and Thermochronology by the ^{40}Ar/^{39}Ar Method (2nd ed.): Oxford, UK, Oxford University Press 288 p.

McLeod, P.J., 1987, The depositional history of the Deer Lodge Basin, western Montana [M.Sc. thesis]: Missoula, Montana, University of Montana, 61 p.

O'Neill, J.M., and Lageson, D.R., 2003, West to east geologic road log: Paleogene Anaconda metamorphic core complex: Georgetown Lake Dam–Anaconda–Big Hole Valley: Northwest Geology, v. 32, p. 29–46.

O'Neill, J.M., Lonn, J., and Kalakay, T.J., 2002, Early Tertiary Anaconda Metamorphic Core Complex, southwestern Montana: Geological Society of America Abstracts with Programs, v. 34, no. 4, p. A-10.

O'Neill, J.M., Lonn, J.D., Lageson, D.R., and Kunk, M.J., 2004, Early Tertiary Anaconda Metamorphic Core Complex, southwestern Montana: Canadian Journal of Earth Sciences, v. 41, p. 63–72, doi:10.1139/e03-086.

Passchier, C.W., and Trouw, R.A.J., 2005, Microtectonics: Berlin; New York, Springer, 366 p.

Parrish, R.R., Carr, S.D., and Parkinson, D.L., 1988, Eocene extensional tectonics and geochronology of the southern Omineca Belt, British Columbia and Washington: Tectonics, v. 7, p. 181–212, doi:10.1029/TC007i002p00181.

Robinson, G.D., Klepper, M.R., and Obradovich, J.D., 1968, Overlapping plutonism, volcanism, and tectonism in the Boulder batholith region, western Montana: Geological Society of America, v. 116, p. 557–576, doi:10.1130/MEM116-p557.

Rutland, C., Smedes, H.W., Tilling, R.I., and Greenwood, W.R., 1989, Volcanism and plutonism at shallow crustal levels: The Elkhorn Mountains volcanic sequence and the Boulder batholith, southwestern Montana, *in* Hyndman, D.W., ed., Cordilleran Volcanism, Plutonism, and Magma Generation at Various Crustal Levels, Montana and Idaho (International Geological Congress, 28th, Field Trip Guidebook T337): Washington, D.C., American Geophysical Union, p. 16–31.

Schmidt, C.J., Smedes, H.W., and O'Neill, J.M., 1990, Syncompressional emplacement of the Boulder and Tobacco Root Batholiths (Montana-USA) by pull-apart along old fault zones: Geological Journal, v. 25, p. 305–318, doi:10.1002/gj.3350250313.

Smedes, H.W., 1962, Lowland Creek volcanics, an upper Oligocene formation near Butte, Montana: The Journal of Geology, v. 70, p. 255–266, doi:10.1086/626818.

Smedes, H.W., and Prostka, H.J., 1972, Stratigraphic Framework of the Absaroka Volcanic Supergroup in the Yellowstone National Park Region: U.S. Geological Survey Professional Paper 729-C, 33 p.

Spear, F.S., Kohn, M.J., and Cheney, J.T., 1999, P-T paths from anatectic pelites: Contributions to Mineralogy and Petrology, v. 134, p. 17–32.

Stockli, D.F., 2005, Application of low-temperature thermochronology to extensional tectonic settings: Reviews in Mineralogy and Geochemistry, v. 58, p. 411–448, doi:10.2138/rmg.2005.58.16.

Tilling, R.I., Klepper, M.R., and Obradovich, J.D., 1968, K-Ar ages and time span of emplacement of the Boulder Batholith, Montana: American Journal of Science, v. 266, p. 671–689, doi:10.2475/ajs.266.8.671.

Vejmelek, L., and Smithson, S.G., 1995, Seismic reflection profiling in the Boulder batholith, Montana: Geology, v. 23, p. 811–814, doi:10.1130/0091-7613(1995)023<0811:SRPITB>2.3.CO;2.

Vogl, J.J., Foster, D.A., Fanning, C.M., Kent, K.A., Rogers, D.A., and Diedesch, T., 2012, Timing of extension in the Pioneer metamorphic core complex with implications for the spatial-temporal pattern of Cenozoic extension and exhumation in the northern U.S. Cordillera: Tectonics, v. 31, TC1008, 22 p.

Wallace, C.A., Lidke, D.J., Elliott, J.E., Desmarais, N.R., Obradovich, D.A., Lopez, D.A., Zarske, S.E., Heise, B.A., Blaskowski, M.J., and Loen, J.S., 1992, Geologic Map of the Anaconda-Pintlar Wilderness and Contiguous Roadless Area, Granite, Deer Lodge, Beaverhead, and Ravalli Counties, Western Montana: U.S. Geological Survey Miscellaneous Field Studies Map MF-1633-C, scale 1:50,000.

Wells, M.L., 1997, Alternating contraction and extension in the hinterlands of orogenic belts: An example from the Raft River Mountains, Utah: Geological Society of America Bulletin, v. 109, p. 107–126, doi:10.1130/0016-7606(1997)109<0107:ACAEIT>2.3.CO;2.

Winegar, R.C., 1971, The petrology of the Lost Creek stock and its relation to the Mount Powell Batholith, Montana [M.S. thesis]: Missoula, University of Montana, 60 p.

Manuscript Accepted by the Society 6 March 2014

Printed in the USA

The Geological Society of America
Field Guide 37
2014

The Yellowstone and Regal talc mines and their geologic setting in southwestern Montana

Sandra J. Underwood*
John F. Childs*
Chad P. Walby*
Helen B. Lynn*
Zachary S. Wall*
Childs Geoscience, Inc., 1700 West Koch St., Suite 6, Bozeman, Montana 59715, USA

Michael T. Cerino*
Barretts Minerals Inc., 8625 Hwy 91 South, Dillon, Montana 59725, USA

Ericka Bartlett*
Imerys Talc, Yellowstone Mine, 280 Johnny Ridge Road, Cameron, Montana 59720, USA

ABSTRACT

We summarize the geologic settings, generalized geology, and inferred conditions of talc formation for two major deposits in southwestern Montana. Imerys Talc operates the Yellowstone Mine in the Gravelly Range. Barretts Minerals Inc., a subsidiary of Minerals Technologies Incorporated, mines talc from two large deposits—the Regal and the Treasure—in the southern Ruby Range. Talc mineralization in southwestern Montana is associated with hydrothermal alteration of Archean dolomitic marbles along faults in the southern margin of the middle Proterozoic Belt Seaway. Conditions of talc formation appear to have varied across the region and probably range from shallow hot spring systems to connate brine circulation pathways in Belt basin sediments. A road log description of the geology along a loop from Bozeman to Dillon, Montana, to visit both the Yellowstone and Regal talc mines accompanies this paper.

INTRODUCTION

Talc is a hydrous magnesium phyllosilicate [$Mg_3Si_4O_{10}(OH)_2$] that may consist of compact masses of <1μm microcrystalline platelets or crystal sizes >100 μm in platy macrocrystalline talc. In addition, highly lamellar talc has individual macroscopic plates similar to mica. Individual talc platelet size is controlled by conditions of mineral formation, and the size of the talc plates affects surface properties. Talc is recognized for its softness (Mohs hardness of 1) and the ease with which it can be carved with a knife.

The talc deposits mined in southwestern Montana produce some of the purest talc in the world (Van Gosen et al., 1998).

*E-mails: sunderwood@childsgeoscience.com; jchilds@childsgeoscience.com; cwalby@childsgeoscience.com; hlynn@childsgeoscience.com; zwall@childsgeoscience.com; mike.cerino@mineralstech.com; ericka.bartlett@imerys.com.

Underwood, S.J., Childs, J.F., Walby, C.P., Lynn, H.B., Wall, Z.S., Cerino, M.T., and Bartlett, E., 2014, The Yellowstone and Regal talc mines and their geologic setting in southwestern Montana, *in* Shaw, C.A., and Tikoff, B., eds., Exploring the Northern Rocky Mountains: Geological Society of America Field Guide 37, p. 161–187, doi:10.1130/2014.0037(08). For permission to copy, contact editing@geosociety.org.

Many of the Montana talc products are of the compact or microcrystalline variety and are in great demand for manufacturing paper, paints and coatings, plastics, rubber, ceramics, and agricultural products. In addition, cosmetic and personal care products, pharmaceuticals, and selected food industry applications are low volume niches for Montana talc. Two companies, Imerys Talc and Barretts Minerals Inc., currently operate mines and mills in southwestern Montana.

OVERVIEW OF REGIONAL GEOLOGY

All of the talc deposits in southwestern Montana (Fig. 1) are hosted by Archean dolomitic marbles that are associated with high-grade metamorphic sequences, including quartzofeldspathic gneiss, garnetiferous kyanite-staurolite mica schist, quartz-rich gneiss, banded iron formation, amphibolite, and many other lithologies. A late Archean tectonothermal event from continental collision is the postulated cause for the oldest amphibolite grade metamorphism, followed by a ca. 1.7–1.8 Ga upper amphibolite facies metamorphic event (Brady et al., 1998). Both the Yellowstone and Regal talc mines are in the Paleoproterozoic suture zone along the northwestern edge of the Archean Wyoming province (Fig. 2).

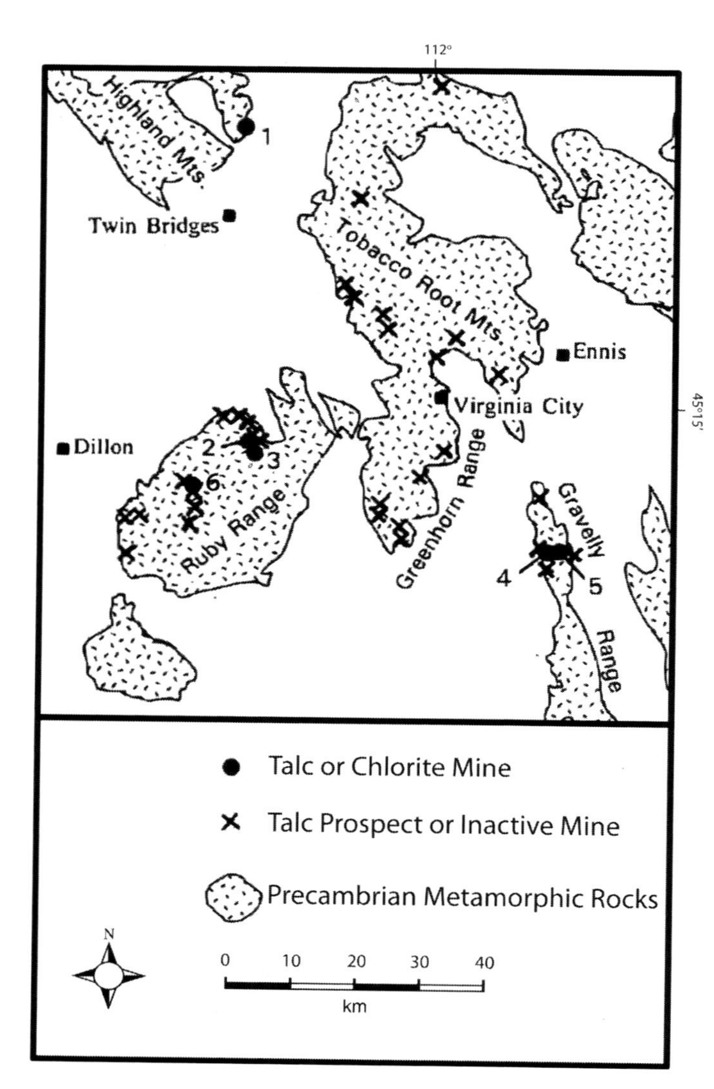

Figure 1. Simplified map for selected chlorite and talc mines in southwestern Montana (after Berg, 1995). 1—Golden Antler Mine (chlorite); 2—Treasure Mine (talc); 3—Beaverhead Mine (talc); 4—Johnny Gulch Mine (talc); 5—Yellowstone Mine (talc); and 6—Regal Mine (talc). X—Other talc and chlorite prospects.

Talc formation in southwestern Montana is often associated with faulting and magmatism that occurred during opening of the Belt basin (ca. 1.4 Ga), by stable isotope studies of selected minerals in and around talc ore bodies, geochronological data, petrological studies, and fluid inclusions in chalcedony (i.e., Anderson et al., 1990; Brady et al., 1998; Gammons and Matt, 2002). If Belt seawater was involved in talc formation, it would require that the Belt basin extended across part of the region uplifted as the Dillon Block. Erosion would have removed the resulting Belt sedimentary rocks from the Ruby Range while preserving Belt basin sediments to the west and southwest. Presently, the extensive exposures of Belt Supergroup sedimentary rocks are found in the Pioneer Mountains 24 km (15 mi) west of Dillon (Vuke et al., 2007), as well as in the Beaverhead Mountains and adjacent ranges in Idaho farther southwest (Burmester et al., 2013).

The geologic setting of the talc deposits is somewhat enigmatic, and textural relationships can be complex in detail (Fig. 3). For example, diabase dikes postulated as the heat engines for the hydrothermal environments where talc formed are not exposed near some of the producing or past producing deposits (i.e., the Treasure and the Beaverhead deposits, respectively). Some talc deposits display thickening and deformation in early fold hinges, e.g., the Regal (Fig. 4), whereas other deposits are in relatively thin marbles and though not within obvious fold hinges, are along major north-to-northwest chloritic fault zones, e.g., the Treasure (Fig. 5) and the Beaverhead. Improving plausible talc formation model(s) must consider these differences.

In the Ruby Range, chlorite appears to have developed in aluminous rocks such as mica schist and quartzofeldspathic gneiss as part of the same process that produced talc in the dolomitic marbles. Berg (1979) described the Nolte chlorite deposit in the Highland Mountains ~48 km (30 mi) north of the southern Ruby Range. Clinochlore, or Mg chlorite, apparently replaced all mineral phases in the quartzofeldspathic gneiss except rutile and zircon. The chlorite at the Nolte deposit, and at least some of the chlorite and associated talc in the southern Ruby Range appears to be associated with north-to-northwest–trending steeply dipping fault zones.

Talc Formation Models for SW Montana

Early models for talc formation in southwestern Montana included a retrograde metamorphic event based on replacement

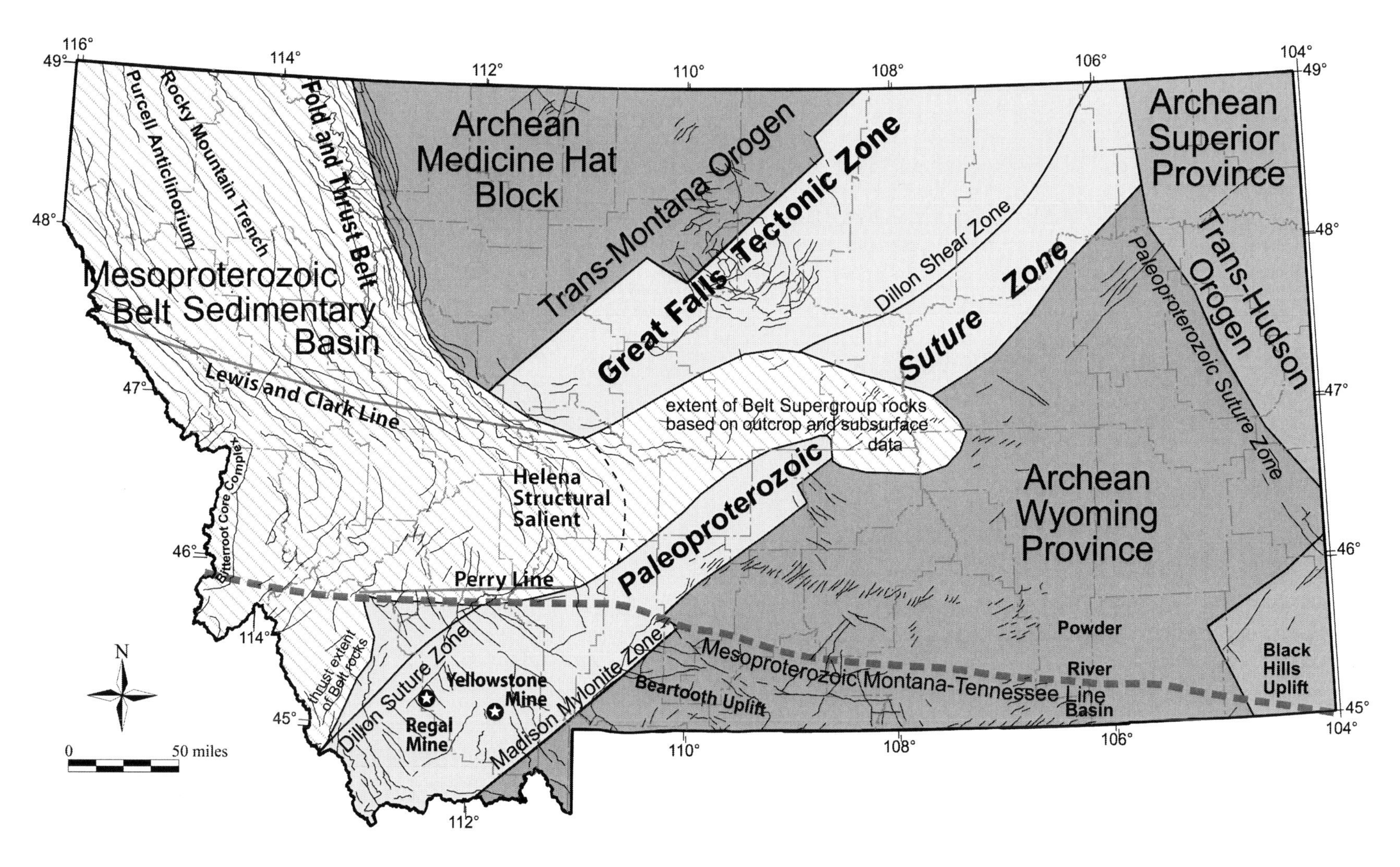

Figure 2. Montana basement rock and tectonic provinces (modified after Vuke et al., 2007). The Yellowstone and Regal Mine localities are each indicated with a star and major faults are delineated with thin gray lines. The Paleoproterozoic fold-and-thrust section southeast of the Dillon suture zone and northwest of the Madison mylonite zone in southwest Montana contains the Dillon Block.

Figure 3. A loose block of ore in the Treasure Mine showing some of the complex relationships between talc and carbonates. The central domain of medium-green chloritic talc is cut by white veinlets <1 cm thick of dolomite or magnesite. These veins terminate at the contact between the chloritic talc and the surrounding coarsely recrystallized white dolomite. The recrystallized dolomite is partially replaced by white coarsely crystalline talc. Hammer and GPS unit provide scale.

of tremolite by talc and biotite by chlorite, and on the restriction of the talc deposits to Archean rocks where radiometric dates had been reset at ca. 1.6 Ga (Berg, 1979). Anderson et al. (1990) suggested a hydrothermal model for the formation of the Beaverhead deposit at ca. 1.4 Ga and, by inference, a similar origin and age for the other talc deposits in the area. Physical-chemical modeling and petrographic evidence developed by Anderson et al. (1990) suggested that the talc was formed in a near-surface hot spring environment and required large volumes of water.

Talc formation in southwestern Montana is probably related to a rifting event, such as the opening of the Belt basin, based on several lines of evidence. Childs (1985) obtained K-Ar age dates of 1.14 and 1.03 Ga for sericite from gneisses adjacent to talc, and Brady et al. (1998) reported a Mesoproterozoic age of 1.36 Ga for sericite associated with talc. Investigations of stable isotopes and Rb-Sr radiometric age dates on Ruby Range talc deposits indicate a middle Proterozoic age for the talc (Childs, 1985; Hurst, 1985).

Northwest-trending Proterozoic diabase dikes, which are common throughout the southern Ruby Range, may have invaded extensional faults related to opening of the Belt basin and are the postulated heat source that drove talc-forming hydrothermal systems (James, 1990). Rose (1984) demonstrated that the trace element geochemistry of the talc, when compared with the unaltered host dolomitic marble, contained elevated Cr and Ni, which could indicate the presence of a mafic or ultramafic component in the talc-forming process. Moreover, Hoy et al. (2000) suggested that diabase sills were emplaced into soft sediment in the northern Belt basin, and that heat loss drove hydrothermal cells. The associated hydrothermal vents or sand boils were presumed coeval with the exposure to metal rich brines that formed the Sullivan massive sulfide deposit and its related tourmaline and chlorite alteration blankets in southern British Columbia.

Talc in the Ruby Range and probably talc at the Yellowstone Mine formed in a near-surface environment. Solid support for this assertion is found in field observations: (1) botryoidal talc (Fig. 6), (2) abundant open vugs in veins (Fig. 7), and (3) thinly banded opaline to chalcedonic quartz as vug linings. Investigators generally agree that the timing and characteristics of talc formation support water-rich environments probably related to the Belt basin opening, but details regarding depth of formation in the "plumbing system" are lacking. Talc is stable at temperatures up to 700 °C and pressures of 10 kbar (Evans and Guggenheim, 1988). When water is the dominant fluid component, talc-forming reactions and talc stability fields easily extend into low temperature and pressure conditions (Anderson et al., 1990). The mass balance calculations by Anderson et al. (1990) permit further conjecture that the type of shallow environment could have been a local sabkha-type setting proximal to the shore of the Belt Seaway. The crystalizing dikes and sills supplied ample amounts of shallow Mg-enriched water for talc formation. Additional SiO_2 and Mg stripped from the Archean basement rocks by deeply circulating, magma-derived hydrous fluids may have enriched Mg content in dolomite and crystallized magnesite domains in the marble and in the subsequent talc formation. A modern example of this type of environment may be the Guaymas Basin of the Gulf of California, where talc formation is documented adjacent to diabase sills at the seawater interface (Lonsdale et al., 1980).

In the Gravelly Range, Gammons and Matt (2002) suggest that deep brines in the Proterozoic Belt basin were instrumental in talc formation at the Yellowstone Mine. They examined fluid inclusions in hydrothermal quartz contained within talc and developed a model that invokes injection of thick mafic sills into the Belt sedimentary pile. The heat forced connate brines out of the bottom of the basin and into the adjacent basement. This hydrothermal engine would have moved fluids along Proterozoic conduits, which are now "growth" faults in the Yellowstone Mine area, and generated chloritic alteration of mafic dikes that had intruded along the faults. Hydrothermal alteration produced talc in the receptive dolomitic marbles in the basement rocks. Temperatures of talc formation at the Yellowstone Mine ranged from 190° to 250 °C at 1–4 kbar, which corresponds to depths of 3.5–14 km (2.2–8.7 mi) (Gammons and Matt, 2002).

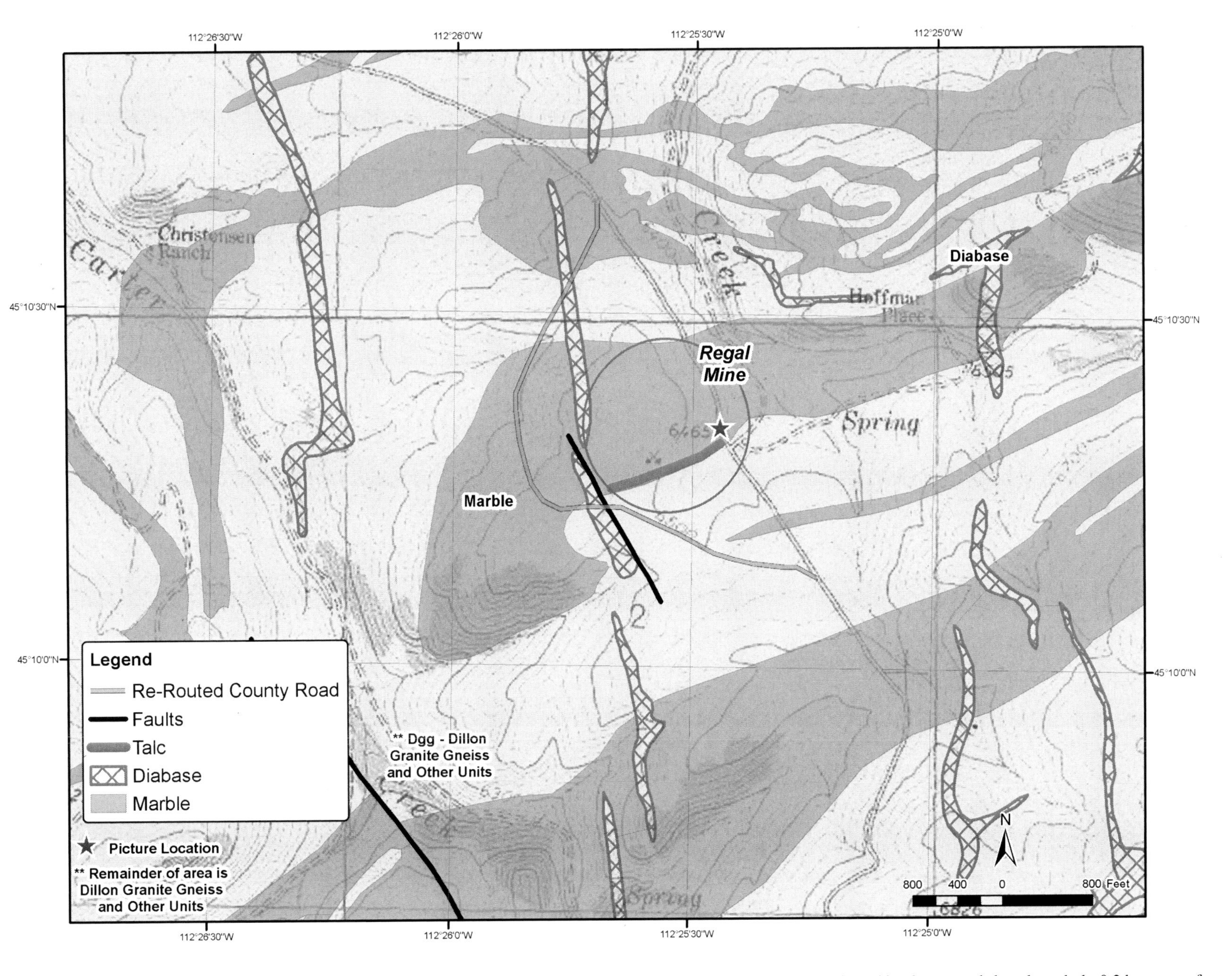

Figure 4. Geologic setting of the Regal Mine, Ruby Range (modified after James, 1990). The original mine locality is indicated by the crossed shovel symbol ~0.2 km west of the original right-of-way for the Sweetwater Road. Expansion of the mine pit necessitated re-routing the road ca. 2006 around a pit within the circle, and the star is the vantage point for Figure 13.

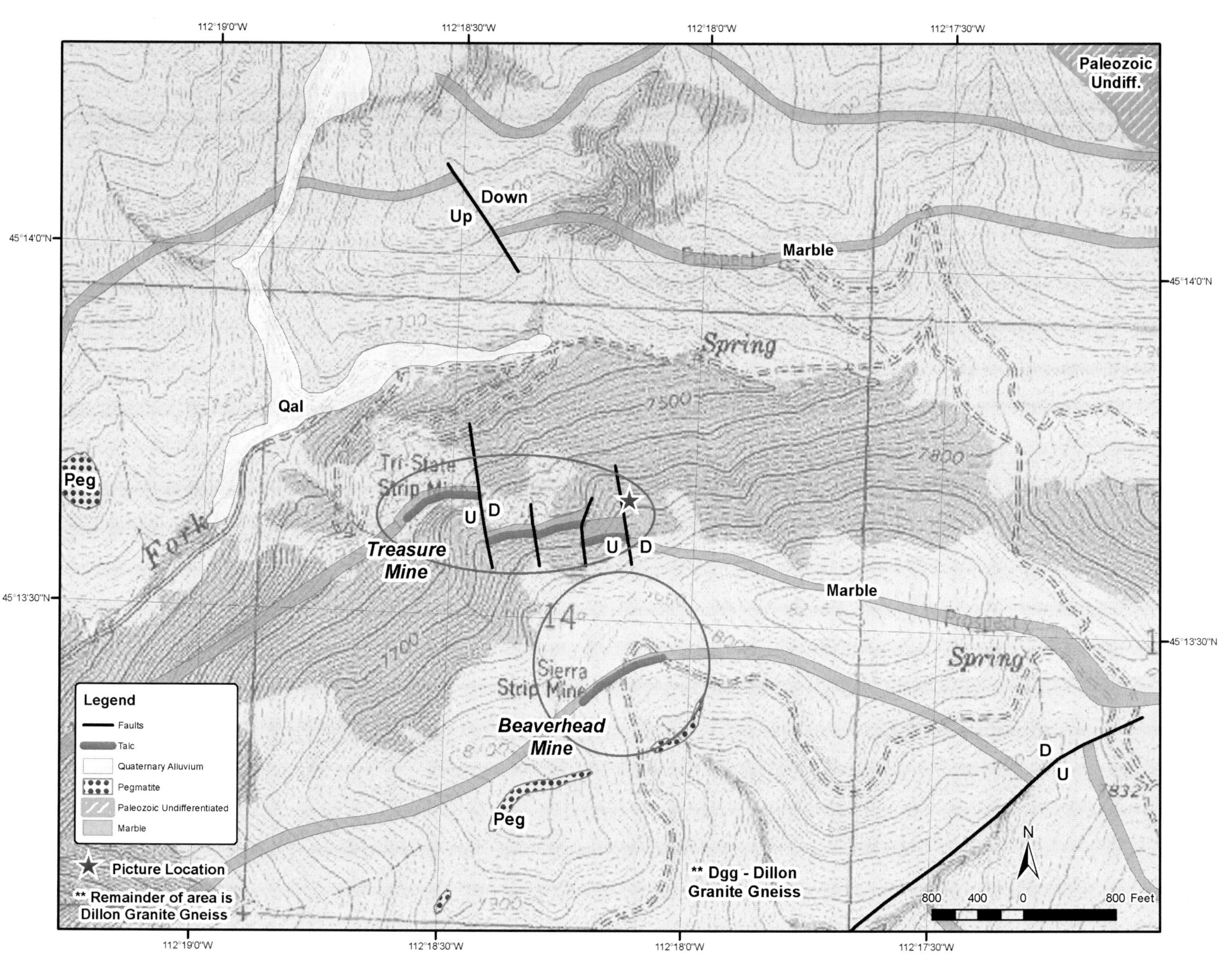

Figure 5. Geologic setting of the Treasure and Beaverhead Mines, Ruby Range (modified after Garihan, 1973). The pre-mining surface geology shown in the map guided placement of the large pits within the indicated oval and circle. The star is the vantage point for Figure 12.

Figure 6. Botryoidal light green talc from the Yellowstone Mine. This texture is strong evidence for open space late in the talc-forming process. Scale is in inches.

Talc Textures

Constant volume replacement of carbonate by talc is indicated by the common occurrence of spectacular coarse dolomite, and possibly magnesite rhomboids, that are partially to completely replaced by talc, while perfectly preserving their rhombic shapes. The Beaverhead and Treasure Mines and the Whitney prospect farther north in the Ruby Range have especially well-developed examples of coarse-grained dolomite pseudomorphically replaced by talc. In addition, talc replaces olivine and diopside formed at amphibolite to granulite metamorphic grades, and tremolite formed during subsequent greenschist-facies metamorphism (Harms et al., 2004). Other evidence of constant volume replacement is found in the quartzofeldspathic gneisses and pegmatites adjacent to the marbles where Mg-chlorite replaces feldspar and mafic minerals and talc replaces quartz ribbons, while the blastomylonitic textures are perfectly preserved (Figs. 8 and 9).

Figure 7. Coarse vugs lined with amethyst, smoky quartz, and locally chalcedonic banded quartz within coarsely recrystallized dolomite and talc. This texture is strong evidence for open space late in the talc-forming process. Samples collected from the Whitney prospect north of the Treasure Mine. Mechanical pencil provides scale.

Evidence for a shallow (epithermal) origin of the talc deposits includes the abundant development of botryoidal textures in the talc, possibly resulting from deposition from a colloidal solution (Fig. 6), and other textural evidence cited above. Veins in the talc and the coarse recrystallized carbonate are lined with euhedral quartz and amethyst crystals (Fig. 7) at the Yellowstone Mine, the Whitney prospect, and many other localities. In order to maintain void space now expressed as vugs and common botryoidal textures in the talc zones and the host marble, talc must have formed at depths shallow enough for the rock to have behaved in a brittle manner. Experimental deformation studies on limestones and marbles (Paterson, 1978) suggest that the brittle-ductile transition in these rocks occurs at 0.3–1.0 kbar or roughly 1–3.5 km (0.6–2.2 mi) in depth. The effects of high heat flow would be expected to further decrease the depths of the brittle-ductile transition.

GRAVELLY RANGE

The Yellowstone Mine, owned and operated by Imerys Talc, is located in the eastern foothills of the Gravelly Range 5.9 km

Figure 8. Loose block of pink quartzofeldspathic gneiss (i.e., Dillon Granite Gneiss) from the Treasure Mine area. Biotite in the gneiss in the upper part of the photo is partially replaced by chlorite but the feldspar and quartz are unaltered. The gneiss below the sharp boundary extending horizontally across the lower part of the block is strongly altered to chlorite and contains quartz veinlets. Individual foliation surfaces can be followed uninterrupted across the alteration boundary. Bird droppings (white elongated blot) are oblique to boundary. Rock hammer provides scale.

Figure 9. Constant volume replacement of quartz. Light green talc has completely replaced the original euhedral quartz crystal extending right to left across the center of the photo at the pencil point and other quartz grains in the left and upper portions of the photo. In this pegmatite, the talc is a pseudomorphic form after quartz, but feldspar remains relatively fresh even where it is immediately adjacent to quartz that is replaced by talc. Fresh quartz is present in this sample just to the left of the pencil point and exemplifies an abrupt alteration boundary. Pegmatite from the Granite Creek area in the southern Tobacco Root Mountains. Scale is in inches.

(3.6 mi) west of the Madison River (Figs. 1 and 10). Operations consist of an open pit mine, sorting and grading plants, and facilities to maintain heavy mining equipment. Continuous open pit mining operations have occurred since the early 1950s. The reserve base will support mining for decades into the future.

The Yellowstone Mine is situated in medium- to high-grade Archean dolomitic marble of the Wyoming province (Vargo, 1990). Low-grade metamorphic and intrusive rocks, which may be as young as Proterozoic in age, are exposed in the mine area and are separated from the marble by a northward-dipping thrust fault (Kellogg and Williams, 2000). The mine area is capped with volcanic tuffs, ashes, and clays of the Pliocene Huckleberry Ridge Tuff (Pritchett, 1993).

Talc ore bodies are hosted in a siliceous dolomitic marble unit that exhibits complex folding and faulting. The marble was structurally thickened by folding and underlies an area 4 km (2.5 mi) along strike and 2.4 km (1.5 mi) wide. Well-developed quartz rods and mullions are found throughout the dolomitic marble. These lineations are parallel to tight F_2 fold axes that have an average orientation of 15°/S35°W. The average strike of foliation is N35°E with dips ranging from 55° to 75° on both limbs of the primary synform. Foliation dips are to the SE on the north limb of the fold and to the NW on the south limb. Gray dolomitic marbles typically have sucrosic textures that range from fine-grained to microcrystalline. Local alteration of the marble by hydrothermal activity produced coarse crystalline carbonate masses stained maroon with hematite.

The primary talc orebody mined at the Yellowstone Mine is known as the Johnny Gulch deposit. This mass of talc is "decoupled" from the host rock, severely deformed, discordant to foliation of the host rock, and likely a replacement of select layer(s) within the massive dolomitic marble. Refolded folds still recognizable within the talc suggest intense deformation within the host dolomitic marble. Evidence of the earlier deformation events is further obfuscated by fault zone brecciation from periodic reactivation.

Of particular interest is the distribution of iron throughout the orebody. Yellowstone talc was formed in a reducing chemical environment with ~1% Fe in the system substituting for Mg^{2+} in octahedral lattice sites. Excess iron in the system combined with sulfur to form pyrite inclusions in the talc, presumably during talc formation. More recently, the pyrite oxidized to limonite and goethite above the water table, with partial mobilization to create rust stain along fracture surfaces. Differential concentrations of iron as FeO_x within the talc zones reveal fold patterns similar to those defined by the larger scale shape of the orebody.

Analysis of drill hole data and surface mapping using 3-D modeling of the talc orebody morphology suggests multiple superposed folding events occurred. Plastic deformation resulting from ductile flow of the carbonate host is indicated by flexures and convolutions in the talc-dolomite contacts. The interference patterns seen in 3-D modeling, i.e., folded hinges of earlier folds, indicate at least three superposed fold sets. The hinge-line of the primary antiform in the deposit plunges to the north. Thickening in the core of the fold nose is responsible for the primary volume of talc ore in the Johnny Gulch deposit.

Three north-south–striking normal faults provide the structural control and the plumbing system for fluids that created the talc mineralization in the Johnny Gulch orebody (Fig. 11). These large faults were invaded by mafic magmas, and periodic reactivation created fluid pathways for talc-forming mineral reactions in the carbonates and chlorite alteration in the mafic intrusive bodies. Multiple generations of carbonate minerals, silica, and to a lesser extent talc, exhibit cross-cutting relationships at the periphery of these structures, indicating multiple fluid flux events (Cerino et al., 2007). The Johnny Gulch orebody consists of eight separate structural domains that are divided by key faults.

Cenozoic reactivation of the western boundary fault dragged the overlying Huckleberry Ridge ash, tuffs, and clays deep onto the footwall along the western boundary fault and related structures. Karst processes created large embayments into the carbonates along the hanging wall side of the western boundary fault. Karst features and offsets in the ash layers along the reactivated growth fault (Fig. 11B) are observable in the pit walls.

The majority of talc mined at the Yellowstone Mine was formed by a constant volume replacement of dolomite by talc. Relict compositional layering marked by minute amounts of graphite is observed to pass from the dolomitic marble host through pods of talc. This textural continuity indicates a constant volume replacement mechanism that could require a

Figure 10. Oblique aerial view of Imerys Yellowstone Mine taken July 2013 looking north-northeast. The Madison River meanders across the upper right-hand portion of the image.

minimum volumetric water/rock ratio of 600 (Anderson et al., 1990). Minor amounts of talc with botryoidal textures exist in the Yellowstone deposit and are indicative of precipitation in open spaces. The botryoidal talc is usually free of iron oxide stain, and it cross-cuts the massive "constant volume" talc. Both the massive and botryoidal varieties are microcrystalline as a result of relatively low temperatures and pressures during talc formation.

Telescoping of the ancient hydrothermal system might explain the significantly different homogenization temperatures in fluid inclusions from the Yellowstone (140 °C) and the Cadillac–Burlington Northern talc deposit (88 °C) located less than one mile to the east in the same marble (Gammons and Matt, 2002). A telescoped hydrothermal system has a steep temperature gradient, such that radically different P-T determinations result depending on which part of the system is tested.

SOUTHERN RUBY RANGE

All of the deposits in the southern Ruby Range share the following characteristics:

- The host rocks are Archean dolomitic marbles.
- Talc appears to preserve older fabrics developed as part of tight folding of the marbles and development of boudinage.
- Talc and chlorite formation was a constant volume process (Figs. 8 and 9).
- The host marble is strongly recrystallized with development of coarse dolomite with or without magnesite.
- Talc replaces the recrystallized coarse-grained dolomitic marble; chlorite is present as an alteration product of the adjacent gneisses and schists (Fig. 8).
- The talc deposits terminate abruptly both along strike (Fig. 12) and perpendicular to the compositional layering.
- The host dolomitic marble in contact with talc tends to be coarsely crystalline and lighter in color, and late silica veinlets are common (Fig. 3).
- Graphite and pyrite are present as accessory minerals in the talc.

Numerous smaller talc deposits are found in the southern Ruby Range both north and south of the Sweetwater Basin, which transects the southern part of the range southeast of Dillon, Montana (Fig. 1). These smaller talc deposits have had varying levels of exploration and development, and some have produced talc from both underground and surface-mining operations.

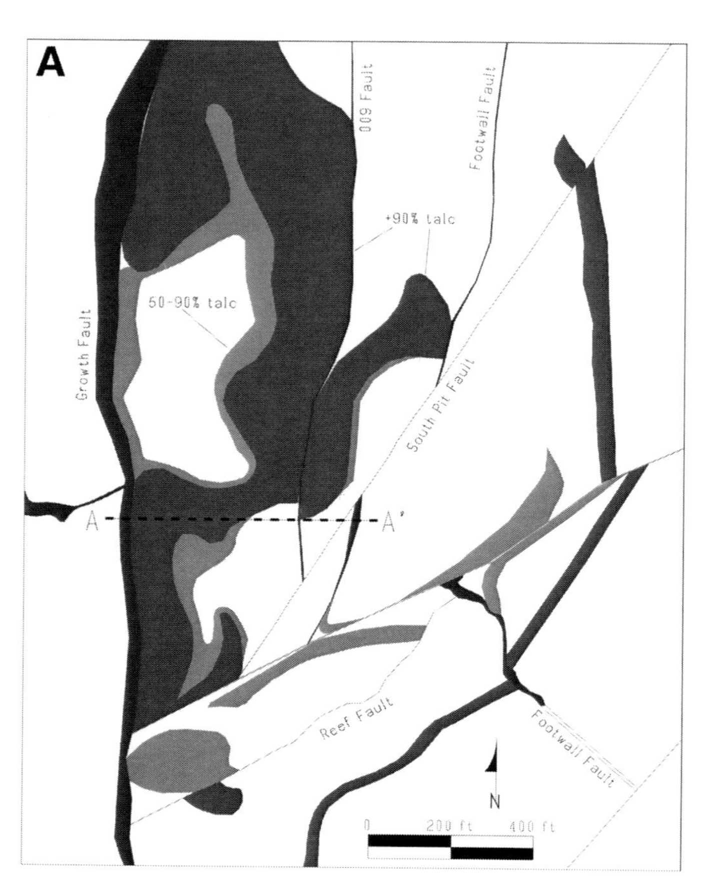

The southern Ruby Range has several major talc deposits—the largest of which are the Regal Mine located ~11 km (7 mi) to the east-southeast of Dillon on the Sweetwater Road (Figs. 4 and 13) and the Treasure Mine and the adjacent Beaverhead Mine in the headwaters of Stone Creek (Figs. 5 and 12). In this chapter, the Treasure Mine includes both the formerly active Treasure Chest Mine and the presently operating Treasure State Mine because they are on the same orebody and have been mined with a single open pit. A major north-south fault termed the "Treasure fault" passes through the Beaverhead and Treasure Mines as well as prospects aligned along the fault farther north. This structure was likely the primary conduit for passage of talc-forming fluids, because strongly chloritized zones in gneiss and schist and talc mineralization in marble coincide with probable fault traces. Barretts Minerals Inc. owns and operates both the Regal and Treasure Mines, and ores are processed at a mill located on a rail line ~13 km (8 mi) south of Dillon, Montana, on the east side of Interstate 15.

Regal Mine

The orebody at the Regal Mine strikes west southwest and dips ~45° to the north. It occupies the footwall of the "Regal" marble, and this marble in the mine area is made up of both limbs of an early isoclinal fold that was subsequently refolded to produce a "fish hook" interference pattern (Fig. 4). The shaft

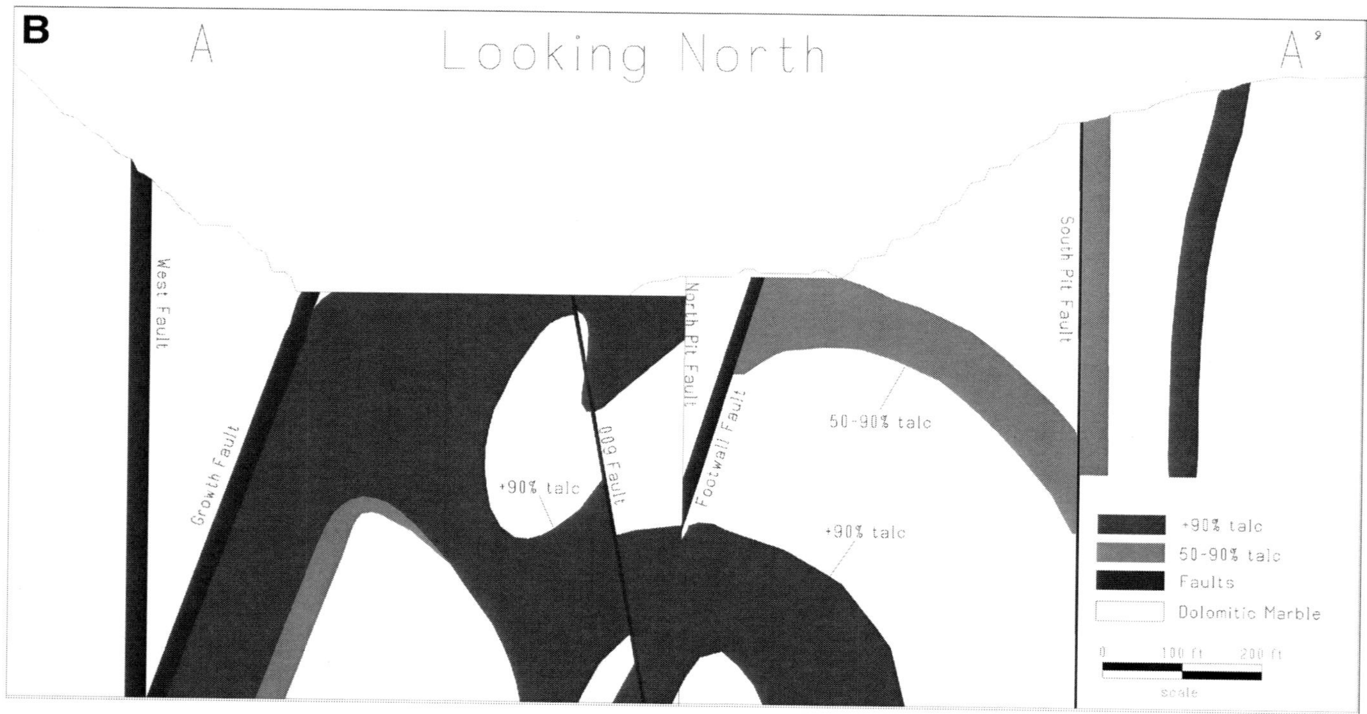

Figure 11. Complex folding and faulting of the Johnny Gulch orebody at the Yellowstone Mine. (A) Plan view geology. (B) Section A–A′ shows geology at mine coordinate 7700N. Faults are labeled and fault zones are black sinuous lines. Dark gray regions are >90% talc, light gray are 50%–90% talc, and white areas are dolomitic marble.

Figure 12. View looking west subparallel to the strike of the Treasure orebody from the location indicated in Figure 5. The marble and contained talc orebody dips to the right at ~40° north. The dark gray rocks outlined by two black lines on the mine face in the center of the photo are brown and green footwall schist and gneiss that are locally intensely altered to green chlorite. The host marble is unaltered where it extends out of the pit to the right in the photo. The host marble visible at the left edge of the photo is strongly altered, but becomes unaltered east of the pit.

of the fish hook in the mine passes to the west where it bends sharply to the south and then back to the east, terminating at the end of the "hook."

The Regal orebody is as much as 61 m (200 ft) thick and is immediately underlain by a strongly chloritized garnetiferous sillimanite mica schist. In most places, the hanging wall of the orebody is a fault contact between the orebody and marble. The hanging wall of the marble in the mine includes interlayered biotite gneiss, amphibolite, quartzofeldspathic gneiss, and pegmatite. Subsidiary talc bodies are found in the marble above the main talc zone. The orebody is terminated on the west by a nearly vertical diabase dike that is ~30 m (100 ft) thick, is locally altered to chlorite, and is the inferred heat source for the hydrothermal cell(s) that produced the talc. The orebody appears to terminate on the east at a major brecciated fault zone with abundant iron and manganese staining, calcite veins, and silicification. This fault strikes north-northwest and is approximately vertical.

To date, only minor talc has been identified by drilling and mapping on the west side of the major diabase dike. The reason for this major difference in talc development east and west of the dike is poorly understood. Thin mafic dikes cut the hanging wall marble and are clearly visible in the west and south wall of the pit. They are probably much younger than the diabase. Potential to expand the talc reserves at the Regal Mine appears to be good.

SUMMARY

The talc deposits in the Gravelly and Ruby Ranges are hosted by Archean dolomitic marbles and are associated with major faults that acted as conduits for hydrothermal fluids thought to be responsible for alteration of marble to talc and of gneisses and schists to chlorite. Mesoproterozoic hydrous fluids essential to talc formation may have originated in or proximal to the Belt basin. A relatively shallow crustal level for talc formation is likely given the abundance of vugs and pseudomorphic mineral replacements, and an open system is needed to satisfy mass balance calculations or fluid inclusion solute compositions. That the talc formed under hot spring or sabhka conditions of the Ruby Range contrasts with hydrothermal alteration mechanism along some peripheral basin faults, where connate brines, magmatic fluids, and meteoric water mixed in the Gravelly Range. Potential for discovery of additional reserves in southwestern Montana mines could be aided with further understanding of the paleoenvironments responsible for talc formation.

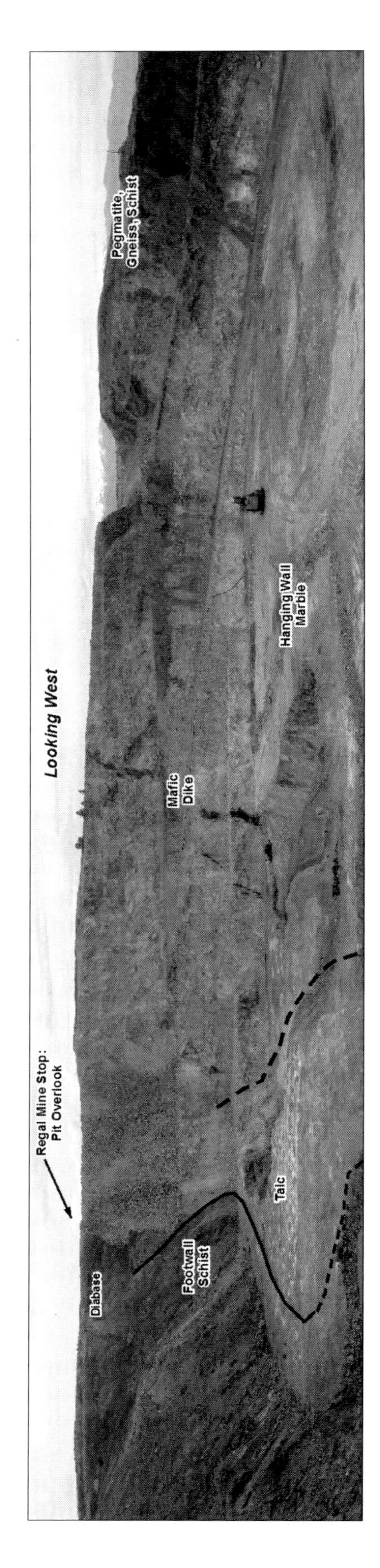

Figure 13. Regal pit view looking west down the fold axis of the re-folded fold from the location indicated in Figure 4. The footwall of the orebody and marble is to the left (south) and the hanging wall of the marble is to the right (north). A thin, continuous mafic dike is visible in the middle of the photograph running from the bottom to the top of the pit wall. The large north-northwest–trending diabase dike shown in map view in Figure 4 is visible to the left (i.e., southwest corner of the pit) in this picture. Pegmatite and gneiss are exposed in the hanging wall on the right side of this picture. Snowy peaks of the Pioneer Mountains are visible where the mine road exits the pit.

ROAD LOG, INTRODUCTION

This road log was generated from a combination of published documents adapted to describe the geology along various segments of the highway routes to, between, and returning from the Yellowstone and Regal Mines in southwest Montana, and are cited below with corresponding road increments. The older road logs were updated and expanded with new geologic information where warranted. The trip length precludes stops other than for tours at the talc mines, but the Archean to Holocene rock exposures and spectacular geologic structures en route are outlined here for geologic interest. The high probability of inclement spring road conditions makes unpaved roads, such as the Sweetwater Road, which traverses the Ruby Range, potentially impassable. However, the Sweetwater Road from Alder to Dillon is a geologically rewarding and scenic road through the Ruby Range, and can provide excellent access to Archean basement and Cenozoic structures.

A route map is provided for reference (Fig. 14). The basic framework of the road log begins with the segment from Bozeman to Norris, Montana, traveling on Montana Highway 84 (McMannis, 1960). From Norris to Ennis along U.S. Highway 287, we updated information from Johns et al. (1981). The road log segment from Ennis to the turnoff of U.S. Highway 287 south of Cameron was created for this trip access to the Yellowstone Mine. From Ennis to Twin Bridges, we again augmented the road log of Johns et al. (1981). The segment that includes Dillon, Whitehall, and Bozeman along Montana Highway 41, Montana Highway 55, and I-90 follows an updated version of the road log from Schmitt et al. (1995). The commentary for the Sweetwater Road from Dillon to the Regal Mine used Sears et al. (1995) as a template.

A schematic depiction of the lithologies and time and/or spatial relationships is presented in Table 1.

ROAD LOG, DETAILED

Mileage	*Directions and Descriptions*
0.0	Leave Montana State University Student Union Building (SUB). WGS84 datum; UTM 0496292 m E; 5056890 m N. From parking area, turn right onto Grant Street. At stop sign, turn left onto 11th Street.
0.3	At stop sign, turn right onto W. Lincoln Street.
0.8	Turn right onto S. 19th Ave.
1.3	At stop light, turn left onto W. College St.
2.0	Turn left on W. Main St. (U.S. Highway 191). The road name switches to Huffine Lane.

Geographic orientation: On the left, at 8 o'clock, is the Gallatin Range, 10 o'clock is the Madison Range, and on the right at 4 o'clock is the Bridger Range.

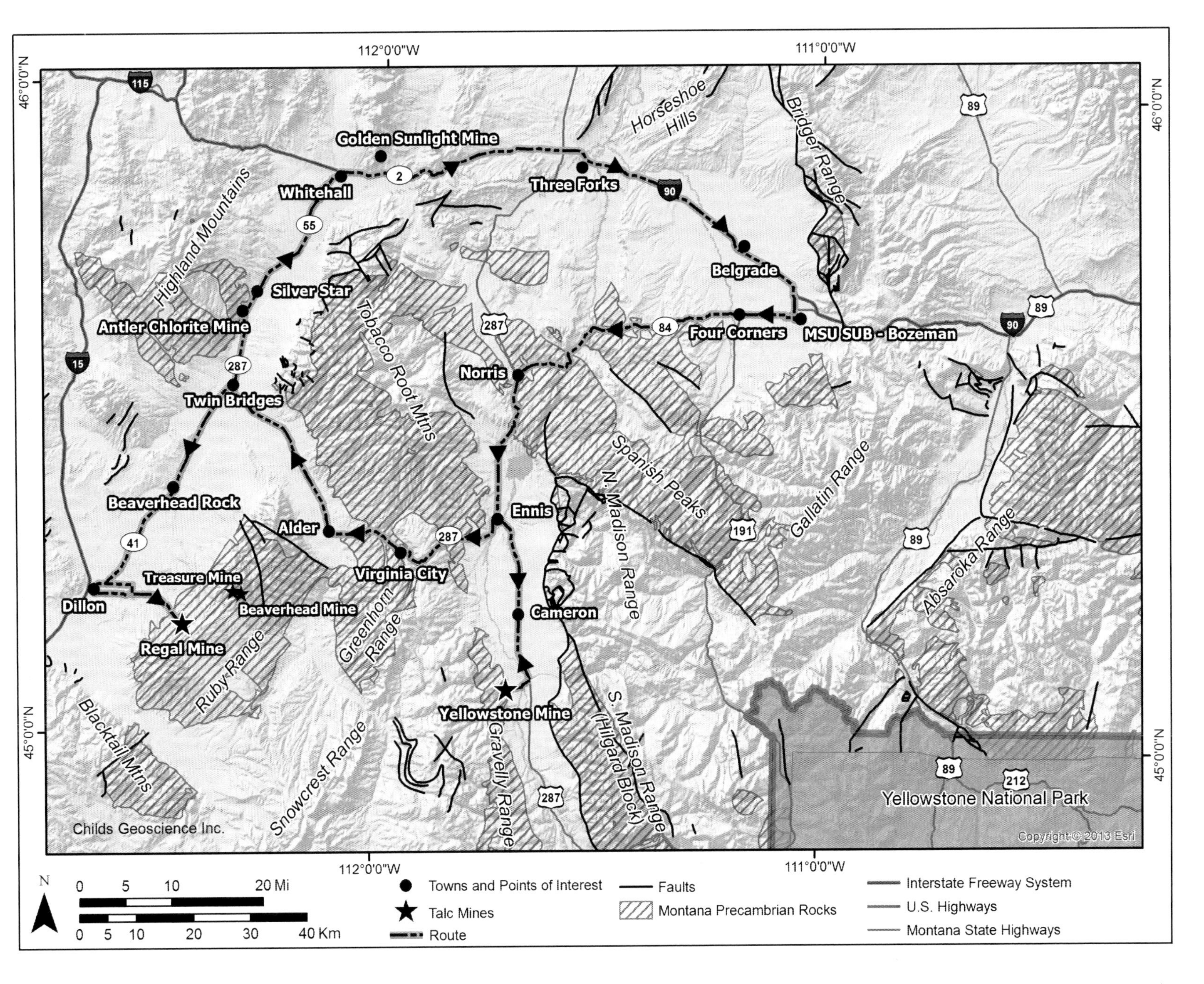

Figure 14. Field trip route map. The Yellowstone Mine (Stop 1) and the Regal Mine (Stop 2) are shown with arrows in the direction of travel. The SUB starting point is on the Montana State University campus.

TABLE 1. SCHEMATIC REFERENCE SECTION OF MAJOR LITHOLOGIC UNITS AND AGES OF OUTCROPS ALONG THE SOUTHWEST MONTANA FIELD TRIP ROUTE FROM BOZEMAN TO DILLON, VIA NORRIS, ENNIS, AND VIRGINIA CITY, AND FROM DILLON TO BOZEMAN VIA WHITEHALL

Eon	Era	Period	Epoch	Thickness (m, avg.)	Sedimentary unit	Igneous unit	Abbreviated lithology	Source
Phanerozoic	Cenozoic	Neogene	Holocene		Recent or active sediments		Recently or actively depositing fluvial, lacustrine, eolian, glacial systems; includes colluvium.	
			Pleistocene				Fresh till of the Pinedale Glaciation, partly weathered till of the Bull Lake Glaciation, and deeply weathered till of a Pre–Bull Lake Glaciation.	J. Montagne, 1989, personal commun.
			Pliocene	3–170		Huckleberry Ridge Tuff	Fine- and coarse-grained facies containing tuffaceous siltstone, conglomeratic sandstone, and brown to gray conglomerate, as well as common exposures of the lithologically varied Six Mile Creek Formation.	Kellogg, 1994; Vuke et al., 2002; McDonald et al., 2012
			Miocene			Basalt flows		
		Paleogene	Oligocene	300	Dunbar Creek Fm.		Tuffaceous siltstone, fine-grained sandstone, poorly sorted coarse-grained arkosic sandstone, and matrix-supported conglomerate beds of the Dunbar Creek Formation (Renova equivalent).	Marvin et al., 1974; Vuke et al., 2002
			Eocene (middle)	300+	Undiff.	Virginia City volcanic field and Absaroka Volcanics	Pale olive to reddish-brown bentonitic, sandy clay and claystone, yellowish-gray coarse-grained argillaceous sand and sandstone, and tuffaceous siltstone and fine-grained volcanic glass sandstone of the Climbing Arrow Formation (Renova equivalent), atop the intercalated gray, fine-grained tuffaceous limestones and conglomerates of the Milligan Creek Formation.	Vuke et al., 2002
			Eocene (early)	—	Red Bluff Fm.	Andesite, latite, and dacite sills	Pale siltstone, sandstone, conglomerate, and brick-red mud and siltstone overlying subrounded to well-rounded, matrix-supported, bouldery diamictite.	Vuke et al., 2002
	Mesozoic	Cretaceous		600	Undiff.	Tobacco Root batholith, Boulder batholith and its satellite plutons (including Hells Creek pluton), Elkhorn Mtn Volcanics	Cretaceous sedimentary rocks include the Sphinx Conglomerate, Beaverhead Conglomerate, Livingston Formation (Sedan equivalent), Everts Formation, Virgelle Sandstone, Eagle Sandstone, Telegraph Creek Formation, Cody Shale, Frontier Formation, Mowry Formation, Muddy Sandstone, Thermopolis Formation, and Kootenai Formation.	Vuke et al., 2002; Kellogg et al., 2007
		Jurassic	Upper	200	Undiff.		Jurassic sedimentary rocks include the Morrison Formation, Swift Formation, Rierdon Formation, and Sawtooth Formation.	J. Montagne, 1989, personal commun.
		Triassic		0–30	Dinwoody		Brown sandy limestone and calcareous shale.	J. Montagne, 1989, personal commun.
	Paleozoic	Permian		0–30	Shedhorn S.S. (Phosphoria equiv.)		Brown to greenish-brown, laminated or thin- to thick-bedded chert, interbedded with oolitic phosphatic sandstone, siltstone, and yellowish-gray dolomitic limestone.	
		Penn.	Upper	40	Quadrant		Pale cream-colored clean washed dolomite-cemented quartz arenite.	J. Montagne, 1989, personal commun.; Kellogg and Williams, 2005
			Lower	88	Amsden Group		Brick-red, reddish-brown, and pink calcareous siltstone.	
		Mississippian		130	Snowcrest Range Group		Dark gray thin- to medium-bedded limestone of the Lombard Formation overlying thin- to medium-bedded, yellow-gray to maroon, friable Kibby Sandstone.	J. Montagne, 1989, personal commun.; Kellogg and Williams, 2005
				135–275	Big Snowy Group		Thin- to thick-bedded, pale olive gray, locally fossiliferous limestone atop red and green-gray shale, calcareous sandstone, and cherty limestone.	
				440	Madison Group		Light gray, locally fossiliferous, thick-bedded to massive limestone of the Mission Canyon Formation atop thin- to medium-bedded limestone and argillaceous limestone of the Lodgepole Formation.	

(Continued)

TABLE 1. SCHEMATIC REFERENCE SECTION OF MAJOR LITHOLOGIC UNITS AND AGES OF OUTCROPS ALONG THE SOUTHWEST MONTANA FIELD TRIP ROUTE FROM BOZEMAN TO DILLON, VIA NORRIS, ENNIS, AND VIRGINIA CITY, AND FROM DILLON TO BOZEMAN VIA WHITEHALL (*Continued*)

Eon	Era	Time units: Period	Epoch	Thickness (m, avg.)	Sedimentary unit	Igneous unit	Abbreviated lithology	Source
Phanerozoic	Paleozoic	Devonian		40–60	Three Forks Fm.		Thin-bedded, yellowish siltstone and silty limestone.	Kellogg and Williams, 2005
Phanerozoic	Paleozoic	Devonian		110	Jefferson Dol. and Maywood Fm.		Thin- to thick-bedded, dark gray to light gray, petroliferous, locally fossiliferous Jefferson Dolomite overlying pale yellow sugary dolostone of the Maywood Formation.	Vuke et al., 2002; Kellogg and Williams, 2005
Phanerozoic	Paleozoic	Ordovician		3	Bighorn Dol.		Scarce occurrences only in the northwestern Madison Range. Thin- to medium-bedded light gray dolostone.	Kellogg and Williams, 2005
Phanerozoic	Paleozoic	Cambrian		40–60	Red Lion Fm. (Snowy Range equiv.)		Thin-bedded tan to gray siliceous dolostone intercalated with orange-tan to reddish-tan cherty ribbons. Lower portion contains intraformational clasts up to 5 mm.	Kellogg and Williams, 2005
Phanerozoic	Paleozoic	Cambrian		30–120	Pilgrim Fm.		Medium-bedded to massive, light gray, locally oolitic dolostone.	
Phanerozoic	Paleozoic	Cambrian		30–75	Park Shale		Greenish-gray to tan, fissile shale.	
Phanerozoic	Paleozoic	Cambrian		50–150	Meagher Fm.		Thin-bedded to massive, light gray micritic limestone.	
Phanerozoic	Paleozoic	Cambrian		30–65	Wolsey Shale		Thin-bedded, greenish-gray, olive, micaceous sandstone, siltstone, and shale.	
Phanerozoic	Paleozoic	Cambrian		15–75	Flathead Formation		Thin- to medium-bedded, medium- to coarse-grained, reddish-brown, tan, and purplish-tan arkosic sandstone. Basal conglomerate contains rounded pebbles of metamorphic rock.	
Proterozoic	Neoprot.			~30		Diabase	Sets of mafic dikes, ferrobasaltic to tholeitic composition.	Kellogg and Williams, 2005; McDonald et al., 2012
Proterozoic	Mesoproterozoic			up to 20,000	Belt Supergroup		A truly venerable assemblage of siliciclastic and carbonate rocks deposited in the Belt Basin from 1.47 to 1.4 Ga. Argillite and quartzite of the Missoula Group overlie the Middle Belt carbonate units, which in turn overlie the argillite and planar- and cross-bedded sericitic quartzite of the St. Regis, Revett, and Burke Formations. At the base of the Belt Supergroup lie the laminated siltite and argillite couplets of the Prichard Formation. LaHood Formation is a basal syntectonic coarse conglomerate along Perry Line and Willow Creek fault.	Schmidt et al., 1988; Ross and Villenueve, 2003; McDonald et al., 2012
Archean				Unknown	Meta-igneous and meta-sedimentary rocks		Quartzofeldspathic gneiss, amphibolite and hornblende-plagioclase gneiss, biotite schist, dolomitic marble, aluminous gneiss and schist, banded iron formation, and ultramafic rocks.	Kellogg and Williams, 2005; McDonald et al., 2012

Bozeman to Four Corners

This road log segment is adapted from Schmitt et al. (1995).

Bozeman was named for John M. Bozeman, who scouted the route from the Oregon Trail near Ft. Fetterman (Douglas), Wyoming, to the gold camps of western Montana. In 1864, he successfully guided fortune seekers over the route from Ft. Laramie. Bozeman was platted in 1864 through his efforts. Early growth of Bozeman was based on the rich agriculture of the Gallatin Valley. In 1893, Montana State College was founded as the state's land-grant institution of higher education, with emphases in agriculture and engineering.

Geologic Setting of the Gallatin Valley

The valley is surrounded by mountains on all sides, and lies at the juncture of four major tectonic provinces including: (1) the south margin of the middle Proterozoic Belt basin; (2) the Sevier fold-and-thrust belt, where the south margin of the Helena salient reactivated Belt basin faults); (3) thick-skinned Laramide deformation; and (4) eastern limit of Neogene "Basin and Range" extensional deformation (Lageson, 1989). Laramide uplifts include the Bridger Range to the east, the Beartooth and Gallatin Ranges to the south, the Spanish Peaks–Madison Range to the southwest, and Tobacco Root Mountains to the west. The north end of the Gallatin Valley is framed by the Horseshoe Hills and Big Belt Mountains, both lying within the Belt basin and Sevier fold-and-thrust belt. Lageson (1989) describes the Bridger Range as "a perched basement wedge of Archean and Proterozoic rocks overlain by steeply east-dipping Paleozoic-Mesozoic strata." In addition, a high-angle normal fault bounds the north flank of the Gallatin Range at the south end of the valley, and several normal faults within the Gallatin Valley are antithetic to range front normal faults.

Between Bozeman and Four Corners, U.S. Highway 191 traverses poorly consolidated Tertiary sediments that fill the Gallatin Valley. Based upon a local anomaly of approximately −30 mGals in gravity data (Kucks, 1999), Schmitt et al. (1995) state that as much as 1525 m (5000 ft) of Tertiary fill is locally present. The crest of the ancestral Laramide Bridger Range lies beneath the valley to the north, having been down-dropped by a listric normal fault(s) that bounds the west flank of the range. Inasmuch as basement rocks are overlain by Tertiary strata along the west margins of the valley, the crest of the ancestral Bridger Range was probably eroded to the basement before being displaced by normal faulting. Thus, the presence of Paleozoic or Mesozoic rocks at depth in this part of the valley is unlikely.

To the south, the Gallatin Range is capped by Eocene volcanic rocks genetically related to the Absaroka and Challis volcanic fields.

Four Corners to Norris

This road log segment is adapted from McMannis (1960).

7.3	Four Corners: Junction of U.S. Highway 191. Landmarks are Korner Club and gas stations. Proceed west to Norris on Montana 84.
8.4	Gallatin River crossing.
9.2	Immediately west of Cottonwood Golf course lies a gravel pit north of the highway. Outcrop of Neogene valley-fill sediments?
11.1	Amsterdam Road on right.
12.7–13.0	Excellent outcrops of Miocene thin-bedded lake sediments, part of the sedimentary fill in the Gallatin Valley (Vuke et al., 2002).
16.0	On right, outcrops of garnetiferous hornblende-biotite gneiss, schist, and amphibolite of the Archean basement.
16.2	Intersection of Camp Creek Road on north with renovated Anceney grain elevator on south side of road and former railroad terminus. This was a center for grain shipment to mainline railroads farther north into the 1980s. The elevator was sold in ~1993 and is now a residence. The ranch that Charles Anceney Sr. established, the Flying D, was purchased by Ted Turner and is where he raises bison.
16.4	Exposures of fine-grained Miocene beds in road-cuts. These strata are unconformable on the Precambrian basement along a surface of strong relief.
17.4	Geographic orientation: At 11 o'clock the crest of the Tobacco Root Mountains is visible. The name, Tobacco Root, is reportedly derived from the habit of local Native Americans, who collected mullein plants in the area and mixed them with kinnikinnick as a substitute for smoking tobacco (Schmitt et al., 1995).
18.8–19.2	Archean gneisses cut by pegmatites are exposed on the right.
21.3	Highway crosses Elk Creek. This valley is the surface topographic expression of one of many northwest-trending oblique reverse faults in the area similar to the fault along Cherry Creek of mile ~26.8.
24.6	Black's Ford Fishing Access to the Madison River and Madison River Road on right. Wildfire swept over the area in June 2012.

High metamorphic grade Archean rocks of the Spanish Peaks region ~10–20 km to the southeast are within the same structural block and experienced the same general geologic history as rocks exposed here (Kellogg, 1994). Mogk et al. (1989) report a U-Pb zircon age from Spanish Peaks quartzofeldspathic gneiss of 3.3–3.2 Ga. Voluminous melts, mostly trondhjemitic and tonalitic, were injected in the Spanish Peaks area at 3.2–3.1 Ga, which is the suggested age of peak metamorphism (Mogk et al., 1989).

25.1	Madison County line.
25.3–25.5	Two old base metal mine portals visible on the left explored veins in Archean gneiss.

26.0 On the right, across the Madison River, is a prominent exposure of Archean quartzofeldspathic gneiss that is a popular rock climbing venue called "Neat Rock Crag" by locals.

26.8 Turn off to Damsel Fly Fishing Access to Madison River and bridge over Cherry Creek. The road crosses a northwest-trending reverse fault similar to the Spanish Peaks fault zone, which offsets Precambrian rocks on the north against Paleozoic rocks to the south in the area immediately east of the highway. West of the Madison River, that same fault zone appears to have recently undergone minor reactivation as a normal fault. Across the river at 2 o'clock is Red Mountain, which is underlain by Tertiary rhyolitic volcanic rocks (rhyolitic vitrophyre, Vuke et al., 2002).

27.9 Cliffs at 3 o'clock across the river are underlain by Archean gneiss at their base and overlain by the Cambrian Flathead Formation.

28.2 Bear Trap Canyon access road on left. The road ends at a trailhead and fisherman's access.

28.4 Highway crosses Madison River.

29.7 Excellent outcrop of migmatitic quartzofeldspathic gneiss and amphibolite cut by steeply dipping pegmatite dikes in cliff on south shore of Madison River.

30.2 To the left, across the river, a rock fall occurred in ~1990, which blocked the Bear Trap Canyon access road, stranding a number of recreationists to the south. Numerous pegmatitic dikes cut the Archean gneiss.

31.1 Parking area for the Warm Springs Creek Fishing Access and mouth of Bear Trap Canyon to south, where the incised Madison River flows from the Madison Dam 16 km (10 mi) south of here. In August 2010, a semi-truck–sized boulder dropped off the canyon wall and onto the concrete dam. The hydroelectric company PPL Montana, who oversees the dam and associated powerhouse, reported some crushed spill gates, but the lower dam structure was deemed sound.

Tertiary Red Bluff conglomerate caps the hill that lies across Warm Springs Creek to the east. For the next several miles, the highway follows the general east-northeast trend of this creek, which flows across Precambrian metamorphic rocks. Many excellent exposures of these rocks are seen in roadcuts and canyon walls. Shoulders on ridges to the south of the road suggest an ancient mature valley was incised by Warm Springs Creek.

31.6 Severe erosion visible on the right was a result of a wildfire in 2012. Continue up Warm Springs Creek through Archean gneisses.

33.7 On the left, Montana Agricultural Experiment Station.

33.9 Bradley Creek Road turnoff to the south and the Lower Hot Springs Mining District. From 1870 to 1948, lode gold was the primary commodity with byproducts of silver, copper, and lead (Kellogg, 1994). Oxidized sulfide-bearing quartz veins cutting Archean gneiss were the typical mining targets. The richest gold-bearing veins in the Lower Hot Springs district (e.g., Boaz and Grubstake Mines) are parallel to a regional set of northwest-striking, northeast-dipping to nearly vertical, Late Cretaceous to early Paleogene reverse faults. The Boaz Mine workings are located at 9 o'clock; a large mine dump is just visible in the upper part of the valley. There are spectacular views of the Tobacco Root Mountains at 12 o'clock. Low hills to the south are underlain by diamictite and fine sediments of the Eocene Red Bluff Formation. Hot spring activity has locally silicified this unit.

Norris to Ennis

This road log segment is adapted from Johns et al. (1981).

36.5 Norris. Junction of U.S. Highway 287 and Montana 84. Turn south up Burnt Creek to McAllister and Ennis. Hills to the west are capped by Miocene basalt, which overlies Miocene siltstone and sandstone (Kellogg, 1994). Miners leaving Virginia City/Alder Gulch following the peak in placer gold production in 1865 located lode gold mines in the Norris area in the late 1860s, and by 1900, load gold mining largely ceased (Montana Abandoned Mine Reclamation, 2014).

39.4 Archean gneiss is visible in roadcuts on the left.

39.7 Roadcut exposes the eastern margin of the Late Cretaceous Tobacco Root batholith, which intrudes Archean gneiss and amphibolite in the central Tobacco Root Mountains. This batholith is a zoned granodioritic pluton (medium-gray, coarse-grained, porphyritic) located at the northwest end of the Spanish Peaks–Gardiner fault system (Schmitt et al., 1995). Left-reverse motion on splays of the Spanish Peaks–Gardiner fault system may have accommodated intrusion of magma into a transtensional pull-apart (Schmidt et al., 1990).

41.9 Summit of Norris Hill. High country to the west is underlain by Archean rocks and the Tobacco Root batholith. The Madison Valley opens to the south.

43.9 Crossing northwest extension of Spanish Peaks fault. This fault is believed to be a continuation of the northwest-trending Gardiner thrust zone, visible at the north entrance to Yellowstone National Park. In this area, the Spanish Peaks fault cannot be readily identified along the road.

Multiple pediment surfaces in this area include the prominent surface that approximately coincides in elevation with that of the Cameron Bench, which lies to the south of Ennis. Fluvial gravels exposed in roadcut to the right (west) are imbricated, indicating flow to the south; beds now dip 10–12 °N.

46.5 McAllister Post Office. Continue south across floodplains of North and South Meadow Creeks, which are slightly incised into the Ennis terrace.
49.8 Highway crosses to higher, slightly older terrace. To the west is a still higher general level named the Cameron Bench, which is ~76 m (250 ft) above the present floodplain of the Madison River. There are excellent views of the Spanish Peaks Wilderness at 8 o'clock, Jack Creek at 10 o'clock, and Fan Mountain at 10:30.
53.0 Junction of U.S. Highway 287 and Montana Highway 287 in Ennis. Bear east into downtown Ennis and south on U.S. Highway 287.

We shall return to this junction following our visit to the Yellowstone Mine, which is ~34 km (21 mi) south. The road log contains information for traveling both directions.

53.5 Rest stop at Lions Club Park on south side of Ennis prior to crossing Madison River.
54.1 Odell Creek bridge. The Gravelly Range, visible at 2 o'clock, lies on the eastern limb of a large southwest-plunging syncline. Phanerozoic sediments exposed in this syncline separate the pre-Belt metamorphic rocks exposed in the Gravelly Range from those exposed in the Greenhorn Range to the west (Johns et al., 1981).
59.8 Ennis airport road (east). The fault-bounded Madison Range is visible on the left, and the Gravelly Range on the right.
62.9 Sphinx Mountain in the Madison Range is at 10 o'clock. A syncline that formed during the Cretaceous Period by progressive crustal loading during thrust faulting was filled with coarse sediments that lithified into conglomerate that caps Sphinx Mountain. Hendrix (2011) provides a nice geology hiking trail guide for up-close inspection of fault surfaces and lithologies of the Helmet and Sphinx Mountain areas.
64.3 Cameron Post Office.
68.5 River terraces are visible to the west as we descend from the Cameron Bench.
70.0 To the right, a prominent extension of the Gravelly Range juts eastward into the Madison Valley. This nose is underlain by Archean dolomitic marble that hosts the talc in the Yellowstone Mine, which is coming into view on the south side of the hills.
71.6 Turn right onto the road leading to McAtee Bridge Fishing Access and the Yellowstone Mine operated by Imerys Talc (signs). An aerial photograph of the mine site (Fig. 10) may aid orientation.
72.3 Cross Madison River.
72.8 Note the rugged Hilgard block, which forms a portion of the southern Madison Range at 7 o'clock. The Yellowstone Mine is located at 11 o'clock.
74.5 The public road goes left; stay right. The old Burlington Northern Talc Pit is at 3 o'clock.
74.6 Turn right to visitor check-in at Imerys Talc Yellowstone Mine office. We will take a site tour led by mine personnel.

STOP 1

The southwest pit overlook is located along the Growth fault (mine name only, not a genetic term), which is stained maroon by hematite, and cuts dolomitic marble. The Growth fault can be traced northward along strike and at the northern end of the structure, stratified deposits of distal Pliocene Huckleberry Ridge volcanic rocks are visible. The North Pit and Footwall faults are visible on the north pit wall at about N30°E and the South Pit fault can be seen along the eastern margin of the pit. As of January 2014, activity in the pit is focused in the southern portion of the talc body, south of section A–A′ in Figure 11A, as well as on upper benches to the north.

74.8 Leave the Yellowstone Mine and retrace route north to Ennis, Montana.
75.0 Recent fault scarps cutting alluvial fans at 2 o'clock formed at the western margin of the Hilgard Block. At 12:30, a large moraine at the mouth of Indian Creek drains west out of the Hilgard Block. Immediately north of this moraine at 12 o'clock is a possible debris flow.
75.3 At 11 o'clock, there is an excellent view of Sphinx Mountain on the right and lower Helmet Peak on the left. Similar to the Sphinx, Helmet Peak is capped by Sphinx Conglomerate and underlain by Cretaceous sedimentary and volcanic rocks.
77.1 Re-cross the Madison River.
77.8 Turn left heading north on U.S. Highway 287.
77.9 Indian Creek.
80.5 Start climb from lower bench up onto the Cameron Bench.
85.1 Cameron Post Office.
86.4 Mill Creek alluvial fan at 3 o'clock. Tan and/or gray Mesozoic and Paleozoic rocks with various dips are visible along the western flank of the Madison Range.
89.1 Descend into the modern drainage incised into Cameron Bench.

Ennis to Twin Bridges via Virginia City

This road log segment is adapted from Johns et al. (1981).

96.5 Junction of U.S. Highway 287 and Montana Highway 287 in Ennis: Turn west on Montana Highway 287 toward Virginia City. At this junction, the road lies at the edge of the Ennis terrace, a recent level of stabilization of the ancestral Madison River, ~15 m (50 ft) above the present floodplain.

98.7 Highway rises to the terrace level below the Cameron Bench. The slightly higher Cameron Bench can be seen to the north.

99.4 Cameron Bench.

100.5 South end of the Tobacco Root Mountains is visible at 12 o'clock, and the northern Gravelly Range at 10 o'clock. This is approximately the nonconformity of Tertiary sediment over Precambrian metamorphic rocks. Pre-Belt schists and gneisses are exposed intermittently in roadcuts and surface exposures.

104.0 Sphinx Mountain scenic turnoff: This observation point provides a magnificent view of the Madison Valley and Madison Range. At the western base of the Madison Range (below Lone and Fan Mountains) is a broad alluvial fan spreading both north and south. This alluvial fan covers older pediment surfaces. No recently formed fault scarps are discernible along this portion of the mountain front, but they are well developed along the range front to the south. About 914 m (3000 ft) of the Sphinx Conglomerate caps Sphinx Mountain.

~105.0 Summit of Ennis Hill: Pegmatite dikes are exposed in Precambrian metamorphic rocks.

105.2 Early Oligocene volcanic flows overlie Precambrian metamorphic rocks. Marvin et al. (1974) suggest a K-Ar age of 34–33 Ma for these volcanic flows.

106.7 Top of grade and commencing down-slope approach to Alder Gulch and Virginia City. A red clay band marking an old erosional surface overlying the volcanic units may no longer be distinct in the roadcut. An overlying andesite flow contains carbonate and zeolite nodules filling vesicles in the flows and the interstices in the breccia.

108.2 The Ruby Range appears at 11 o'clock. Cenozoic faulting has cut this range into three distinct blocks trending southwest to northeast. These blocks are a succession of half-grabens; the faults that bound each block trend west-northwest, roughly orthogonal to the northeast trend of basin and range faulting in the region.

110.7 Don't forget to check out the Brewery Follies live comedy show.

110.8 Virginia City: Madison County Courthouse on left (south), Boothill to right on Skyline ridge. The Virginia City placer deposits were discovered in May 1863 by prospectors panning gravels in Alder Gulch. Within a year, there was an estimated population of 10,000 people in the area and the Virginia City Mining District (VCMD), along with numerous subdistricts (Granite Creek, Fairweather, Highland, Browns Gulch, Pinegrove, Summit, and Barton Gulch) was established. The placer deposits in the area became the richest single stream placers in the United States. The last major dredging operation was shut down in 1933. In total, over 2.6 million ounces of gold and 350,000 ounces of silver were recovered from placer operations in the VCMD between 1863 and 1963 (Barnard, 1993). Lode deposits contributed another 170,000 ounces of gold and >2.4 million ounces of silver. Figures for base metal production (Cu, Pb, and Zn) were not accurately recorded.

110.9 Jackson Street intersection: To reach the original discovery site for Alder Gulch placers along Alder Creek, drive 1.2 km south (Jackson Street becomes Alder Gulch Road). Bill Fairweather and others discovered gold at this site on 26 May 1863. The U.S. Grant flotation mill is located ~0.4 km (0.25 mi) up (south) Alder Gulch from this point.

111.1 Bummer Dan's bar, the highest placer bar on the west edge of town, was one of the fabulously rich deposits of Alder Gulch. Bummer Dan, aka Dan McFadden, a ne'er-do-well who was caught stealing a pie, was given an ultimatum that he was to work the unclaimed ground or get out of town. With due persuasion he was left scratching at the gravels, but the gleam of gold soon turned perfunctory efforts to feverish labor.

A large placer gold-bearing cavity in volcanic rocks reportedly lies adjacent to the highway. Initially, the placer was too deep 21 m (69 ft) to be recovered during early dredging operations in 1896–1897. However, when the pocket was eventually mined, possibly between July 1935 and June 1937 by more powerful dredges, $1 million in gold was reportedly recovered (Lyden, 1948). The Virginia City district still has intermittent production of gold and silver from both placer and lode deposits.

Placer gravel spoils banks are visible in Alder Gulch. Tertiary volcanic rocks are exposed in a roadcut.

112.3 Nevada City: This is the site at which George Ives, the first of the criminal gang known as the "Innocents" was convicted and hanged, paying the supreme penalty in 1863 for murdering Nicholas Thiebalt for the sum of $200 in gold dust (www.virginiacitymt.com/Nevada.asp). The Nevada City Hotel on the right has a two-story outhouse.

112.7 Browns Gulch enters Alder Gulch from the south: Five men reportedly extracted $30,000 in gold nuggets in 11 working days from this drainage early in the Gold Rush. A placer operation started in 1967 after test drilling gave encouraging shows from gravel beneath a clay "false bedrock." This operation has been recently reclaimed.

114.5 Granite Creek: Junction City, a flourishing mining town of the 1860s, was probably located a short distance up the gulch. The roadcut contains Archean metamorphic rock including metaquartzite, veined amphibolite, gneiss, biotite schist, hornblende schist, and pegmatite.

114.9 McNeal Gulch: The roadcut just ahead exposes Archean metasediments. Important lithologies are amphibolite, gneiss, and schist.

115.3 Placer operation on the left.

115.4 Water Gulch: The ridge-forming metaquartzite or vein quartz of the Archean sequence contains locally massive rose quartz.

116.2 The highway descends through Archean metamorphic rocks with exposures of thin marbles.

117.3 Northern Ruby Mountains at 12 o'clock: Pre-Belt metamorphic rocks in the core of northern Ruby Range are flanked by Paleozoic and Mesozoic rocks.

119.0 Garnet USA has a mill facility within the ~early 1900s placer dredge tailings on the north side of the road. The plant requires a ready water source for the concentration of almandine alluvial garnet product. A hardrock mine is planned in Archean garnetiferous gneiss from a location ~6 km (3.7 mi) to the southeast.

Entering the Ruby River Valley.

119.6 To the right is an area of dredge tailings. These extensive piles of gravel were deposited by the stackers of the large dredges, which began operation during the late 1890s and continued until 1922. One source estimates the value of gold extracted by this method at about $30 million. The largest of these dredges cost half a million dollars and handled 7645 m^3 of gravel per day.

119.9 Alder Post Office: Stay right and continue on Montana Highway 287. To the west is an excellent view of the Ruby Range. The Tobacco Root Mountains are to the north. A north-trending fault parallels the eastern front of the Ruby Range. However, the scarp is covered by alluvial fans. Talc from the Beaverhead Mine in the southern Ruby Range (Cyprus Industrial Minerals Company) was shipped overseas for processing from a railroad spur in Alder.

121.9 Turn off to Laurin is on the left. Copper Mountain (iron and copper deposits) is at 2 o'clock (fringe of trees on mountaintop below skyline). Refolded folds defined by Archean dolomitic marbles (Cordua, 1973) similar to those hosting the talc deposits at the Yellowstone and Regal Mines are found throughout the Tobacco Root Mountains to the right. Several talc prospects up California, Harris, and Bivens Creeks were mined by small-scale operations.

122.0 Turn right on the road to California Creek and Mill Gulch Station to reach a lunch stop with good exposures of Archean marbles and other lithologies. Proceed east for 0.3 km (0.2 mi) to where the California Creek road intersects on the left, but continue straight ahead past the IMOC Lumber building for 0.5 km (0.3 mi) to a tee in the road. Turn left on Mill Gulch Road and continue 0.15 km (0.1 mi). The road bends sharply to the right. The lunch stop is in an open valley 1.4 km (0.9 mi) farther up the road.

Pull off on the road to the right for lunch. Please restrict your lunch traverses to the south side of the road because the land north of the road is private property. While enjoying lunch, a stroll to the southeast rewards the hungry geologist with excellent grassroots outcrops of Archean quartzofeldspathic gneiss, and farther east, grassroots outcrops of light gray, dolomitic to calcitic marble and minor hornblende gneiss and other lithologies are found. The regionally metamorphosed amphibolite- to granulite-grade marbles seen in these exposures contrast sharply with the strongly recrystallized marbles at the talc mines we are visiting today.

Investigation of the rocks and minerals in hand specimen: The marble is fine to medium grained, has light green calcsilicate and gray quartzitic interlayers and pods, and contains disseminated grains of diopside and olivine. Graphite and phlogopite are common. Iron oxides are abundant on fracture and layering surfaces and as weathering products of diopside and olivine grains. The calcsilicate layers contain phlogopite, diopside, calcite, tremolite, graphite, and other fine grained phases. Calcite veins are common.

After lunch, we will retrace our route 2.4 km (1.5 mi) back to Montana Highway 287 and turn north, resuming our mileage log northward toward Sheridan, Montana.

With this side trip, the mileage log should be incremented 3.0 miles.

122.6 Alder Creek: The highway is built on Tertiary sediments.

123.4 California Creek.

124.9 The Highland Mountains are at 12 o'clock.

125.2 Robber's Roost Historical Marker, formerly Pete Daly's place: The building served as a stage stop and the second floor once served as a dance hall. When a lower partition was removed recently on the first floor separating the bar from the living quarters, several pounds of lead bullets were recovered from the wall, which apparently was used as a target by the more boisterous patrons. The stage stop served as an incubator where Henry Plummer

(1832–1864), notorious sheriff of Bannack, and his criminal gang—the "Innocents"—hatched many of their plots. Some of the loot was reported to have been cached on the property.

127.1 Horse Creek road: The low foothills on the right are metamorphic rocks.

130.1 Sheridan Post Office: Mill Creek drainage from Tobacco Root Mountains to the east. A number of productive gold-bearing quartz veins occur in a marble layer of the Archean metamorphic rocks in the Mill Creek area. One recent producer is the Red Pine Mine near the head of Indian Creek.

132.3 Mine workings in upper slopes to the left of rounded Mount Baldy at 3 o'clock are part of the Tidal Wave mining district that produced gold and silver.

136.4 Wet and Dry Georgia Creeks road turnoff. Continue on Montana Highway 287.

138.6 At 3 o'clock, light gray–colored cliffs of Mississippian Madison Limestone are visible.

139.7 Twin Bridges. Turn west toward Dillon at junction with Montana Highway 41.

139.8 Just across the bridge, on the west bank of the Beaverhead River is the Jessen Park rest area, which is open for the summer, usually beginning on Memorial Day weekend.

Twin Bridges to Dillon

This road log segment is adapted from Schmitt et al. (1995).

150.9 Point of Rocks Cemetery on the right (west): The route traverses the outer floodplain of the Beaverhead River. Low hills visible to the left and right above the floodplain are composed of the Tertiary "Bozeman Group," consisting of Eocene(?) through Pliocene, poorly consolidated, basin-fill sediments. From here, one has a good view to the east of the north end of the Ruby Range, the Ruby Valley, and the south end of Tobacco Root Mountains.

152.3 A warm spring is on the west side of the road on Beaverhead River floodplain.

152.9 Crossing Beaverhead River: Mississippian carbonates to the west form Beaverhead Rock. A good section of the Tertiary Bozeman Group strata is visible to the west, unconformably overlying Paleozoic and Triassic strata.

153.7 Turnout for Beaverhead Rock historic site: Beaverhead Rock was named by Native Americans and first recorded by the Lewis and Clark expedition on 10 August 1805. However, the real Beaverhead Rock, as recognized by locals, is south of Dillon at the entrance to the Beaverhead River Canyon and is composed of Tertiary volcanic rocks. That beaver is best viewed traveling north on I-15.

Overview of the north flank of the collapsed Laramide Blacktail-Snowcrest uplift: The Blacktail-Snowcrest uplift trends northeast and is basement cored. It is bounded along its southeastern margin by the Snowcrest-Greenhorn thrust system and related sub-Snowcrest thrust. The remnant high portions of the Blacktail-Snowcrest uplift form the southern portion of the Blacktail Mountains immediately south of Dillon, Montana, and the adjacent Snowcrest Range to the south. The central portion of this Laramide-style uplift has been down-dropped by Tertiary normal faulting to form the Sage Creek basin.

Ruby Range (to the east): According to Tysdal (1976), the Ruby Range consists of Precambrian metamorphic rocks, which include dolomitic marble, quartzofeldspathic gneiss, hornblende-rich gneiss, sillimanite schist, and minor banded iron formation, overlain by Phanerozoic sedimentary rocks. The Archean dolomitic marbles that host the talc deposits in the Dillon area are a component of this basement complex. Mid- to late Tertiary sediments overlap older rocks along the range margins.

The central part of the Ruby Range comprises a series of northwest-plunging, basement-cored Laramide folds associated with high-angle reverse faults. The present margins of the uplift, which are buried beneath Quaternary alluvial fans, could be some reactivated Laramide high-angle reverse faults inverted to accommodate Tertiary extension. Normal faults here have been active since Oligocene time.

153.9 Madison-Beaverhead County line: Traveling on Tertiary conglomerate and sandstone.

155.0 To the west lie the high peaks of the Pioneer Mountains cored by the Cretaceous Pioneer batholith. McCartney Mountain is in the middle distance.

158.6 Crossing Stone Creek.

The highway crosses the top of a pediment surface, above the Beaverhead River floodplain to the west. This upland surface has been moderately dissected by intermittent streams flowing down to the Beaverhead River. Loess deposits are locally exposed adjacent to the road. Roadcuts expose sections of tuffaceous Tertiary sandstone and siltstone with resistant calcareous interbeds, which are overlain by matrix-supported channel conglomerate containing abundant Proterozoic quartzite cobbles and other mixed lithologies. The cobbles and pebbles are likely reworked from conglomerates in the Beaverhead Group. Beaverhead River floodplain lies to the west.

166.5 Entering Dillon on Montana Highway 41: Airport road lies to the east. We are traveling south on the upper floodplain of the Beaverhead River (to west in the forested area); this is a Quaternary surface that has been cut laterally into Tertiary strata by the Beaverhead River. Water treatment lakes are north of roadway and armory on south.

167.3 Intersection with I-15 Interchange.

167.5 Junction of Montana Highways 91 and 41: Continue south into Dillon.

167.7 Turn east-southeast on Johnson Ave./Vine St. Turn south on Vine St. before reaching the entrance to the Safeway parking lot.

168.1 Turn east on Kentucky Ave. At the eastern city limit 0.6 km (0.4 mi), Kentucky Ave. becomes Sweetwater Road (Rd 206). The road starts as a two-lane blacktop, but shortly after passing the cemetery (mile 169.3) becomes a two-lane improved gravel road. Inquire locally about conditions.

Dillon to Regal Talc Mine

This road log segment is supplemented from entries by Sears et al. (1995).

170.4 Sweetwater Estates subdivision: Dump material from the Treasure talc mine is visible on the horizon at 12 o'clock. The Sweetwater Road proceeds due east out of Dillon, across the East Bench, a pediment cut on Miocene Six Mile Creek Formation. The Six Mile Creek Formation here includes thin tephra beds, an indurated conglomerate, and garnetiferous sands. The conglomerate is 3–5 m (10–16 ft) thick near the northwest margin of the East Bench. The clasts are well-rounded and imbricated, indicating fluvial transport to the northeast. Pebbles include veined chert and jasper-bearing pink feldspathic quartzite, suggesting a central Idaho provenance. The conglomerate passes southeast into poorly sorted, angular, feldspathic gravel made up of metamorphic and igneous clasts. The clasts were derived from the Ruby Range Archean metamorphic rocks and Eocene volcanics, and the deposit laps onto the source rocks. The Sweetwater Road crosses onto Archean bedrock ~16 km (10 mi) out of Dillon.

171.6 Canal crossing.

172.4 Intersection with Carter Creek Road. Continue east on Sweetwater Road.

173.9 The Regal Mine dumps are visible at 1 o'clock. Forested hills at 3 o'clock mark the location of the Mineral King talc prospect. The Smith-Dillon talc mine is out of sight on the south side of these hills.

174.7 On the right is the talc ore transfer facility for the Regal Mine.

177.9 The Regal Mine is visible at 12 o'clock. Near the turnoff to the Christensen Ranch, Sears et al. (1995) report a basin-bounding, Miocene age fault, down to the northwest, passing through this area. The location of the fault is based on gravity data. The bedrock surface near Sweetwater Pass is an exhumed, dissected, erosional surface of Eocene age; it is overlain by an Eocene lava flow dated at 41.5 Ma (K-Ar whole rock) just north of Sweetwater Road.

179.6 Entering Regal Mine property: Proceed straight through underpass.

179.9 Turn right, through gate, into Regal Mine. Office is ahead at 10 o'clock.

180.0 Arrive at the Regal Mine office and check-in. Mine personnel will provide site tour.

STOP 2

From the pit overlook, the Regal orebody is visible directly east in the southwest corner of the Regal pit. The orebody strikes west and dips ~45° north, to the left as viewed from the overlook. A major north-northwest diabase dike, a potential heat source for talc formation at the Regal, is in subcrop below the overlook. The Regal marble is the host for the talc ore. The marble is deformed into a hook-shaped, refolded fold hinge (Fig. 4) that is west-southwest of the overlook and is not visible from this point. Where the rerouted Sweetwater Road curves on a ridge southwest of the mine property, the road is on the refolded marble of the hook fold, yet the subtle expression of the hook cannot be fully appreciated. Vegetation on the undisturbed land surrounding the mine is a good indicator of lithology.

The return segment from the Regal Mine may follow the Sweetwater Road to the intersection with the Carter Creek Road. Turn north on Carter Creek Road. At the intersection with Nissen Lane, a left turn will allow us to reach Montana Highway 41 north of Dillon. The total distance from the mine to Twin Bridges is 60 km (37.3 mi). The road log for this trip resumes at Twin Bridges.

Twin Bridges to Bozeman

This road log segment is adapted from Schmitt et al. (1995).

Reset odometer at the intersection of Montana Highways 41 and 287.

Mileage	*Directions and Descriptions*
0.0	Turn north on Montana Highway 41 toward Whitehall at the main intersection in downtown Twin Bridges.

The Rochester Mining District is 21 km (13 mi) west in the Highland Mountains.

Tobacco Root Mountains to the east: A spectacular sloping alluvial fan complex lies at the base of the Tobacco Root Mountains. The fan complex is moderately dissected at the top.

The Tobacco Root Mountains form the eastern boundary of the Jefferson River Valley. They extend for 55 km (34.5 mi) southward to Virginia City. South of Virginia City, the general structural trend continues and forms the Greenhorn and Gravelly Ranges. The west flank of the Tobacco Root Range is characterized by a west-dipping Laramide thrust system that has been mostly down-dropped on a normal fault system of similar trend. The rise of the range along the normal fault has tilted the block eastward.

3.6 Historic point: Meriwether Lewis camped here when scouting ahead of William Clark in the Jefferson River valley. In this area, the Jefferson River hugs the west side of the valley, probably the result of depositional "crowding" by the young alluvial fan complex along the east side of valley.

Low on the hills to the west, dissected Tertiary and Quaternary alluvial and pediment gravels are preserved along the southeast flank of the Highland Mountains. These older deposits are mostly unconsolidated sediments of deeply weathered, granitic clasts in a sandy matrix, commonly capped by eolian silt (Ruppel et al., 1993).

Highland Mountains to the west: Most faults in the Highland Mountains strike northwest, except for the northeast-striking range-front normal fault on the east flank of the range and the east-west Camp Creek fault in the south-central part of the range. The Camp Creek fault dips gently northward and places Belt Supergroup strata (Newland and LaHood Formations) over the basement rocks to the south. Schmidt and O'Neill (1982) interpreted the Camp Creek fault to be a reactivated middle Proterozic normal fault.

Rugged canyons on the left dissect Archean crystalline rock. Chloritic alteration in Cottonwood Canyon along major west-northwest–trending faults was explored as a possible source of commercial chlorite.

4.5 The Highland Mountains host several plutons of granodioritic (e.g., Rader Creek pluton) and monzogranitic (e.g., Hells Canyon pluton) composition that intrude a faulted metamorphic basement complex containing wedges of Phanerozoic strata. The Hells Canyon pluton is cut by Hells Canyon to the west. It is a southern satellite intrusion of the Cretaceous Boulder batholith, which hosts the Butte porphyry copper-molybdenum system ~48 km (30 mi) to the north.

6.9 Crossing Jefferson River. To the far east, coalescing alluvial fans occur along base of the Tobacco Root Mountains.

7.7–7.9 The highwall of the reclaimed open-pit Golden Antler mine is just visible west of the highway above the road. This mine produced high-quality magnesian chlorite from an orebody discovered by local prospector Bob Nolte. Cyprus Industrial Minerals and Luzenac America mined chlorite for ceramic products from this mine in the ~1990s. Chlorite formation was localized along a series of northwest-trending high-angle faults and fractures (O'Neill, 1995). Sericitic alteration of Precambrian basement rocks presently exposed in the Highland Mountain gneiss dome produced magnesium-rich chlorite (clinochlore) via complete chemical replacement reactions (Berg, 1983). Veins of pure chlorite as much as 8 m (25 ft) thick in a zone of alteration ~80 m (260 ft) wide were described by Berg (1983).

9.3–9.5 Intensely deformed Paleozoic rocks are juxtaposed against Cretaceous granodiorite along the Green Campbell thrust.

10.0 Silver Star Hot Springs. This hot springs may be associated with a north-northeast–trending normal fault parallel to range-front faults that exist along the west side of Jefferson Valley.

A roadcut just south of Silver Star exposes the Pennsylvanian Quadrant Formation, which is overlain by porphyritic andesite, probably equivalent to the Elkhorn Mountains volcanics. The town of Silver Star is located along the southern exposed edge of the main Boulder batholith. The batholith intrudes a south- and west-dipping sequence of Cambrian through Pennsylvanian strata. The mines of the Silver Star district were developed in Paleozoic carbonates adjacent to the batholith and in veins in Precambrian metamorphic rocks. The principal metals sought were gold, silver, lead, and copper. In the heyday of mining in the 1870s, Silver Star was the only town between Helena and Virginia City.

10.3 In the town of Silver Star, Lloyd Harkins' private mining museum on the west side of the highway is adorned with a man cage, huge compressor wheels from underground mines at Butte, and a caboose. A road entering the highway (on the west) is from the Silver Star mining district, where the reactivated Victoria gold mine ships high-grade copper oxide ores to China and gold ore to the Golden Sunlight Mine.

North of town, the highway traverses Quaternary alluvium and talus deposits that were shed off the Highland Mountains to the west. To the east, Paleozoic and Mesozoic strata crop out on the west and northwest flanks of the Tobacco Root Mountains.

11.9 Possible fault scarps cutting the alluvial fans are visible (if the lighting is favorable) at 3 o'clock along the fault-controlled western range front of the Tobacco Root Mountains. Point of Rocks Hot Spring, at the northern end of the Tobacco Root Mountains, is located at ~1:30.

14.3 Junction of Montana Highways 41 and 55. Continue straight (Montana Highway 55) toward Whitehall.

16.4 Silver Bow County line.

16.7 Jefferson County line.

16.7 Road to Waterloo, Montana, takes off on the right. At 3 o'clock, one can see flat irons of Paleozoic carbonate rocks at the base of the Tobacco Root Mountains. The dark shrub dominating these slopes is mountain mahogany, which grows prolifically on carbonate rocks.

18.3 Fish Creek Bridge: Boulders in alluvium adjacent to Fish Creek are interpreted to have been catastrophically deposited in middle to late Pleistocene time in response to the breaching or breaking of a glacially dammed lake along the upper reaches of Fish Creek in the Highland Mountains (O'Neill, 1995). The east flank of the Highland Mountains lies to the west (left) and the west flank of Tobacco Root Mountains lies to the east (right). Traveling on the outer floodplain of the Jefferson River, the road traverses a Quaternary pediment surface. This surface has been dissected to form the current valley floor and floodplain of the Jefferson River (to the east). As we approach Whitehall, the road skirts the eastern margin of a large outcrop of the Miocene Six Mile Creek Formation (low hills to the north).

18.5 Large dumps visible on the skyline at 1 o'clock are from the Golden Sunlight Mine (Barrick Gold Corp.) at the south end of Bull Mountain.

Bull Mountain is a horst composed of Belt Supergroup rocks covered by Late Cretaceous Elkhorn Mountains volcanics farther north. Mineralization occurs within a large (213 m diameter) hydrothermal breccia pipe, which likely formed in an epithermal environment above a molybdenum porphyry system, based upon the increase in potassic alteration and molybdenite mineralization with depth (Foster and Chadwick, 1990). The breccia pipe is believed to have been contemporaneous with Late Cretaceous latite porphyry magmatism. Lamprophyre dikes and small hypabyssal intermediate mafic intrusions (with mantle-derived chemical signatures) may be similar in age to the dominant silicic intrusive and extrusive igneous rocks (DeWitt et al., 1996). Gold is hosted by the latite, and open-pit mining began in the early 1980s. The Golden Sunlight orebody is one of many porphyry-related systems that occur within the Great Falls tectonic zone (O'Neill and Lopez, 1985), a long lived, deep-seated zone of crustal weakness that appears to have controlled late Mesozoic and early Cenozoic magmatism in Idaho and Montana. The Golden Sunlight Mine is the subject of another field trip included in this field guide volume (Oyer et al.).

26.0 Crossing Pipestone Creek.

26.2 Junction with Montana Highway 2. Turn east onto Montana Highway 55 to I-90.

26.6 At stop sign in downtown Whitehall, turn north on Montana Highway 55 and continue through Whitehall.

27.2 Turn east on I-90 ramp (toward Bozeman).

28.2 At 11 o'clock, a prominent ridge is underlain by a Late Cretaceous latite sill that intrudes Proterozoic Belt sedimentary rocks and forms part of the intrusive complex at the Golden Sunlight Mine.

30.4 Tertiary sediments on left are part of the valley-fill marginal to the Bull Mountain block. Tertiary/Quaternary debris flows farther north along the range front are derived from erosion of the Golden Sunlight breccia pipe and contain enough gold-mineralized clasts to have been mined in recent years.

31.1 Triangular valley fill on left is reclaimed dump material from the Golden Sunlight gold mining operation.

32.2 Buttress of a mill tailings dam visible at 9 o'clock is currently being used by the Golden Sunlight Mine.

34.2 Cardwell and Boulder (Exit 256). Continue eastward to Bozeman. Poorly consolidated Neogene strata lie to the north adjacent to long straightaway in highway.

35.0 Boulder River crossing and base of Doherty Mountain Pass: This area is the north end of the Tobacco Root Mountains. Doherty Mountain (to the north) is a large, north-plunging anticline cored by Belt Supergroup rocks (LaHood Formation), and flanked by steeply dipping Cambrian and younger strata. The sedimentary strata are intruded by greenish-black weathering dikes and sills of probable Late Cretaceous age.

46.9 Crest of Doherty Summit between Three Forks and Whitehall: This broad saddle is capped by the "Ballard gravels," a deposit of locally occurring, late Tertiary and/or Quaternary(?) well-rounded, cobble-sized sediment. This saddle may mark the ancestral drainage of the Jefferson River, or some other river, prior to late Tertiary and/or Quaternary extension and basin excavation. Alternatively the Ballard gravels may represent a remnant of a regionally extensive blanket of late Cenozoic, high-elevation pediment gravel.

49.7 Jefferson-Broadwater County line.

52.6 Helena (Exit 274). Continue eastward to Bozeman.

53.2 Gallatin-Broadwater County line. Jefferson River crossing.

54.0 Three Forks, Montana (Exit 278): The early settlement of Gallatin City, founded in 1862 and situated east of the confluence of the Madison, Jefferson, and Gallatin Rivers, was moved in 1882 to a location a mile to the southwest (a locality north of I-90 and marked as Oldtown on modern maps) and named Three Forks. In 1912, the community was relocated to its present site.

Gallatin City, the first settlement in Gallatin County, was platted in 1862 on the west bank of the Missouri River, opposite the mouth of the Gallatin River. The first cabin was built by Frank Dunbar. People bought lots under the impression that the site would become a head of navigation for steamers coming up the Missouri. However, no boats came up the river, because of the Great Falls and other falls and rapids downstream. The town site was abandoned and moved to the east bank of the Madison River. At this site stood several cabins and stores, a grist mill, Dunbar's hotel, and a horse-racing track. The hotel still stands today although rooms are not available.

The confluence of the Jefferson, Madison and Gallatin Rivers (named by Lewis and Clark in 1805) to form the Missouri River lies three miles to the north. After traveling up the Missouri, Lewis and Clark camped at the confluence (in present-day Missouri Headwaters State Park) and decided to take the Jefferson Fork to continue their exploration of a route to the Pacific. Their party was guided by a Shoshoni woman named Sacajawea. Her French Canadian husband, Toussaint Charbonneau, was the interpreter for the party. She had been captured by the Minnetaree subgroup of the Dakota Nation in this region five years earlier and, on the trip west, recognized familiar landmarks.

At Trident, which is ~0.8 km farther north of the confluence, Holcim (U.S.) Inc. operates the Trident cement plant. Holcim is an international company that operates 12 cement plants in the United States with 1800 employees. The Trident cement plant was built in 1910, and some of the current 74 employees are third-generation employees.

The mine is in Paleozoic carbonate rocks in the Trident syncline, which is bounded by the Green thrust on the east and the Trident thrust on the west. The cement plant obtains lime from Madison Group limestones at the plant site, shale from nearby, and iron ore from Radersburg and White Sulfur Springs. These ingredients are crushed, ground, blended (in slurry form), and fed into a rotary kiln, where the mixture is burned to partial fusion. The sintered product is ground with gypsum into Portland cement and shipped.

56.5 Madison River crossing.

58.1 Madison River bluffs to the south consist of Tertiary tuffaceous and calcareous siltstones, sandstones, and ash beds. Originally called the "Bozeman lake beds" by A.C. Peale (1896), these cliffs now constitute the Bozeman Group, defined as "... Tertiary fluvial, eolian, and lacustrine rocks which accumulated in the basins of western Montana after the Laramide orogeny" (Robinson, 1963). Kuenzi and Fields (1971) identified two lithostratigraphic units in the Bozeman Group, the Renova and Six Mile Creek Formations. The lower cliffs are mostly lacustrine deposits equivalent to the Eocene–Oligocene Renova Formation, overlain by dominantly fluviatile deposits of the late Miocene Six Mile Creek Formation (Hackett et al., 1960).

A few miles to the south, these Tertiary cliffs were used by early Native Americans as a "buffalo jump" mass kill site (Madison Buffalo Jump State Park at Exit 283).

63.6 Logan-Trident (Exit 283): The limestone cliff directly across the Gallatin River at Logan is the type section of the Madison Group (Sando and Dutro, 1974). Peale (1893) proposed the name "Madison Formation" in the Three Forks area of southwest Montana for carbonates underlain by shales of the Three Forks Formation and overlain by sandstones of the Quadrant Formation, although he never specified a type section. Sloss and Hamblin (1942) reviewed and synthesized previous work on the Madison Group and proposed the stratigraphic nomenclature now used by most geologists throughout Montana (Sando and Dutro, 1974). The Madison was divided into the Lodgepole and Mission Canyon Formations. Sando and Dutro (1974) established a detailed type-section description of the Madison Group at Logan, with a reference set of fossils permanently housed at the U.S. National Museum in Washington, D.C.

67.5 Manhattan (Exit 288): The Horseshoe Hills are well-exposed to the north across the Gallatin River, consisting of northwest-dipping arkose and shale of the middle Proterozoic LaHood Formation (in the bluffs along the river), overlain by a Phanerozoic succession that crops out farther north in the hills. Here, I-90 runs approximately parallel to the old southern margin of the Belt basin, variously known as the Perry line (Winston, 1986), Willow Creek fault (Robinson, 1963; Harrison et al., 1974), Central Park fault (Hackett et al., 1960), and/or the southwest Montana transverse fault zone (Schmidt and O'Neill, 1982).

The Perry line is a major cross-strike structural discontinuity in the Northern Rocky Mountains that is generally delimited by middle Proterozoic normal faults that were reactivated as right-oblique thrust faults during Cretaceous and Paleogene contractional orogenesis (Sevier and Laramide orogenies). As such, the Perry line is one of several fault zones in western Montana that have structurally inverted the Belt basin. During the middle Proterozoic, arkosic debris was eroded from Archean metamorphic highlands across southwest Montana (Dillon Block) and deposited in the Belt basin. These strata are preserved as the LaHood Formation along Montana Highway 2 near the old Lahood station in the Jefferson River Canyon. Prominent northwest-dipping exposures in the Horseshoe Hills north of the river are from east to west: Cambrian (Pilgrim Limestone), Devonian (Maywood, Jefferson, Three Forks, Sappington Formations) and Mississippian (Lodgepole and Mission Canyon Formations)—principally carbonates, except for the Sappington Sandstone.

The structure of the Horseshoe Hills consists of a series of northeast-trending, tight, sigmoidal folds with steep NW-dipping axial surfaces, and right-oblique thrust faults. The Horseshoe Hills were originally mapped by Verrall (1955). Parts of the hills have been subsequently remapped in great detail by field camp students at Montana State University over the years. Lageson (1992) has interpreted the structure of the Horseshoe Hills as being the result of transverse lateral ramping along the old Perry line, coupled with megascopic dextral simple shear.

72.7 Gallatin River crossing.

75.2 Belgrade-Amsterdam (Exit 298). The Bridger Range is well exposed to the east, and Ross Pass is the prominent saddle on the skyline in the middle of the range. Ross Pass is the trace of the Pass Fault, a reactivated middle Proterozoic fault that juxtaposes the Belt Supergroup (LaHood Formation) in the core of the northern Bridger Range and the Archean basement complex south of this fault (e.g., southern Bridger Range). Cambrian strata nonconformably overlie the entire range. The topographic crest of the Bridger Range consists of steeply east-dipping to overturned Mississippian carbonate rocks of the Madison Group, which strike the length of the range. These Phanerozic strata represent the steep east limb of the ancestral (Laramide) Bridger Range (Lageson, 1989).

79.9 Gravel pit north of the highway is in thick Quaternary alluvium of the Gallatin Valley. The bulk of the gravel is composed of rounded cobbles of andesite and dacite derived from the Eocene volcanic rocks of the Gallatin Range to the south.

84.7 Take Exit 305 N. 19th Avenue, Bozeman, south to return to Montana State University.

ACKNOWLEDGMENTS

We thank Dick Berg and Bob Lankston for timely and constructive reviews. David Crouse at Imerys Talc provided valuable input on the Yellowstone Mine. Management personnel at both Imerys Talc and Barretts Minerals Inc. generously shared information and provided access to their properties.

REFERENCES CITED

Anderson, D.L., 1987, Timing and mechanism of formation of selected talc deposits in the Ruby Range, southwestern Montana [M.Sc. thesis]: Bozeman, Montana State University, 90 p.

Anderson, D.L., Mogk, D.W., and Childs, J.F., 1990, Petrogenesis and timing of talc formation in the Ruby Range, southwestern Montana: Economic Geology and the Bulletin of the Society of Economic Geologists, v. 85, p. 585–600, doi:10.2113/gsecongeo.85.3.585.

Barnard, F., 1993, District scale zoning pattern, Virginia City, Montana, USA, *in* Reprint 93-29, 96th National Western Mining Conference of the Colorado Mining Association (Denver), 18 p.

Berg, R.B., 1979, Talc and Chlorite Deposits in Montana: Butte, Montana Bureau of Mines and Geology Memoir 45, 66 p.

Berg, R.B., 1983, New chlorite in an old Montana gold district: Mining Engineering, v. 35, p. 347–350.

Berg, R.B., 1995, Geology of western U.S. talc deposits, *in* Tabilio, M., and Dupras, D.L., eds., 29th Forum on the Geology of Industrial Minerals: Proceedings: Long Beach, California Department of Conservation, Division of Mines and Geology Special Publication 110, p. 69–79.

Brady, J.B., Cheney, J.T., Rhodes, A.L., Vasquez, A., Green, C., Duvall, M., Kogut, A., Kaufman, L., and Kovaric, D., 1998, Isotope geochemistry of Proterozoic talc occurrences in Archean marbles of the Ruby Mountains, Southwest Montana, U.S.A.: Geological Materials Research, Mineralogical Society of America, v. 1, no. 2, p. 1–41.

Burmester, R.F., Lonn, J.D., Lewis, R.S., and McFaddan, M.D., 2013, Toward a Grand Unified Theory for stratigraphy of the Lemhi Subbasin of the Belt Supergroup, *in* Lewis, R.S., Garsjo, M.M., and Gibson, R.I., eds., 38th Annual Field Conference, Belt Symposium V: Northwest Geology, v. 42, p. 1–20.

Cerino, M.T., Childs, J.F., and Berg, R.B., 2007, Talc in southwestern Montana, *in* Thomas, R.C., and Gibson, R.I., eds., Introduction to the Geology of the Dillon Area: Northwest Geology, v. 36, p. 9–22.

Childs, J.F., 1985, Radiometric date on sericite associated with talc formation: Internal Report to Cyprus Industrial Minerals Company, 2 p.

Cordua, W.S., 1973, Precambrian geology of the southern Tobacco Root Mountains, Madison County, Montana [Ph.D. dissertation]: Bloomington, Indiana University, 258 p.

DeWitt, E., Foord, E.E., Zartman, R.E., Pearson, R.C., and Foster, F., 1996. Chronology of Late Cretaceous Igneous and Hydrothermal Events at the Golden Sunlight Gold-Silver Breccia Pipe, Southwestern Montana: U.S. Geological Survey Bulletin 2155, 48 p.

Evans, B.W., and Guggenheim, S., 1988, Talc, pyrophyllite, and related minerals, *in* Bailey, S.W., ed., Hydrous Phyllosilicates: Washington, D.C., Mineralogical Society of America, Reviews in Mineralogy 18, p. 225–294.

Foster, F., and Chadwick, T., 1990, Relationship of the Golden Sunlight Mine to the Great Falls tectonic zone, *in* Moye, F.J., ed., Geology and Ore Deposits of the Trans-Challis Fault System/Great Falls Tectonic Zone: Guidebook for the Fifteenth Annual Field Conference, Tobacco Root Geological Society, p. 77–81.

Gammons, C.H., and Matt, D.O., 2002, Using fluid inclusions to help unravel the origin of hydrothermal talc deposits in southwest Montana: Northwest Geology, v. 31, p. 41–53.

Garihan, J.M., 1973, Geology and talc deposits of the central Ruby Range, Madison County, Montana [Ph.D. dissertation]: University Park, Pennsylvania State University, 282 p.

Hackett, O.M., Visher, F.N., McMurtrey, R.G., and Steinhilber, W.L., 1960, Geology and Ground-Water Resources of the Gallatin Valley, Gallatin County, Montana: U.S. Geological Survey Water-Supply Paper 1482, 282 p.

Harms, T.A., Brady, J.B., Burger, H.R., and Cheney, J.T., 2004, Advances in the geology of the Tobacco Root Mountains, Montana, and their implications for the history of the northern Wyoming province, *in* Brady, J.B., Burger, H.R., Cheney, J.T., and Harms, T.A., eds., Precambrian Geology of the Tobacco Root Mountains, Montana: Geological Society of America Special Paper 377, p. 227–243.

Harrison, J.E., Griggs, A.G., and Wells, J.D., 1974, Tectonic Features of the Precambrian Belt Basin and Their Influence on Post-Belt Structures: U.S. Geological Survey Professional Paper 866, 15 p.

Hendrix, M.S., 2011, Geology Underfoot in Yellowstone Country: Missoula, Montana, Mountain Press Publishing, 302 p.

Hoy, T., Anderson, D., Turner, R.J.W., and Leitch, C.H.B., 2000, Tectonic, magmatic and metallogenic history of the early synrift phase of the Purcell Basin, Southeastern British Columbia, *in* Lydon, J.W., ed., The Geological Environment of the Sullivan Deposit, British Columbia: Geological Association of Canada Special Publication No. 1, p. 32–60.

Hurst, R.W., 1985, Rb/Sr analyses of talc, dolomite and calcite: Internal Report to Cyprus Industrial Minerals Company, 2 p.

James, H.L., 1990, Precambrian Geology and Bedded Iron Deposits of the Southwestern Ruby Range, Montana: U.S. Geological Survey Professional Paper 1495, 39 p.

Johns, W.M., Berg, R.B., and Dresser, H.W., 1981, First day geologic road log Part 4. Three Forks to Twin Bridges via U.S. Highways 10 and 287 and State Highway 287, *in* Tucker, T.E., ed., Montana Geological Society Field Conference and Symposium Guidebook to Southwest Montana: Billings, Montana, p. 388–392.

Kellogg, K.S., 1994, Geologic Map of the Norris Quadrangle, Madison County, Montana: U.S. Geological Survey, Geologic Quadrangle GQ-1738, scale 1:24,000, 1 sheet.

Kellogg, K.S., and Williams, V.S., 2000, Geologic Map of the Ennis 30′ × 60′ Quadrangle, Madison and Gallatin Counties, Montana, and Park County, Wyoming: U.S. Geological Survey Geologic Investigation Series Map I-2690, scale 1:100,000, 1 sheet, 16 p. text.

Kellogg, K.S., and Williams, V.S., 2005, Geologic Map of the Ennis 30′ × 60′ Quadrangle, Madison and Gallatin Counties, Montana, and Park County, Wyoming: Montana Bureau of Mines and Geology Open-File Report MBMG 529, scale 1:100,000, 1 sheet, 27 p. text.

Kellogg, K.S., Ruleman, C.A., and Vuke, S.M., 2007, Geologic Map of the Central Madison Valley (Ennis Area) Southwestern Montana: Montana Bureau of Mines and Geology Open-File Report MBMG 543, scale 1:50,000, 1 sheet, 19 p. text.

Kucks, R.P., 1999, Bouguer gravity anomaly data grid for the conterminous US, http://mrdata.usgs.gov/services/gravity?request=getcapabilities&service=WMS&version=1 .1.1 (accessed 18 March 2014). This is part of a larger work: Phillips, J.D., Duval, J.S., and Ambroziak, R.A., 1993, National geophysical data grids; gamma-ray, gravity, magnetic and topographic data for the conterminous United States: U.S. Geological Survey Digital Data Series DDS-9, http://pubs.er.usgs.gov/publication/ds9.

Kuenzi, W.D., and Fields, R.W., 1971, Tertiary stratigraphy, structure, and geologic history, Jefferson basin, Montana: Geological Society of America Bulletin, v. 82, p. 3373–3394, doi:10.1130/0016-7606(1971)82[3373:TSSAGH]2.0.CO;2.

Lageson, D.R., 1989, Reactivation of a Proterozoic continental margin, Bridger Range, southwestern Montana, *in* French, D.E., and Grabb, R.F., eds., Geologic Resources of Montana, Volume II: Montana Geological Society Field Conference Guidebook, p. 279–298.

Lageson, D.R., 1992, Structural analysis of the Horseshoe Hills transverse fold-thrust zone, Gallatin County, Montana: A preliminary report, *in* Elliot, J.E., ed., Guidebook for the Red Lodge-Beartooth Mountains-Stillwater area: Tobacco Root Geological Society Seventeenth Annual Field Conference: Northwest Geology, v. 20/21, p. 117–124.

Lonsdale, P.F., Bischoff, J.L., Burns, V.M., Kastner, M., and Sweeney, R.E., 1980, A high-temperature hydrothermal deposit on the seabed at a gulf of California spreading center: Earth and Planetary Science Letters, v. 49, p. 8–20, doi:10.1016/0012-821X(80)90144-2.

Lyden, C.J., 1948, Gold Placers of Montana: Butte, Montana Bureau of Mines and Geology, 120 p. (reprinted in 1987).

Marvin, R.F., Wier, K.L., Mehnert, H.H., and Merritt, V.M., 1974, K-Ar ages of selected Tertiary igneous rocks in southwestern Montana: Isochron-West, no. 10, p. 17–20.

McDonald, C., Elliott, C.G., Vuke, S.M., Lonn, J.D., and Berg, R.B., 2012, Geologic Map of the Butte South 30′ × 60′ Quadrangle, Southwestern Montana: Montana Bureau of Mines and Geology Open-File Report MBMG 622, scale 1:100,000, 1 sheet.

McMannis, W.J., 1960, Exit geologic road log Ennis to Bozeman via state highways 287 and 289, *in* Campau, D.E., and Anisgard, H.W., eds., West Yellowstone–Earthquake Area, Billings Geological Society, 11th Annual Field Conference: Billings, Montana, p. 312–313.

Mogk, D.W., Mueller, P.A., Weyand, E., and Wooden, J.L., 1989, Tectonic and geochemical mixing in the middle crust—Evidence from the Archean basement of the northern Madison Range, Montana: Geological Society of America Abstracts with Programs, v. 21, no. 6, p. A183.

Montana Abandoned Mine Reclamation, 2014, Department of Environmental Quality of the Official Montana State Government Website: www.deq.mt.gov/abandonedmines/linkdocs/117tech.mcpx (accessed March 2014).

O'Neill, J.M., 1995, Early Proterozoic geology of the Highland Mountains, southwestern Montana, and field guide to the basement rocks that compose the Highland Mountain gneiss dome, *in* Mogk, D.W., ed., Field Guide to Geologic Excursions in Southwest Montana: Northwest Geology, v. 24, p. 85–97.

O'Neill, J.M., and Lopez, D.A., 1985, Character and regional significance of the Great Falls tectonic zone of east-central Idaho and west-central Montana: American Association of Petroleum Geologists Bulletin, v. 69, p. 437–447.

Oyer, N., Childs, J., and Mahoney, J.B., 2014, this volume, Regional setting and deposit geology of the Golden Sunlight Mine: An example of responsible resource extraction, *in* Shaw, C.A., and Tikoff, B., eds., Exploring the Northern Rocky Mountains: Geological Society of America Field Guide 37, doi:10.1130/2014.0037(06).

Paterson, M.S., 1978, Experimental Rock Deformation—The Brittle Field: Berlin, Springer-Verlag, 254 p.

Peale, A.C., 1893, The Paleozoic Section in the Vicinity of Three Forks, Montana: U.S. Geological Survey Bulletin No. 110, 56 p.

Pritchett, K., 1993, Huckleberry Ridge Tuff of the Madison valley, southwest Montana: Northwest Geology, v. 22, p. 57–75.

Robinson, G.D., 1963, Geology of the Three Forks Quadrangle, Montana: U.S. Geological Survey Professional Paper 370, 140 p.

Rose, A., 1984, Geochemical methods of exploration for Beaverhead-type talc deposits: Internal Report to Cyprus Industrial Minerals Company, Talc Division, 51 p.

Ross, G.M., and Villeneuve, M., 2003, Provenance of the Mesoproterozoic (1.45 Ga) Belt basin (western North America): Another piece in the pre-Rodinia paleogeographic puzzle. Geological Society of America Bulletin, v. 115, no. 10, p. 1191–1217, doi:10.1130/B25209.1.

Ruppel, E.T., O'Neill, J.M., and Lopez, D.A., 1993, Geologic Map of the Dillon 1° × 2° Quadrangle, Idaho and Montana: U.S. Geological Survey Miscellaneous Geologic Investigation Series Map I-1803-H, scale 1:250,000, 1 sheet.

Sando, W.J., and Dutro, J.T., 1974, Type Sections of the Madison Group (Mississippian) and Its Subdivisions in Montana: U.S. Geological Survey Professional Paper 842, 22 p.

Schmidt, C.J., and O'Neill, J.M., 1982, Structural evolution of the southwest Montana transverse zone, *in* Powers, R.B., ed., Geologic Studies of the Cordilleran Thrust Belt: Denver, Colorado, Rocky Mountain Association of Geologists, v. 1, p. 167–180.

Schmidt, C.J., O'Neill, J.M., and Brandon, W.C., 1988, Influence of Rocky Mountain foreland uplifts on the development of the frontal fold and thrust belt, southwest Montana, *in* Schmidt, C.J., and Perry, W.J., eds., Interaction of the Rocky Mountain Foreland and the Cordilleran Thrust Belt: Geological Society of America Memoir 171, p. 171–202.

Schmidt, C.J., Smedes, H.W., and O'Neill, J.M., 1990, Syncompressional emplacement of the Boulder and Tobacco Root batholiths (Montana-USA) by pull-apart along old fault zones: Geological Journal, v. 25, no. 3–4, p. 305–318, doi:10.1002/gj.3350250313.

Schmitt, J.G., Haley, J.C., Lageson, D.R., Horton, B.K., and Azevedo, P.A., 1995, Sedimentology and tectonics of the Bannack-McKnight Canyon-Red Butte Area, Southwest Montana: New perspectives on the Beaverhead Group and Sevier Orogenic Belt, *in* Mogk, D.W., ed., Field Guide to Geologic Excursions in Southwest Montana: Northwest Geology, v. 24, p. 245–313.

Sears, J.W., Hurlow, H., Fritz, W.J., and Thomas, R.C., 1995, Late Cenozoic disruption of Miocene grabens on the shoulder of the Yellowstone Hotspot track in southwest Montana: Field guide from Lima to Alder, Montana, *in* Mogk, D.W., ed., Field Guide to Geologic Excursions in Southwest Montana: Northwest Geology, v. 24, p. 201–219.

Sloss, L.L., and Hamblin, R.H., 1942, Stratigraphy and insoluble residues of Madison Group (Mississippian) of Montana: American Association of Petroleum Geologists Bulletin, v. 26, p. 305–335.

Tysdal, R.G., 1976, Geologic Map of Northern Part of Ruby Range, Madison County, Montana: U.S. Geological Survey Miscellaneous Geologic Investigation Series Map I-951, scale 1:24,000, 1 sheet.

Van Gosen, B.S., Berg, R.B., and Hammarstrom, J.M., 1998, Map Showing Areas with Potential for Talc Deposits in the Gravelly, Greenhorn, and Ruby Ranges and the Henry's Lake Mountains of Southwestern Montana: U.S. Geological Survey Open-File Report 98-224-B, scale 1:250,000, 1 sheet.

Vargo, A.G., 1990, Structure and petrography of the Prebeltian rocks of the north-central Gravelly Range, Montana [M.S. thesis]: Fort Collins, Colorado State University, 157 p.

Verrall, P., 1955, Geology of the Horseshoe Hills area, Montana [Ph.D. dissertation]: Princeton, New Jersey, Princeton University, 260 p.

Vuke, S.M., Lonn, J.D., Berg, R.B., and Kellogg, K.S., 2002, Preliminary Geologic Map of the Bozeman 30′ × 60′ Quadrangle, Southwestern Montana: Montana Bureau of Mines and Geology Open-File Report MBMG 469, scale 1:100,000, 1 sheet, 39 p. text.

Vuke, S.M., Porter, K.W., Lonn, J.D., and Lopez, D.A., 2007, Geologic Map of Montana—Information Booklet, Montana Bureau of Mines and Geology: Geologic Map 62D, scale 1:500,000, 2 sheets, 73 p. text.

Winston, D., 1986, Sedimentation and tectonics of the middle Proterozoic Belt basin, and their influence on Phanerozoic compression and extension in western Montana and northern Idaho, *in* Peterson, J., ed., Tectonics and Sedimentation in the Rocky Mountain Region: American Association of Petroleum Geologists Memoir 41, p. 87–118.

Manuscript Accepted by the Society 27 February 2014

The Geological Society of America
Field Guide 37
2014

Glacial and Quaternary geology of the northern Yellowstone area, Montana and Wyoming

Kenneth L. Pierce*
U.S. Geological Survey, 2327 University Way, Box 2, Bozeman, Montana 59715, USA

Joseph M. Licciardi*
Department of Earth Sciences, University of New Hampshire, Durham, New Hampshire 03824, USA

Teresa R. Krause*
Cathy Whitlock*
Department of Earth Sciences, Montana State University, Bozeman, Montana 59717, USA

ABSTRACT

This field guide focuses on the glacial geology and paleoecology beginning in the Paradise Valley and progressing southward into northern Yellowstone National Park. During the last (Pinedale) glaciation, the northern Yellowstone outlet glacier flowed out of Yellowstone Park and down the Yellowstone River Valley into the Paradise Valley. The field trip will traverse the following Pinedale glacial sequence: (1) deposition of the Eightmile terminal moraines and outwash 16.5 ± 1.4 ^{10}Be ka in the Paradise Valley; (2) glacial recession of ~8 km and deposition of the Chico moraines and outwash 16.1 ± 1.7 ^{10}Be ka; (3) glacial recession of 45 km to near the northern Yellowstone boundary and moraine deposition during the Deckard Flats readjustment 14.2 ± 1.2 ^{10}Be ka; and (4) glacial recession of ~37 km and deposition of the Junction Butte moraines 15.2 ± 1.3 ^{10}Be ka (this age is a little too old based on the stratigraphic sequence). Yellowstone's northern range of sagebrush-grasslands and bison, elk, wolf, and bear inhabitants is founded on glacial moraines, sub-glacial till, and outwash deposited during the last glaciation. Floods released from glacially dammed lakes and a landslide-dammed lake punctuate this record.

The glacial geologic reconstruction was evaluated by calculation of basal shear stress, and yielded the following values for flow pattern in plan view: strongly converging—1.21 ± 0.12 bars (*n* = 15); nearly uniform—1.04 ± 0.16 bars (*n* = 11); and strongly diverging—0.84 ± 0.14 bars (*n* = 16). Reconstructed mass balance yielded accumulation and ablation each of ~3 km^3/yr, with glacial movement near the equilibrium line altitude dominated by basal sliding.

Pollen and charcoal records from three lakes in northern Yellowstone provide information on the postglacial vegetation and fire history. Following glacial retreat,

*E-mails: kpierce@usgs.gov; joe.licciardi@unh.edu; teresa.krause@msu.montana.edu; whitlock@montana.edu.

Pierce, K.L., Licciardi, J.M., Krause, T.R., and Whitlock, C., 2014, Glacial and Quaternary geology of the northern Yellowstone area, Montana and Wyoming, *in* Shaw, C.A., and Tikoff, B., eds., Exploring the Northern Rocky Mountains: Geological Society of America Field Guide 37, p. 189–203, doi:10.1130/2014.0037(09).

sparsely vegetated landscapes were colonized first by spruce parkland and then by closed subalpine forests. Regional fire activity increased significantly with the development of closed subalpine forests as a result of increased fuel biomass and warmer summers. Warm dry conditions prevailed at low elevations during the early Holocene, as indicated by the presence of steppe and open mixed conifer forest. At the same time, closed subalpine forests with low fire frequency were present at higher elevations, suggesting relatively wet summer conditions. Douglas fir populations expanded throughout northern Yellowstone in the middle Holocene as a result of effectively drier conditions than before, and a decline of mesophytic plant taxa during the late Holocene imply continued drying, even though fire frequency decreased in recent millennia.

INTRODUCTION

This one-day trip originates in and returns to Bozeman, Montana, on 22 May 2014 (Fig. 1). The first part of the trip examines the Quaternary geology north of Yellowstone National Park within the Paradise Valley, and the second part examines the Quaternary geology in northern Yellowstone National Park. Much of the material in this guide is derived from a 2003 International Association for Quaternary Research (INQUA) guidebook (Pierce et al., 2003).

This field guide focuses on glacial geology and paleoecology. Previous field guides of the Yellowstone River Valley include those of Locke et al. (1995) and Montagne and Locke (1989). Non-technical overviews of Yellowstone National Park are by Good and Pierce (1996) and Smith and Siegel (2000). Christiansen (2001) extensively describes the volcanic geology of the park and Pierce (1979) describes the glacial geology of the northern Yellowstone region. You may wish to obtain detailed maps of the bedrock and surficial geology of Yellowstone National Park (U.S. Geological Survey, 1972a, 1972b).

GLACIAL OVERVIEW

Under full glacial conditions (~17–15 ka; Fig. 2), the northern Yellowstone outlet glacier was formed by convergence from multiple sources. Sources, from east to west, were (1) glaciers from the Beartooth uplift and the Absaroka Range, (2) ice flowing over the Washburn Range from the eastern Yellowstone Plateau, (3) ice flowing from the Yellowstone Plateau between the Washburn and Gallatin Range, and (4) ice from the Gallatin Range (Fig. 2). From the northern Yellowstone National Park boundary, this outlet glacier flowed 65 km (40 miles) down the Yellowstone River Valley to the Eightmile terminal moraines (Fig. 1).

Pleistocene glacial flow patterns for the greater Yellowstone glacial system were complex. Glaciation was initiated in the mountains that surround the Yellowstone Plateau and flowed onto the plateau. Glaciers then built up on the plateau to ~1000 m (3000 ft) thickness and flowed outward down the major valleys that drain Yellowstone National Park (Fig. 2). During deglaciation, the plateau ice cap stagnated, and glaciers from the adjacent mountains flowed into terrain previously occupied by the plateau ice cap (Pierce, 1979). This pattern of buildup, full glacial conditions, and recessional changes resulted in shifts by up to 180° in direction of glacial flow and the transport of glacial erratics. Changing flow patterns temporarily dammed glacial lakes, including lakes on the Yellowstone Plateau. Outlet glaciers augmented by ice from the Yellowstone Plateau probably culminated later than typical mountain-valley glaciers, apparently because of both the interval needed to build up the plateau ice cap and its self-amplifying nature once established.

Cosmogenic ages for the last glacial maximum are oldest for the Clarks Fork of the Yellowstone drainage along the eastern periphery of the greater Yellowstone glacial system (18.8 ± 0.9 ^{10}Be ka; Fig. 2) that headed in the Beartooth uplift (Licciardi and Pierce, 2008). Cosmogenic ages are of intermediate age (16.5 ± 1.4 ^{10}Be ka) for terminal moraines on the north side of the greater Yellowstone glacial system that resulted from the combined glacial flow from the Yellowstone Plateau and surrounding mountains. Terminal moraine ages are youngest for the outer Jenny Lake moraines (14.6 ± 0.7 ^{10}Be ka). These are buried by younger outwash from terminal moraines on the south margin of the greater Yellowstone glacial system. This pattern suggests that the center of mass of the greater Yellowstone glacial system migrated southwest through time, as diagramed by the green symbols that show the change in the inferred center of mass of the Yellowstone glacial system at 19 ka, 16 ka, and 14 ka (Fig. 2). These centers become younger to the southwest, inferred to result from orographic moisture supplied by winter storms moving northeastward up the Snake River Plain. As the center of mass migrated southwestward, this placed the eastern part of the source area in a precipitation or snow shadow.

Bull Lake and Pinedale glaciations are generally correlated with marine oxygen isotope stages (OIS) 6 and 2, respectively. In the Rocky Mountains, Bull Lake terminal moraines are typically 5%–10% farther down valley from their glacial source areas than Pinedale moraines. However, for the Greater Yellowstone glacial system (Fig. 2, red and black lines), Bull Lake terminal moraines to the west, southwest, and south of Yellowstone are much farther down valley than Pinedale moraines, whereas to the north and locally to the east, Pinedale glaciers have overridden Bull

Lake terminal moraines. One explanation is that on the trailing and subsiding margin of the Yellowstone hotspot, the Bull Lake glaciation occurred on higher landscapes than did the Pinedale. Thus, Bull Lake glaciation extended much farther than did the Pinedale. Whereas on the leading and uplifting margin of the Yellowstone hotspot, Pinedale glaciation occurred on higher, uplifting landscapes and thus Pinedale glaciation extended farther than did the Bull Lake and commonly overrode Bull Lake moraines (Pierce and Morgan, 1992). For Jackson Hole along the southern margin of the Greater Yellowstone Glacial System (Fig. 2), the much greater extent of the Bull Lake glaciation compared to the Pinedale glaciation is not explained by emplacement of rhyolite flows in Yellowstone after the Bull Lake glaciation.

Calculation of basal shear stress was used to evaluate the reconstruction of the northern Yellowstone glacial system (Fig. 3; Pierce, 1979). For the vast majority of modern glaciers, empirical observations indicate that basal shear stress is between 0.5 and 1.5 bars (Nye, 1952; Patterson, 1969). Basal shear stress was highest in areas of extending flow (convergent in plan view) and lowest in areas of compressing flow (divergent in plan view). Values of basal shear stress for the reconstructed northern Yellowstone outlet glacier and its source areas are as follows (Pierce, 1979):

strongly converging flow 1.21 ± 12 ($n = 15$),
uniform flow 1.04 ± 0.16 ($n = 11$),
strongly diverging flow 0.84 ± 0.14 ($n = 16$).

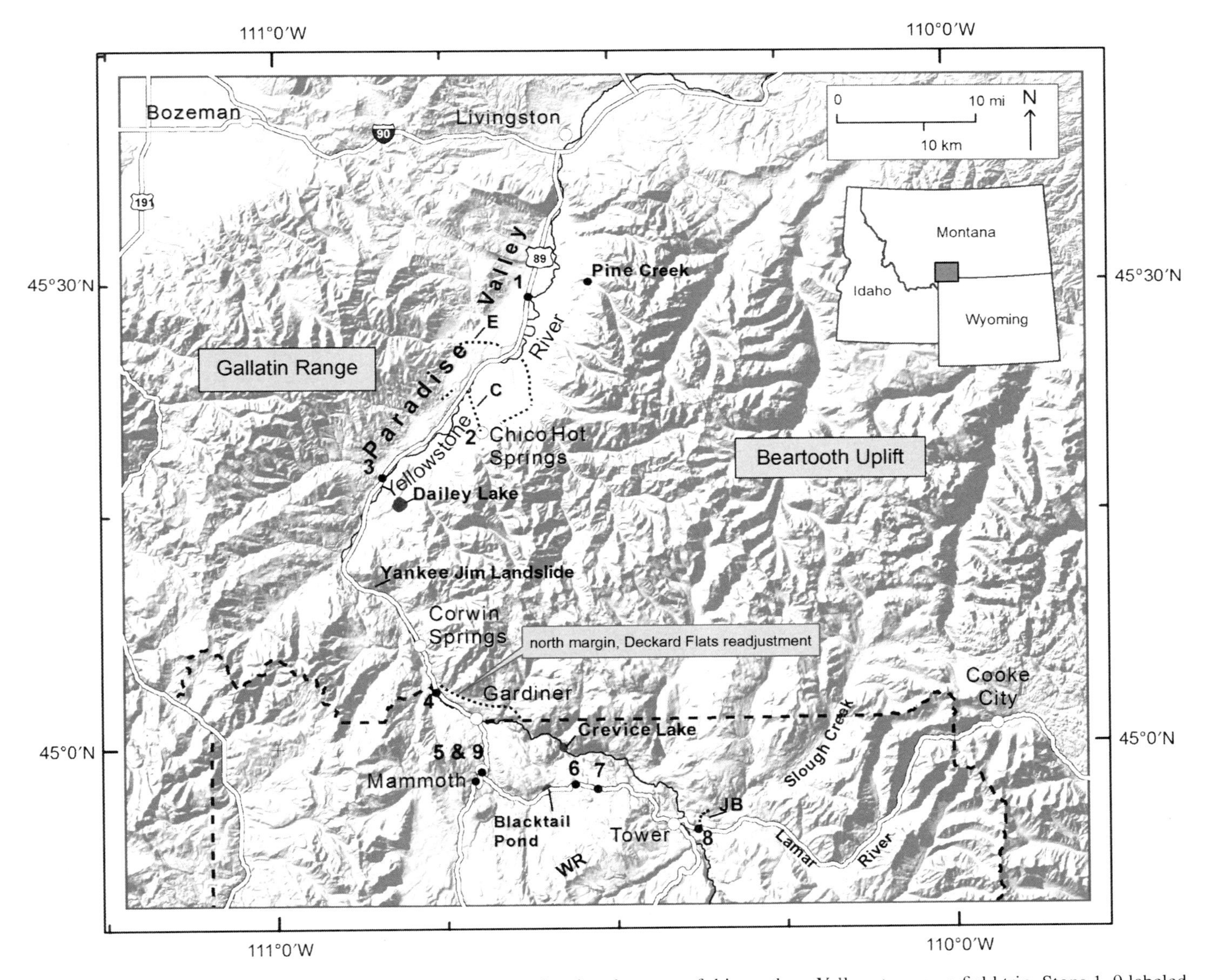

Figure 1. Shaded relief map of the northern Yellowstone area showing the route of this northern Yellowstone area field trip. Stops 1–9 labeled. E—Eightmile moraines; C—Chico moraines; JB—Junction Butte moraines. Black dashed line is northern boundary of Yellowstone National Park. WR—Washburn Range which extends south of this map area.

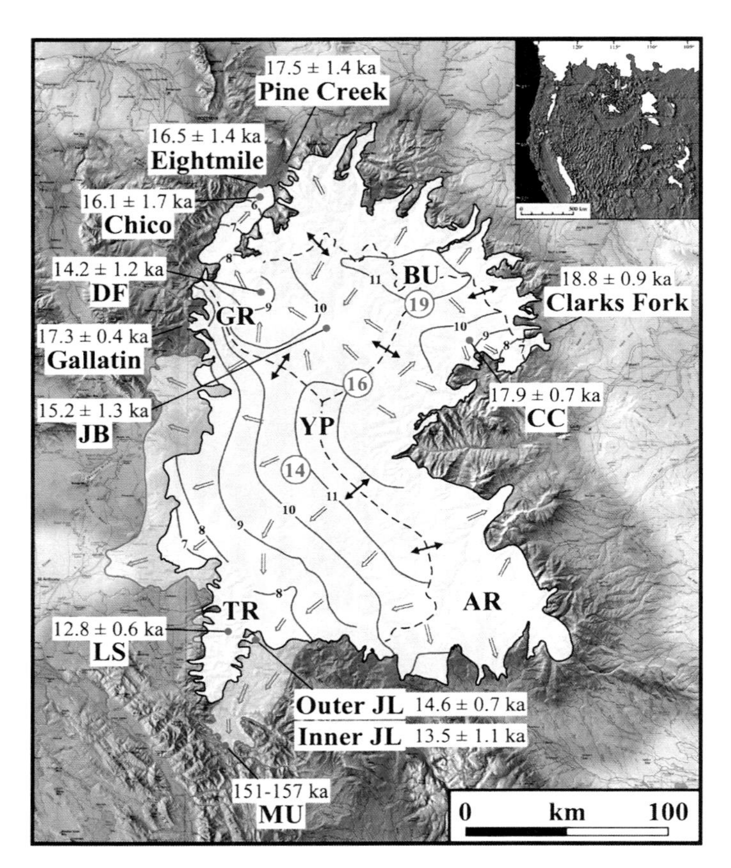

Figure 2. The greater Yellowstone glacial system. Light shaded area bounded by black line indicates the area covered during the Pinedale glaciation. Blue lines are contours in thousands of feet on reconstructed Pinedale glacier surface. Heavy black lines with double-pointed arrows indicate main ice divides. Bull Lake glaciation beyond the Pinedale glaciation is bounded by red line. Green dots are locations of cosmogenic ages in thousands of years. Green circles with numbers 19, 16, and 14 diagrammatically show migration of center of mass to southwest through time (in thousands of years). AR—Absaroka Range; BU—Beartooth uplift; CC—Crandall Creek; DF—Deckard Flats; GR—Gallatin Range; JB—Junction Butte; JL—Jenny Lake; LS—Lake Solitude; MU—Munger glaciation in southern Jackson Hole that correlates with Bull Lake glaciation; TR—Teton Range; YP—Yellowstone Plateau. After Licciardi and Pierce (2008).

The basal shear stress of the reconstructed northern Yellowstone glacial system yields values between 0.5 and 1.5 bars (Fig. 3), suggesting the reconstruction is reasonable (Pierce, 1979). An alternate reconstruction (Richmond, 1969, 1986; U.S. Geological Survey 1972a) yielded values of basal sheer stress that are unreasonably low—0.05–0.15 and 0.0–0.1 bars (Pierce, 1979).

Mass balance for the northern Yellowstone glacier was also estimated (Fig. 4). The equilibrium-line altitude was inferred to be the same as the glaciation limit (2835 m; 9300 ft). The area-altitude distribution of the reconstructed northern Yellowstone outlet glacier, with an equilibrium line altitude of 9300 ft, yielded an accumulation area ratio of 0.75. Three specific net balance estimates (Fig. 4) were based on modern continental glaciers that are thought to be similar to the Pleistocene northern Yellowstone outlet glacier and its accumulation area (Pierce, 1979). The "best" estimate yielded accumulation of 2.8 km^3/yr and ablation of 3.3 km^3/yr, which comes reasonably close to balancing at a value of ~3 km^3/yr.

Basal sliding is important, but difficult to quantify. For a cross section of the lower part of the northern Yellowstone outlet glacier, discharge due to basal sliding was estimated by subtracting flow within the glacier from total discharge (Pierce, 1979). The total discharge was estimated based on mass balance (similar to that shown in Fig. 4). The discharge based on flow within the glacier was calculated based on flow-law equations (Patterson, 1969), and this amount was subtracted from the total discharge. This remaining discharge suggests that discharge was dominated (~78%?) by basal sliding (Pierce, 1979, model C).

PALEOECOLOGY AND PALEO-FIRE OVERVIEW

Paleoecologic research has addressed a variety of topics in the Greater Yellowstone region, including the influence of park management policies on watershed stability in the last century (Engstrom et al., 1991), the vegetation history (Barnosky, 1984; Whitlock, 1993; Whitlock and Bartlein, 1993; Huerta et al., 2009; Mumma et al., 2012; Krause and Whitlock, 2013), and the fire history (Millspaugh and Whitlock, 1995; Millspaugh et al., 2000, 2004; Huerta et al., 2009; Whitlock et al., 2012). The diatom record from Yellowstone Lake offers an important example of biotic evolution on Quaternary time scales (Theriot et al., 2006). Studies of stable isotopes of oxygen and carbon have been undertaken at low resolution in some sites, and water chemistry data are available (National Park Service, 1994; Theriot et al., 1997; Dean, 2006). Other research in the park includes analyses of late Holocene vertebrate assemblages (Hadly, 1996, 1999), tree-ring records of fire and climate (Littell, 2002), past beaver activity (Persico and Meyer, 2009), testae amoebae in wetlands (Booth et al., 2003), and fossil beetles (Elias, 1997).

During the last glacial maximum, nonglaciated valleys and exposed ridges at the margins of the Yellowstone ice field supported tundra communities and isolated populations of Engelmann spruce, subalpine fir, whitebark pine, lodgepole pine, and possibly Douglas fir (Gugger and Sugita, 2010; Mumma et al., 2012). Warming at 17,000 cal. yr B.P. is evidenced by rapid ice recession as well as by a sequence of plant migrations into newly deglaciated regions. In most areas, tundra communities with birch, aspen, willow, and juniper developed initially, and this vegetation was replaced by subalpine parkland communities of Engelmann spruce, whitebark pine, and subalpine fir by ~13,000 cal. yr B.P. (Whitlock, 1993; Huerta et al., 2009; Mumma et al., 2012; Krause and Whitlock, 2013). In central Yellowstone, nonforested communities persisted until ~11,000 cal. yr B.P. and transitioned directly to closed lodgepole pine forest as a result of nutrient-poor rhyolitic substrates (Whitlock, 1993). The Younger Dryas cold interval (12,900–11,500 cal. yr B.P.) is not evident as a distinctive vegetation reversal in the Yellowstone region;

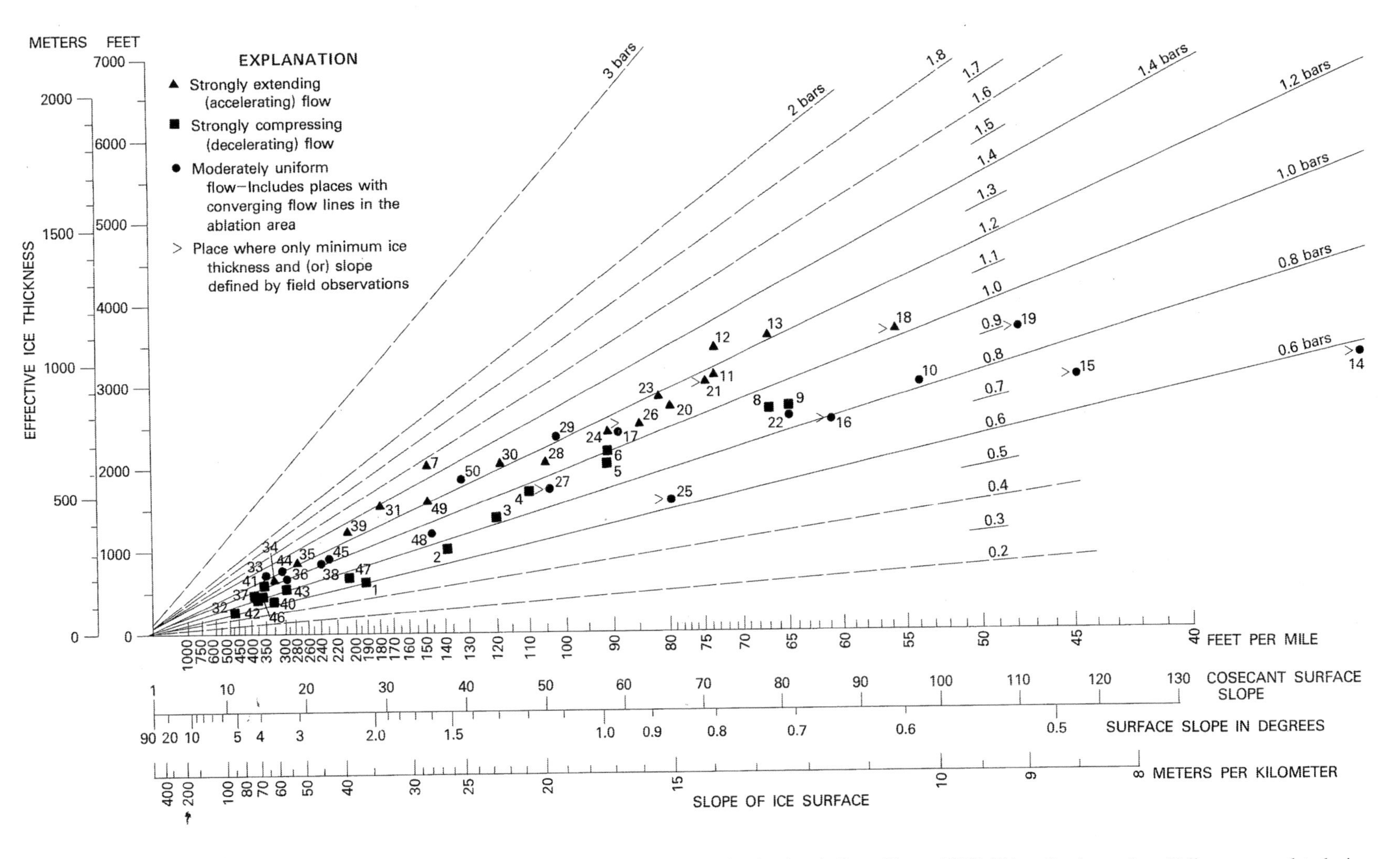

Figure 3. Basal shear stress for the reconstructed northern Yellowstone outlet glacier and ice feeding into it (from Pierce, 1979). Values for the northern Yellowstone outlet glacier and its source areas are within the 0.5–1.5 bar value for the normal range of glaciers.

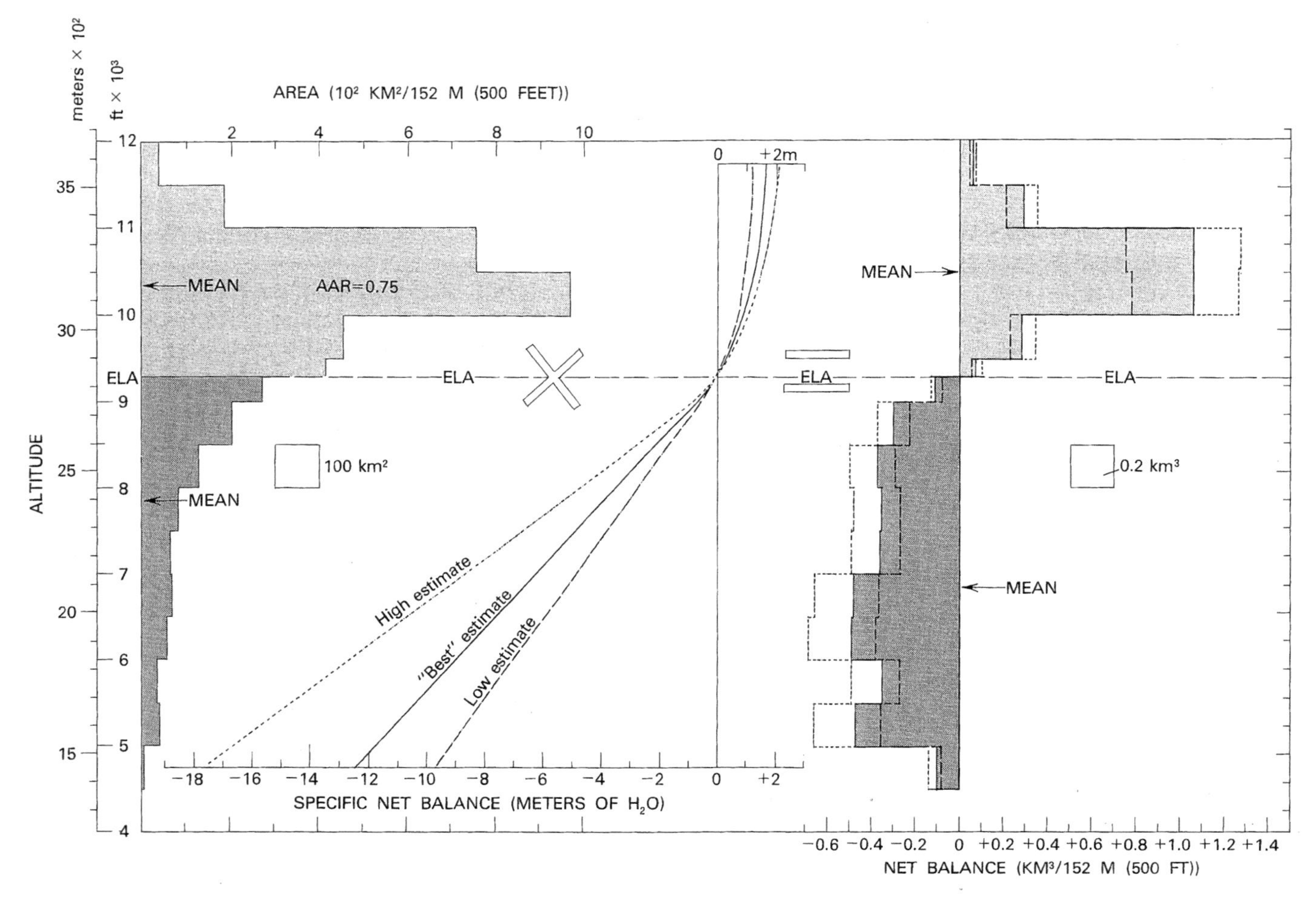

Figure 4. Reconstructed mass balance of the northern Yellowstone outlet glacier and the ice feeding into it (from Pierce, 1979). AAR—accumulation area ratio; ELA—equilibrium line altitude.

however, a slowing of reforestation at mid-elevations (~2000 m; 6500 ft) suggests a prolonged cool period following deglaciation and lasting until 11,500 cal. yr B.P. (Krause and Whitlock, 2013). Between 12,000 and 11,000 cal. yr B.P., closed subalpine forests replaced parkland vegetation at most sites, and fire activity increased significantly (Whitlock, 1993; Huerta et al., 2009; Krause and Whitlock, 2013). The early Holocene (~10,000–7000 cal. yr B.P.) featured warmer-than-present summer conditions, but levels of effective moisture varied from north to south and with elevation (Whitlock and Bartlein, 1993). In summer-wet (high summer precipitation relative to annual precipitation) areas of the northern Yellowstone region, growing season moisture was sufficient to support closed pine and pine and/or juniper forests and fire activity was relatively low (Millspaugh et al., 2004; Huerta et al., 2009; Whitlock et al., 2012; Krause and Whitlock, 2013). In contrast, summer-dry (low summer precipitation relative to annual precipitation) areas in central and southern Yellowstone National Park and Grand Teton National Park featured high fire activity and open forests of Douglas fir and steppe communities at low to mid-elevations and lodgepole pine forest at high elevations (Whitlock, 1993). Geographic differences in effective moisture were less pronounced in the middle and late Holocene. In northern Yellowstone, present-day steppe and montane conifer forests of lodgepole pine and Douglas fir established in response to drying conditions (Millspaugh et al., 2004; Huerta et al., 2009), while in central and southern Yellowstone, mixed forests of Engelmann spruce, lodgepole pine, and subalpine fir developed as conditions became wetter. Lodgepole pine forest persisted on nutrient-poor rhyolitic substrates in central Yellowstone National Park throughout the Holocene, although the charcoal data suggest high fire activity consistent with drier-than-present summer conditions followed by decreased fires after 6000 cal. yr B.P. (Whitlock, 1993; Millspaugh et al., 2000).

DESCRIPTION OF STOPS

From Bozeman, Montana, drive east on I-90 to Livingston, where you will turn south on U.S. 89 and drive ~20 km (12.4 miles) to Mallards Rest Fishing Access. Park on bench at the same level as the highway before the road goes down to the river.

Please note that latitudes and longitudes for all stops are given in the WGS 84 system.

Stop 1. Mallards Rest Fishing Access
(45.4860° N, 110.6218° W)

This stop is on Pinedale outwash that heads into the Eightmile moraines of the northern Yellowstone outlet glacier (Fig. 5). The branching of the paleo-outwash channels is remarkably well preserved, as shown on aerial photographs (Fig. 5). This outwash fan is 60 m (200 ft) above the current Yellowstone River at the Eightmile terminal moraines, 18 m (60 ft) above the river at this stop, and only 3–6 m (10–15 ft) above the present river on this side of the Livingston canyon. The outwash terrace had a steeper gradient than the present Yellowstone River, thus explaining the convergence downstream of 60 m (200 ft) in 18 km (11 miles).

On the east side of the valley is the Beartooth uplift that supported local valley glaciers (Montagne and Locke, 1989). The Pinedale age Pine Creek moraines have a cosmogenic age of 17.5 ± 1.4 ^{10}Be ka (Licciardi and Pierce, 2008; Huss et al., 2012), whereas Bull Lake moraines are probably in the ~140 ka age range.

Near the base of the range, at sites such as Barney Creek straight across the valley from here, is the trace of the Emigrant active fault with scarps 5–8 m (15–25 ft) high and 3–6 m (10–20 ft) post-glacial offset (Personius, 1982). This may result from a single, very large surface faulting event, or 2–3 events very closely spaced in time (Personius, 1982). This fault is on the leading edge of the Yellowstone hotspot (Pierce and Morgan, 1992).

Turn left (south) back onto U.S. 89; after 7 km (4.3 miles) turn left (east) onto Mill Creek Road and after crossing the Yellowstone River, turn right (southwest) onto old U.S. highway 89. The road then climbs up onto outwash of Chico age (ca. 16.1 ka). Note the deep kettle on the right where an ice block of Eightmile age (ca. 16.5 ka) was buried by outwash gravel of Chico age. Turn left (southeast) onto road to Chico Hot Springs and note Chico moraines that form the high ridges to the right (northeast). Just before Chico Hot Springs, turn right (north) and climb up onto the Chico outwash terrace, and then onto the Chico moraine complex.

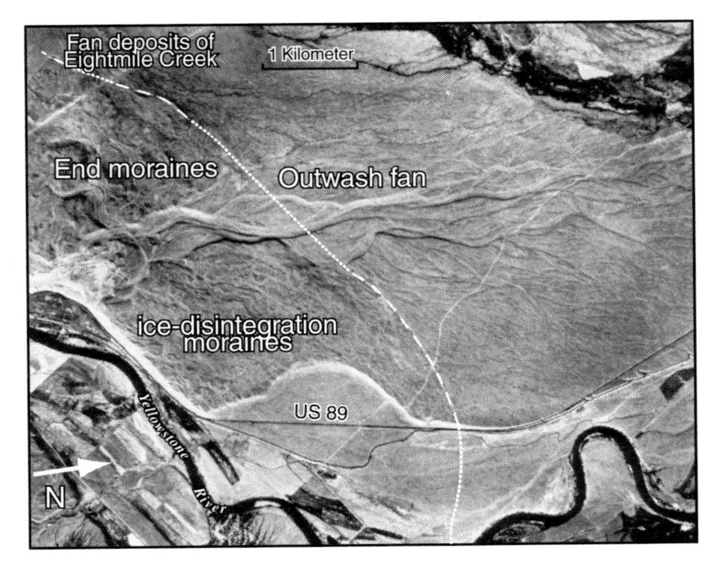

Figure 5. Pinedale end moraines and outwash fan of the northern Yellowstone outlet glacier, showing preservation of detailed braided channel pattern (from Pierce, 1979; photograph by U.S. Geological Survey taken in 1949).

Stop 2. Chico Moraines and Chico Outwash
(45.3402° N, 110.6967° W)

The Chico moraines (Figs. 1, 2) of the northern Yellowstone outlet glacier are bounded on the north by the Chico melt-water channel. Note the succession of outwash terraces that line the melt-water channel now occupied by the Chico road. Cosmogenic ages on boulders on the Chico moraines average 16.1 ± 1.7 ^{10}Be ka. Eightmile moraines average 16.5 ± 1.4 ^{10}Be ka (Fig. 2) and 16.6 ± 1.3 ^{3}He ka (helium cosmogenic age for olivine, see Licciardi and Pierce, 2008). It is problematic to note that these cosmogenic ages are younger than those based on radiocarbon dating of deglaciation in Yellowstone (Licciardi et al., 2001; Pierce et al., 2003).

The Eightmile moraines (Fig. 2) are well expressed just to the north on the other side of the Chico melt-water channel, and from there, they arc around to their terminus 7 km (4.5 miles) beyond the Chico moraines. Across Paradise Valley, ice-marginal channels 50 m (150 ft) deep were eroded into bedrock at the Eightmile and Chico ice margins. Walter Weed (1893) observed these channels and noted their close relation to glaciation, but questioned if there was enough time for them to have been cut by ice-marginal streams.

To fit the Bull Lake–Pinedale paradigm, Horberg (1940) thought irregular stony deposits in the inner valley of the Yellowstone River demonstrated a Pinedale age for the Chico moraines in contrast to the Eightmile moraines on the terrace seen at Stop 1, which he considered to be Bull Lake in age. But this inner valley deposit is flood gravel rather than a morainal deposit because of the northeast alignment of excessively drained ridges and the gravelly, but not boulder-studded, highs. In contrast to normal, flat-topped terraces, the surface of these flood deposits has spoon-shaped forms; the lows are concave-up spoon forms and the highs are convex-up spoon forms. Montagne (Montagne and Locke, 1989) inferred a Pinedale age for both advances based on the kettled Chico outwash; that interpretation is strengthened by the cosmogenic dating results.

Return to U.S. 89 at community of Emigrant, turn left (southeast), and drive 11.4 km (7 miles) to highway rest stop on southeast side of road.

Stop 3. Highway Rest Stop and Dailey Lake
(45.2949° N, 110.8347° W)

Across the Yellowstone River is the basalt of Hepburns Mesa, the age of which is 2.2 Ma (Smith et al., 1995), consistent

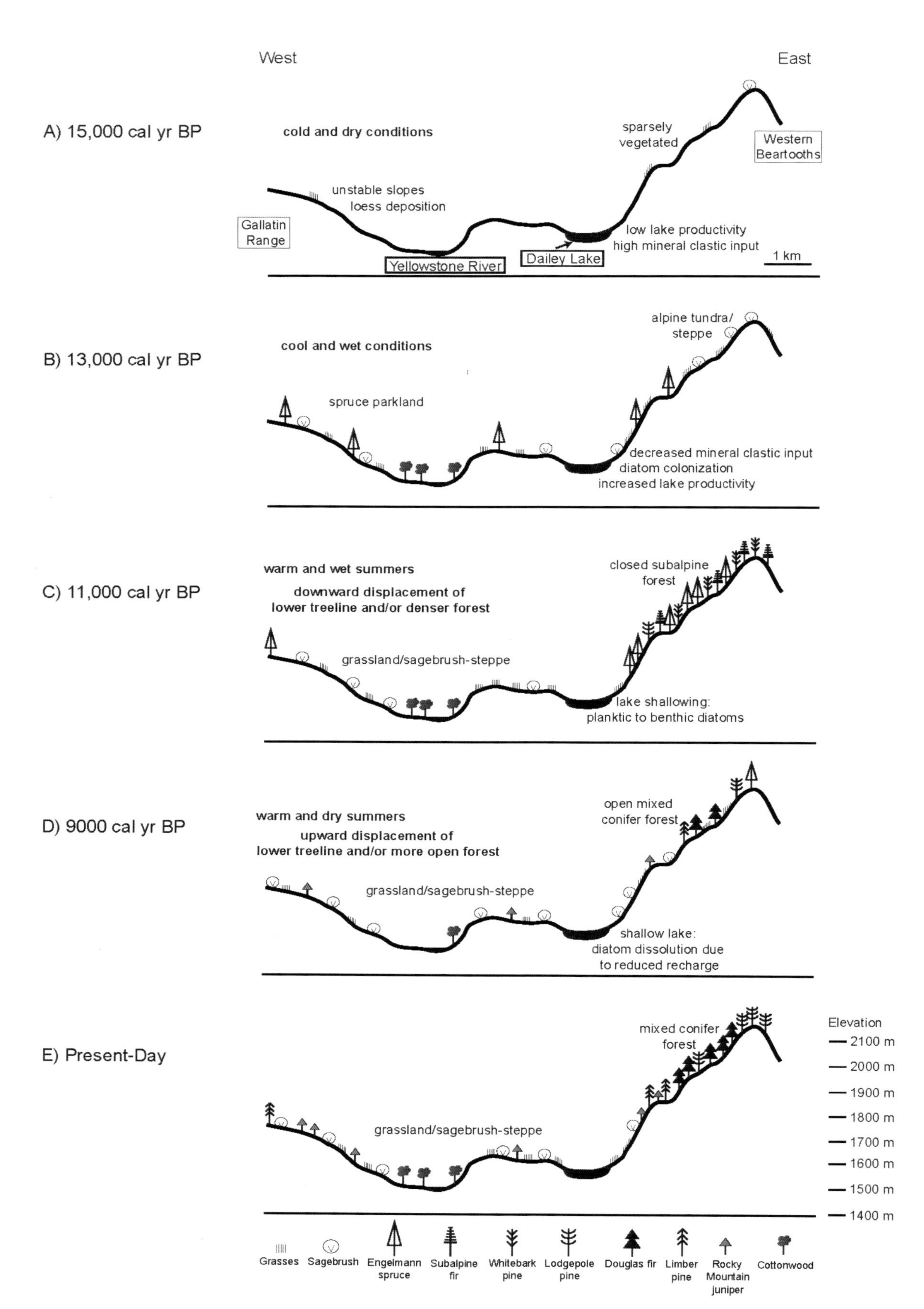

Figure 6. Schematic reconstruction of vegetation and limnologic changes at Dailey Lake during the late-glacial/early Holocene transition. The ages given here are within the pollen-zone interval, and are not the start of that interval.

with the start of the Yellowstone volcanic field. Olivine phenocrysts from this basalt have $^3He/^4He$ ratios >21 R_A (Mark Kurz and Joe Licciardi, 2007, written commun.). Ratios above 8 R_A (eight times the ratio in the atmosphere) are commonly thought to indicate a deep mantle source, and thus support a deep mantle source for the Yellowstone hotspot. The surface of this basalt is striated and polished by the Pinedale glaciation. Locally, this basalt forms cliffs that were used by Native Americans as buffalo jumps to kill bison. Beneath the basalt are white tuffaceous beds that dip 8–10 degrees into the Emigrant fault and are Barstovian (ca. 15 Ma) in age (Burbank and Barnosky, 1990).

Dailey Lake is on the southeast side of the basalt bench just south of this stop (Fig. 1) and occupies a shallow trench carved by the northern Yellowstone outlet glacier. Because of its proximity to the Eightmile and Chico moraines, the lake-sediment record from Dailey Lake provides one of the longest reconstructions of postglacial environmental change in the Yellowstone region. Pollen, charcoal, diatom, geochemical, and lithologic data were used to reconstruct postglacial terrestrial and limnologic development from ice retreat ca. 16,000 cal. yr B.P. to the early Holocene insolation maximum (Fig. 6). Following glacial retreat, the slopes surrounding Dailey Lake were sparsely vegetated and supported pioneering shrubs and forbs. Slopes were unstable and loess deposition occurred at this time. As summer conditions warmed and slopes stabilized, spruce parkland was established at 13,400 cal. yr B.P., and the near synchronous colonization of diatoms marks the onset of productive conditions within the lake itself. Planktic diatom assemblages beginning at 13,100 cal. yr B.P. indicate rapid warming, and this transition was followed by the development of a closed subalpine forest upslope of Dailey Lake at 12,200 cal. yr B.P. With continued sedimentation and warming in the early Holocene, the lake shallowed significantly, leading to a loss of diatom preservation and the establishment of an open mixed conifer forest of Douglas fir, lodgepole pine, and Engelmann spruce.

Continue south on U.S. 89; ~3 km (2 miles) after crossing the Yellowstone River, note the scarp of the Emigrant fault that offsets alluvial fan on the left (Personius, 1982; Ruleman, 2002). The road is on sub-angular flood boulders from catastrophic release of a lake dammed by an early postglacial landslide in Yankee Jim canyon (Fig. 1; Good, 1964). This lake is dated by tan lake silts near Corwin Springs that contain charcoal with an age of 12,492 ± 168 cal. yr B.P. (10,531 ± 53 ^{14}C years, WW5486). Continue on U.S. 89 to Corwin Springs, turn right (west) and cross Yellowstone River on new bridge, and turn left (south) onto gravel road. Enter Yellowstone National Park when crossing Reese Creek and continue 1.3 km (0.8 miles) to where mid-channel flood bar deposits with giant ripples are on left.

Stop 4. Giant Ripples
(45.0551° N, 110.7659° W; WGS84)

This mid-channel flood bar (Fig. 7) has "giant ripples" spaced ~15 m (50 ft) apart and up to 2 m (6 ft) high with boulders up to 1.5 m (5 ft) in diameter. Cosmogenic ages on the boulders average 13.4 ± 1.2 ^{10}Be ka. The flood deposits are in the position where moraines of the Deckard Flats "readjustment" (DF, 14.2 ± 1.2 ka, Fig. 2) had originally been deposited here on the valley floor and are still well represented on the 180-m (600-ft)-high bench on the east side of the valley. The mid-channel flood bar and the bouldery deposits on the other side of the road consist of reworking of these moraines into flood deposits.

The alluvial fan of Reese Creek (located along the park boundary 1.3 km [0.8 miles] to the northwest of here; Fig. 1) provides evidence of at least two floods 45–60 m (150–200 ft) deep. First, an alluvial fan was deposited and then a flood down the Yellowstone River Valley eroded the fan front and left a "flood-ripped" fan front. The younger fan was then built and, later, another flood undercut and eroded the fan front (Pierce, 1979). These floods were probably from release of glacially dammed lakes upstream, most likely from a lake dammed in the Lamar Valley by the Slough Creek glacier.

The type area of the Deckard Flats moraine complex is 4.9 km (3 miles) east of Gardiner (Fig. 1). Large boulders in the moraines of the Deckard Flats "readjustment" date to 14.2 ± 1.2 ^{10}Be ka (DF on Fig. 2). The Deckard Flats position is termed a "readjustment" rather than an advance because it represents the time when the ice cap on the Yellowstone Plateau no longer contributed to the northern Yellowstone outlet glacier, and the glaciers from adjacent mountains (Beartooth uplift, Absaroka Range, and Gallatin Range) readjusted to this loss and established a stable ice margin that can be traced over much of the northern Yellowstone area. In full-glacial time, the outlet glacier was ~1070 m (3500 ft) thick at Gardiner and terminated ~65 km (40 miles) down valley, whereas during the Deckard Flats it was ~300 m (1000 ft) thick at Gardiner and terminated 7 km (4 miles) down valley.

At this position, there is probably a scour basin that extends as much as 150 m (500 ft) below the Yellowstone River (Fig. 8; Pierce et al., 1991). From this viewpoint, we can see how the erosion of the Yellowstone Valley can be dated by volcanic units. A set of basalt flows dating from the start of Yellowstone volcanism ca. 2.1 Ma and resting on Yellowstone River gravel can be seen to the north ~370 m (1200 ft) above the Yellowstone River (Fig. 8). Lower on the valley wall, the 0.6 Ma Undine Falls Basalt (Quf) also rests on Yellowstone River gravels here ~150 m (500 ft) above the Yellowstone River. This basalt underlies the travertine bench and Deckard Flats bench (Christiansen, 2001). These basalts, as well as rhyolite tuffs, date the incision of the Yellowstone Valley (Fig. 8).

Continue on the gravel road to Gardiner, enter the park through the Roosevelt Arch, and continue up to Mammoth.

Stop 5. Mammoth
(44.9760° N, 110.6999° W)

Rest and lunch stop. The conical hills are thermal kames. Across the road from the visitor center are sinkholes in the postglacial travertine.

Figure 7. Mid-channel flood bar with giant ripples. Photograph by John Shelton.

From Mammoth, drive 12.5 km (7.8 miles) east toward Tower Junction and turn right onto paved road and park in paved parking area. Take the trail south past the large glacial erratic named "Frog Rock" to the moraine crest overlooking Blacktail Deer Plateau.

Stop 6. Blacktail Deer Plateau
(44.9577° N, 110.5652° W)

The Blacktail Deer Plateau extends from Stop 6 (Fig. 1) to the southeast. The plateau is underlain by moraines of Deckard Flats age (14.2 ± 1.2 ^{10}Be ka) that extend ~5 km (3 miles) south and west from this viewpoint. The successively lower morainal benches, locally including kame gravel, define about 10 recessional ice margins. Eolian deposition of soil and snow makes for a loamier and moister soil on the northeast sides of ridges, whereas the south sides are drier due to greater solar insolation, wind exposure, and erosion of fine soil.

One way to appreciate the influence of geologic processes on the vegetation is to note that the Deckard Flats moraines form extensive sagebrush-grasslands. If glaciation had not occurred, rhyolite would have been at the surface and the region would have probably been covered with lodgepole pine forest. If rhyolite had not been emplaced, then andesite such as on the Washburn Range to the south would have been at the surface and form a mosaic of meadows and spruce-fir forests. These substrate differences have exerted long-term influences on the vegetation that are evident in the paleoecological record within the region (Whitlock, 1993).

Blacktail Pond lies within a pitted (kettled) glacial outwash terrace that becomes a meltwater channel alongside the present road from Blacktail Pond west for 2.5 km (1.5 miles) to Lava Creek (Pierce, 1979). The pond has been the site of many paleoecological studies (Gennett and Baker, 1986; Huerta et al., 2009; Krause and Whitlock, 2013). A multiproxy study of pollen, charcoal, geochemical, and stable isotope data clarifies the ecological changes during the late-glacial/early Holocene transition (Fig. 9; Krause and Whitlock, 2013). Prior to 11,500 cal. yr B.P., cool conditions supported alpine tundra and spruce parkland. Carbonate δ^{18}O data indicate a step-like change to warm summer conditions at 11,500 cal. yr B.P., and this shift facilitated a transition from spruce parkland to closed lodgepole pine forest and increased fire activity. After 8200 cal. yr B.P., Douglas fir populations expanded and the forest became more open. The dominance of lodgepole pine and sagebrush pollen in the middle Holocene suggests that conifers grew on the upland rocky areas, and sagebrush-steppe was present in the valley. Further decline of spruce and birch in the last 4000 years implies that drying continued through the late Holocene and fire-episode frequency remained relatively high until 2000 cal. yr B.P. (Huerta et al., 2009).

Blacktail Deer Creek descends into the Black Canyon of the Yellowstone River and joins the river near Crevice Lake, which lies ~300 m (1000 ft) lower in elevation than Blacktail Pond. The limited surface area, conical bathymetry, and deep water (31 m [100 ft]) of Crevice Lake create oxygen-deficient conditions in the hypolimnion and preserve annually laminated sediment (varves) for much of the record. Cores from Crevice Lake were analyzed for pollen, geochemistry, mineralogy, diatoms, and stable isotopes to gain a better understanding of the Holocene history of seasonal climate variation (Fig. 10; Whitlock et al., 2008, 2012). The proxy data suggest wet winters, protracted springs, and warm, effectively wet summers in the early Holocene and less snowpack, cool springs, and warm, dry summers in the middle Holocene. The shift from effectively wet summer conditions in the early Holocene to progressively drier conditions in the late Holocene reflects changes in both winter and summer precipitation. A high-resolution reconstruction of late Holocene conditions registers multi-decadal changes in winter, spring, and summer conditions in the last 2650 years (Whitlock et al., 2008). Particularly short springs, dry summers, and dry winters occurred during the Roman Warm Period (~2000 cal. yr B.P.) and Medieval Climate Anomaly (1200–800 cal. yr B.P.), and present-day conditions of long springs and mild summers were established during the Little Ice Age (~1300–1850 AD).

Return to main road and turn right and proceed 3.2 km (2 miles) to Phantom Lake parking area.

Stop 7. Phantom Lake Ice-Marginal Channel
(44.9554° N, 110.5289° W)

This ice-marginal channel, as deep as 60 m (200 ft) and more than 1.9 km (1.2 miles) long, was cut into volcanic bedrock of the Absaroka Supergroup during Pinedale glacial recession. A large portion of the paleo–Yellowstone River flowed alongside the recessional glacier at a position ~360 m (1200 ft) above the Yellowstone River. It is one of several such ice-marginal channels formed alongside the retreating Pinedale glacier. Phantom Lake

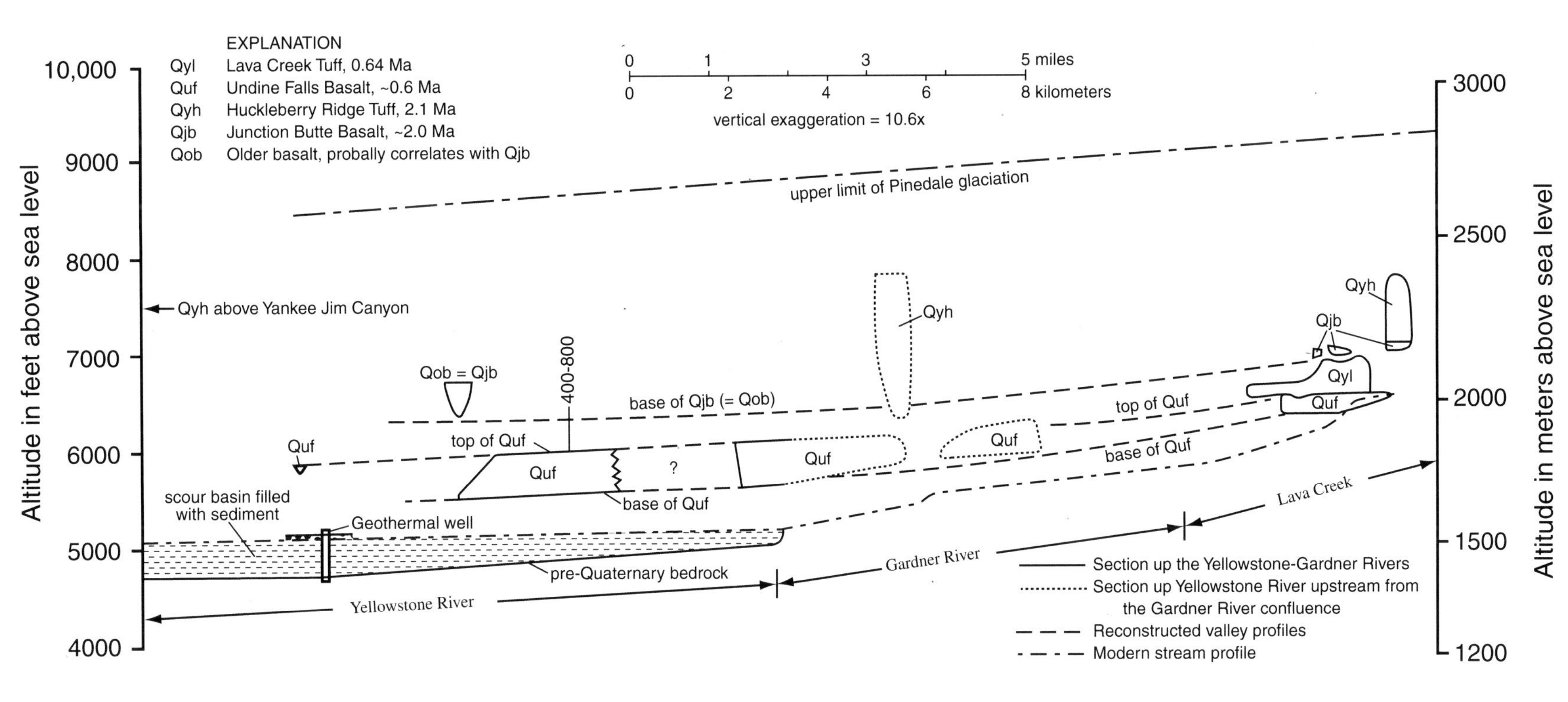

Figure 8. Valley deepening in Gardiner-Mammoth area based on incision of dated volcanic units (from Pierce et al., 1991). K-Ar ages are from Obradovich (1992).

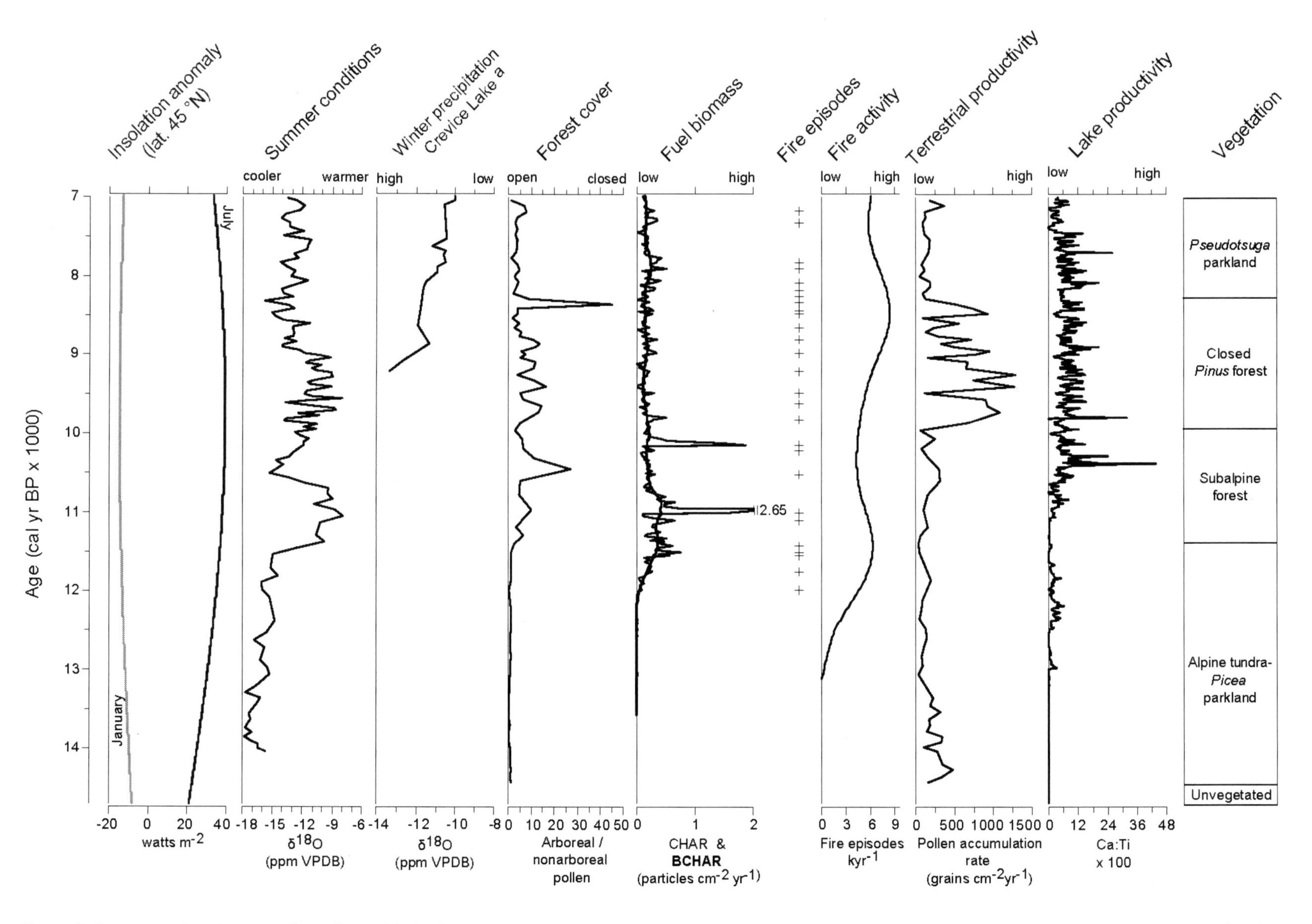

Figure 9. Summary of environmental proxies at Blacktail Pond during the late-glacial/early Holocene transition plotted against January and July insolation anomalies (from Krause and Whitlock, 2013). [a]Crevice Lake plot after Whitlock et al. (2012). CHAR—charcoal accumulation rate; BCHAR—background charcoal accumulation rate; VPDB—Vienna Pee Dee Belemnite.

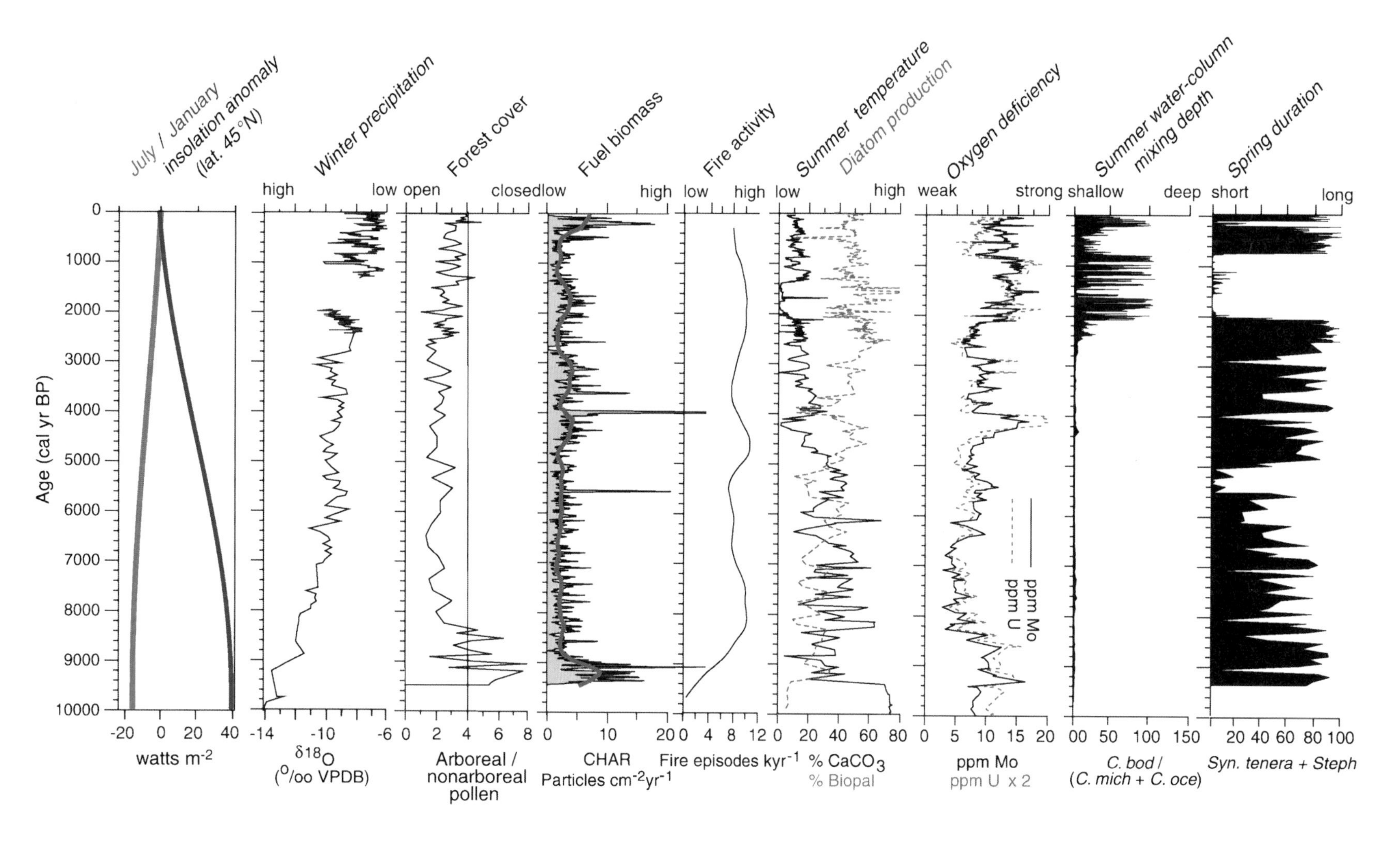

Figure 10. Summary of environmental proxies at Crevice Lake over the last 9400 cal. yr B.P. plotted against January and July insolation anomalies (from Whitlock et al., 2012). CHAR—charcoal accumulation rate; Biopal—biogenic opal; *C. bod—cyclotella bodanica*; *C. mich—Cyclotella michiganiana*; *C. oce—Cyclotella ocellata*; *Syn. tenera—Synedera tenera*; *Steph—Stephanodiscus*; VPDB—Vienna Pee Dee Belemnite.

resides in the low-gradient channel and is dammed on its downstream margin by a small alluvial fan built in postglacial time.

Continue east to Tower Junction, then turn left (north) on the Northeast Entrance Road. From bridge across Yellowstone River continue 2 km (1.3 miles) to the large parking area on the left side of the road just west of Junction Butte moraines (JB on Fig. 1). Walk 150 m (500 ft) east to the edge of intermittent pond in glacial moraines with very large boulders.

Stop 8. Junction Butte Moraines (Turnaround for Field Trip)

(44.9128° N, 110.3854° W)

In the post–Deckard Flats glacial recession, the Junction Butte moraines are the most prominent and occupy a position typically referred to as late Pinedale in the Rocky Mountains (Figs. 1, 2). Large boulders of Precambrian crystalline rocks stud their surface, and permanent and seasonal ponds occupy common depressions. The Junction Butte moraines date 15.2 ± 1.3 ^{10}Be ka, a little too old based on the stratigraphic sequence. A large glacier flowing down Slough Creek deposited these moraines and blocked the Lamar Valley upstream from Slough Creek. The till has enough fine-grained sediment for good water-holding capacity and adequate nutrients for good plant growth. The mineral soil is well covered by the vegetation. The water level in the ponds has diminished since the 1970s as a result of prolonged winter drought. Paleoecological data suggest that droughts of this magnitude or greater characterized the Medieval Climate Anomaly, leading to more fires and reduced beaver activity (Millspaugh et al., 2004; Persico and Meyer, 2009).

Turn around and retrace road back to Mammoth. In Mammoth if time allows, drive or walk to viewpoint up gravel road that starts directly behind and north of Mammoth Hotel (44.9774° N, 110.7014° W).

Stop 9 (Optional). Mammoth Overview

(44.9793° N, 110.6995° W)

The bench north of the Mammoth Hotel is the Deckard Flats margin of glaciers originating to the southwest in the Gallatin Range. The bench is kame gravel and includes boulders of travertine. The conical hills to the south and east are inferred to be thermal kames formed where hot springs melted cavities in the ice that filled with kame gravel. To the north above the Deckard Flats bench is a grassy, rounded hill of sub-glacial till also bearing erratics of travertine. North in the Yellowstone River Valley near Gardiner, the full-glacial northern Yellowstone outlet glacier was ~1070 m (3500 ft) thick and created glacial streaming from northward flow. Under full-glacial conditions, ice from three points of the compass converged in this area to form the northern Yellowstone outlet glacier. Mt. Everts to the east exposes readily erodible, Cretaceous shale and sandstone on its west face. The southern part of Mt. Everts is capped by 2.1 Ma Huckleberry Ridge Tuff, with local, small valley fills of ca. 640 ka Lava Creek Tuff. The upland on top of Mt. Everts is glacially scoured with ridges and lakes, forming a diverse wildlife habitat of lush meadows bordered by bands of forest.

Hydrothermal activity at the Mammoth terraces is always changing. This is the sequence of travertine deposits in the Mammoth area: (1) Terrace Mountain, the highest travertine near Mammoth Hot Springs, rests on 2.1 Ma Huckleberry Ridge Tuff and has a U–Th age of 406 ± 30 ka (Pierce et al., 1991). Pinedale glacial erratics rest on this travertine, which was scoured by northward-flowing ice that reached an altitude of almost 2750 m (9000 ft) on Sepulcher Mountain 5.5 km (3.4 miles) to the northwest. (2) About ~2.5 km (1.5 miles) southwest of here and above the active terraces is the inactive and forested Pinyon terrace, which is postglacial and has U–Th ages near 10 ka (Pierce et al., 1991). (3) The lowest travertine terraces are postglacial. A deep research hole was drilled in the 1960s in the main terrace and a U–Th age of 7.72 ± 0.88 ka was obtained on travertine from a depth of 72.8 m (239 ft).

If you have driven the road up to here, you must continue on this old entrance road down to Gardiner. If you walked up, you can walk the road back down to Mammoth.

REFERENCES CITED

Barnosky, C.W., 1984, Late Miocene vegetational and climatic variations inferred from a pollen record in northwest Wyoming: Science, v. 223, p. 49–51, doi:10.1126/science.223.4631.49.

Booth, R.K., Zygmunt, J.R., and Jackson, S.T., 2003, Testate amoebae as Paleoclimatic Proxies in Rocky Mountain Peatlands: A Case Study in the Greater Yellowstone Ecosystem: University of Wyoming–National Park Service 28th Annual Report, p. 85–96.

Burbank, D.W., and Barnosky, A.D., 1990, The magnetochronology of Barstovian mammals in southwestern Montana and implications for the initiation of Neogene crustal extension in the northern Rocky Mountains: Geological Society of America Bulletin, v. 102, p. 1093–1104, doi:10.1130/0016-7606(1990)102<1093:TMOBMI>2.3.CO;2.

Christiansen, R.L., 2001, The Quaternary and Pliocene Yellowstone Plateau Volcanic Field of Wyoming, Idaho, and Montana: U.S. Geological Survey Professional Paper 729-G, 145 p.

Dean, W.E., 2006, Characterization of Organic Matter in Lake Sediments from Minnesota and Yellowstone National Park: U.S. Geological Survey Open-File Report 2006-1053, 40 p.

Elias, S.A., 1997, Reconstructing Yellowstone's climate history: Yellowstone Science, v. 5, p. 9–16.

Engstrom, D.R., Whitlock, C., and Fritz, S.C., 1991, Recent environmental changes inferred from the sediments of small lakes in Yellowstone's northern range: Journal of Paleolimnology, v. 5, p. 139–174, doi:10.1007/BF00176875.

Gennett, J.A., and Baker, R.G., 1986, A late Quaternary pollen sequence from Blacktail Pond, Yellowstone National Park, Wyoming, U.S.A.: Palynology, v. 1, p. 61–71, doi:10.1080/01916122.1986.9989303.

Good, J.D., 1964, Prehistoric landslide in Yankee Jim Canyon, Park County, Montana [abs.]: Geological Society of America Special Paper 82, p. 327–328.

Good, J.D., and Pierce, K.L., 1996, Interpreting the Landscapes of Grand Teton and Yellowstone National Parks, Recent and Ongoing Geology: Moose, Wyoming, Grand Teton National History Association, 58 p.

Gugger, P.F., and Sugita, S., 2010, Glacial populations and postglacial migration of Douglas-fir based on fossil pollen and macrofossil evidence: Quaternary Science Reviews, v. 29, p. 2052–2070, doi:10.1016/j.quascirev.2010.04.022.

Hadly, E.A., 1996, Influence of late-Holocene climate on northern Rocky Mountain mammals: Quaternary Research, v. 46, p. 298–310, doi:10.1006/qres.1996.0068.

Hadly, E.A., 1999, Fidelity of terrestrial vertebrate fossils to a modern ecosystem: Palaeogeography, Palaeoclimatology, Palaeoecology, v. 149, p. 389–409, doi:10.1016/S0031-0182(98)00214-4.

Horberg, L., 1940, Geomorphic problems and glacial geology of the Yellowstone Valley, Park County, Montana: The Journal of Geology, v. 48, p. 275–303, doi:10.1086/624886.

Huerta, M.A., Whitlock, C., and Yale, J., 2009, Holocene vegetation-fire-climate linkages in northern Yellowstone National Park, USA: Palaeogeography, Palaeoclimatology, Palaeoecology, v. 271, p. 170–181, doi:10.1016/j.palaeo.2008.10.015.

Huss, E.G., Laabs, B.J.C., Leonard, E.M., Licciardi, J.M., Plummer, M.A., and Caffee, M.W., 2012, Pace of glacial retreat and limits on paleoclimate conditions for the Pine Creek Glacier, Montana, during the Pinedale Glaciation, American Geophysical Union, 2012 Fall Meeting, abstract C53A-0823.

Krause, T.R., and Whitlock, C., 2013, Climate and vegetation change during the late-glacial/early-Holocene transition inferred from multiple proxy records from Blacktail Pond, Yellowstone National Park, USA: Quaternary Research, v. 79, p. 391–402, doi:10.1016/j.yqres.2013.01.005.

Licciardi, J.M., and Pierce, K.L., 2008, Cosmogenic exposure-age chronologies of Pinedale and Bull Lake glaciations in greater Yellowstone and the Teton Range, USA: Quaternary Science Reviews, v. 27, p. 814–831, doi:10.1016/j.quascirev.2007.12.005.

Licciardi, J.M., Clark, P.U., Brook, E.J., Pierce, K.L., Kurz, M.D., Elmore, D., and Sharma, P., 2001, Cosmogenic ^{3}He and ^{10}Be chronologies of the northern outlet glacier of the Yellowstone ice cap, Montana, USA: Quaternary Research, v. 29, no. 12, p. 1095–1098.

Littell, J.S., 2002, Determinants of fire regime variability in lower elevation forests of the northern greater Yellowstone ecosystem [M.S. thesis]: Bozeman, Montana State University, 121 p.

Locke, W.W., Clarke, W.D., Elliott, J.E., Lageson, D.R., Mokt, D.W., Montagne, J., Schmidt, J.G., and Smith, M., 1995, The middle Yellowstone valley from Livingston to Gardiner, Montana: A microcosm of northern Rocky Mountain geology: Northwest Geology, v. 24, p. 1–65.

Millspaugh, S.H., and Whitlock, C., 1995, A 750-year fire history based on lake sediment records in central Yellowstone National Park: The Holocene, v. 5, p. 283–292, doi:10.1177/095968369500500303.

Millspaugh, S.H., Whitlock, C., and Bartlein, P.J., 2000, Variations in fire frequency and climate over the last 17,000 years in central Yellowstone National Park: Geology, v. 28, p. 211–214, doi:10.1130/0091-7613(2000)28<211:VIFFAC>2.0.CO;2.

Millspaugh, S.H., Whitlock, C., and Bartlein, P., 2004, Postglacial fire, vegetation, and climate history of the Yellowstone-Lamar and Central Plateau provinces, Yellowstone National Park, *in* Wallace, L., ed., After the Fires: The Ecology of Change in Yellowstone National Park: New Haven, Connecticut, Yale University Press, p. 10–28.

Montagne, J., and Locke, W.W., 1989, Trip 7 road log, Cenozoic history of Yellowstone Valley between Livingston and Gardiner, Montana, *in* French, D.E., and Grabb, R.F., eds., Geologic Resources of Montana, Guidebook 1989 Field Conference: Montana Geological Society, v. II [road logs], p. 502–521.

Mumma, S.A., Whitlock, C., and Pierce, K.L., 2012, A 28,000 year history of vegetation and climate from Lower Red Rock Lake, Centennial Valley, southwestern Montana: Palaeogeography, Palaeoclimatology, Palaeoecology, v. 326–328, p. 30–41, doi:10.1016/j.palaeo.2012.01.036.

National Park Service, 1994, Baseline Water Quality Data, Inventory and Analysis, Yellowstone National Park: Fort Collins, Colorado, National Park Service, Water Resources Division, Technical Report NPS/NRWRD/NRTR-94/22, 941 p.

Nye, J.F., 1952, The mechanics of glacier flow: Journal of Glaciology, v. 2, p. 82–93.

Obradovich, J.D., 1992, Geochronology of Late Cenozoic Volcanism of Yellowstone National Park and Adjoining Areas, Wyoming and Idaho: U.S. Geological Survey Open-File Report 82-408, 45 p.

Patterson, W.S.B., 1969, The Physics of Glaciers: New York, Pergamon Press, 250 p.

Persico, L.P., and Meyer, G., 2009, Holocene beaver damming, fluvial geomorphology, and climate in Yellowstone National Park, Wyoming: Quaternary Research, v. 71, p. 340–353, doi:10.1016/j.yqres.2008.09.007.

Personius, S.F., 1982, Geologic setting and geomorphic analysis of the Deep Creek fault, upper Yellowstone Valley, south-central Montana [unpublished M.S. thesis]: Bozeman, Montana, Montana State University, 77 p.

Pierce, K.L., 1979, History and Dynamics of Glaciation in the Northern Yellowstone National Park Area: U.S. Geological Survey Professional Paper 729-F, 91 p.

Pierce, K.L., and Morgan, L.A., 1992, The track of the Yellowstone hot spot: Volcanism, faulting and uplift, *in* Link, P.K., Kuntz, M.A., and Platt, L.W., eds., Regional Geology of Eastern Idaho and Western Wyoming: Geological Society of America Memoir 179, p. 1–53.

Pierce, K.L., Adams, K.D., and Sturchio, N.C., 1991, Geologic setting of the Corwin Springs Known Geothermal Resource area–Mammoth Hot Springs area in and adjacent to Yellowstone National Park, *in* Sorey, M.L., ed., Effects of Potential Geothermal Development in the Corwin Springs Known Geothermal Resource Area, Montana, on the Thermal Features of Yellowstone National Park: U.S. Geological Survey Water Resources Investigations Report 91-4052, p. C1–C37.

Pierce, K.L., Despain, D.G., Whitlock, C., Cannon, K.P., Meyer, G., Morgan, L., and Licciardi, J.M., 2003, Quaternary geology and ecology of the Greater Yellowstone area, *in* Easterbrook, D.J., ed., Quaternary Geology of the United States: INQUA 2003 Field Guide Volume: Reno, Nevada, Desert Research Institute, p. 313–344.

Richmond, G.M., 1969, Development and stagnation of the last Pleistocene icecap in the Yellowstone Lake Basin, Yellowstone National Park, U.S.A.: Eiszeitalter und Gegenwart, v. 20, p. 196–203.

Richmond, G.M., 1986, Stratigraphy and chronology of glaciations in Yellowstone National Park, *in* Sibrava, V., Bowen, D.Q., and Richmond, G.M., eds., Quaternary Glaciations in the Northern Hemisphere: Quaternary Science Reviews, v. 5, p. 83–98.

Ruleman, C.A., 2002, Quaternary tectonic activity within the northern arm of the Yellowstone Tectonic Parabola and associated seismic hazards, southwest Montana [unpublished M.S. thesis]: Bozeman, Montana, Montana State University, 157 p.

Smith, M., Lageson, D., Heatherington, A., and Harlan, S., 1995, Geochronology, geochemistry, and isotopic systematics of the basalt of Hepburns Mesa, Yellowstone River Valley, Montana: Geological Society of America Abstracts with Programs, v. 27, no. 4, p. 56.

Smith, R.B., and Siegel, L.J., 2000, Windows into the Earth, the Geologic Story of Yellowstone and Grand Teton National Parks: New York, Oxford University Press, 242 p.

Theriot, E.C., Fritz, S.C., and Gresswell, R., 1997, Long-term limnological data from the larger lakes of Yellowstone National Park: Arctic and Alpine Research, v. 29, p. 304–314, doi:10.2307/1552145.

Theriot, E.C., Fritz, S.C., Whitlock, C., and Conley, D.J., 2006, Late Quaternary rapid morphological evolution of an endemic diatom in Yellowstone Lake, USA: Paleobiology, v. 32, p. 38–54, doi:10.1666/02075.1.

U.S. Geological Survey, 1972a, Surficial Geologic Map of Yellowstone National Park: U.S. Geological Survey Miscellaneous Investigations Map 710, scale 1:125,000.

U.S. Geological Survey, 1972b, Bedrock Geologic Map of Yellowstone National Park: U.S. Geological Survey Miscellaneous Geologic Investigations Map I-711, scale 1:125,000.

Weed, W.H., 1893, The Glaciation of the Yellowstone Valley North of the Park: U.S. Geological Survey Bulletin 104, 41 p.

Whitlock, C., 1993, Postglacial vegetation and climate of Grand Teton and southern Yellowstone National parks: Ecological Monographs, v. 63, p. 173–198, doi:10.2307/2937179.

Whitlock, C., and Bartlein, P.J., 1993, Spatial variations of Holocene climatic change in the Yellowstone region: Quaternary Research, v. 39, p. 231–238, doi:10.1006/qres.1993.1026.

Whitlock, C., Dean, W., Rosenbaum, J., Stevens, L., Fritz, S., Bracht, B., and Power, M., 2008, A 2650-year-long record of environmental change from northern Yellowstone National Park based on a comparison of multiple proxy data: Quaternary International, v. 188, p. 126–138, doi:10.1016/j.quaint.2007.06.005.

Whitlock, C., Dean, W.E., Fritz, S.C., Stevens, L.R., Stone, J.R., Power, M.J., Rosenbaum, J.R., Pierce, K.L., and Bracht-Flyr, B.B., 2012, Holocene seasonal variability inferred from multiple proxy records from Crevice Lake, Yellowstone National Park, USA: Palaeogeography, Palaeoclimatology, Palaeoecology, v. 331–332, p. 90–103, doi:10.1016/j.palaeo.2012.03.001.

Manuscript Accepted by the Society 27 February 2014

Printed in the USA